SPRINGER
LAB MANUAL

Springer-Verlag Berlin Heidelberg GmbH

Maria R. Micheli   Rodolfo Bova   (Eds.)

Fingerprinting Methods Based on Arbitrarily Primed PCR

With 53 Figures

Springer

Dr. Maria Rita Micheli

Dipartimento di Biologia Cellulare e Molecolare
Università di Perugia
Via Pascoli
06123 Perugia, Italia

Dr. Rodolfo Bova

Dipartimento di Medicina Sperimentale
e Scienze Biochimiche
Università di Roma "Tor Vergata"
Via Tor Vergata
00133 Roma, Italia

Library of Congress Cataloging-in-Publication Data

Fingerprinting methods based on arbitrarily primed PCR / Maria R.
  Micheli, Rodolfo Bova (eds.).
      p.   cm.
      Includes bibliographical references and index.

    1. Polymerase chain reaction--Laboratory manuals. 2. DNA
fingerprinting--Laboratory manuals.   I. Micheli, Maria R., 1954–
    II. Bova, Rodolfo, 1956–
  QP606.D46F56   1996
  574.87'328--DC20   96–30117

ISBN 978-3-642-47812-3   ISBN 978-3-642-60441-6 (eBook)
DOI 10.1007/978-3-642-60441-6

©Springer-Verlag Berlin Heidelberg 1997
Originally published by Springer-Verlag Berlin Heidelberg New York in 1997
Softcover reprint of the hardcover 1st edition 1997

The use of general descriptive names, registered names, trademarks, etc. in this publication does not imply, even in the absence of a specific statement, that such names are exempt from the relevant protective laws and regulations and therefore free for general use.

Product liability: The publisher cannot guarantee the accuracy of any information about dosage and application thereof contained in this book. In every individual case the user must check such information by consulting the relevant literature.

Cover design: Design & Production GmbH, Heidelberg
Typesetting: Mitterweger Werksatz GmbH, Plankstadt
SPIN 10506236   27/3137 5 4 3 2 1 0 - Printed on acid-free papier

Preface

Before the polymerase chain reaction (PCR) was established, DNA fingerprinting technology had relied for years on restriction fragment length polymorphism (RFLP) analysis, a very efficient technique but quite laborious and not suitable for high throughput applications. Since its development, PCR has provided a new and powerful tool for DNA fingerprinting. According to the earlier PCR-based approaches, fingerprints are generated through the selective amplification of hypervariable loci such as mini- or microsatellites.

More recently, a novel PCR-based strategy involving the use of arbitrary primers to amplify random genomic DNA fragments has been developed. In a short time the random amplification methodology has become very popular due to its ability to easily and rapidly generate polymorphic markers using very small amounts of starting DNA, independently of any prior knowledge of the target DNA sequence. This feature makes random amplification technology a tool potentially useful in many areas of genetic research such as gene mapping, individual and strain identification, population genetics and phylogenetics.

The same strategy has been successfully extended to the comparative analysis of RNAs from different sources in order to detect and clone differentially expressed genes. This RNA fingerprinting approach provides substantial advantages compared to the previously available methods, i.e., differential and subtractive hybridization: it is easier and faster, requires only minimal amounts of RNA and allows the simultaneous comparison of many samples detecting both up-regulated and down-regulated messages. Thanks to this feature, RNA fingerprinting has rapidly become the method of choice for detecting and isolating differentially expressed genes.

This manual is intended to be a working tool for researchers interested in applying random amplification fingerprinting methodology, independently of the specific area of biology or medicine in which they are involved. Detailed protocols for each step of both DNA and RNA fingerprinting procedures are provided and, whenever possible, alternative strategies are made available. Methods for isolation and characterization of the amplification products are also included. Finally a few specific research applications of particular interest, which provide further experimental procedures, are reported. These applications span a wide range of different fields thus illustrating the high versatility of random amplification techniques.

Perugia, Autumn 1996   MARIA R. MICHELI
                                               RODOLFO BOVA

Contents

Further Protocols for Amplicon Visualization

Marker Isolation and Characterization

Research Applications

List of Contributors

Yojiro Anzai
Nippon Roche Research Center
200 Kajiwara
Kamakura 247
Japan

Rosa Arribas
Institut de Recerca Oncologica
Autovia Castelldefels km 2,7
L'Hospitalet
08907 Barcelona
Spain

Marco Bazzicalupo
Dipartimento di Biologia
Animale e Genetica
Università di Firenze
Via Romana 17
50125 Florence
Italy

Douglas E. Berg
Department of Molecular
Microbiology and Genetics
Washington University School of
Medicine
St. Louis, MO 63110
USA

Maja Bévort
Department of Growth and
Reproduction
Rigshospitalet
Blegdamsvej 9
2100 Copenhagen
Denmark

Thomas C.G. Bosch
Zoologisches Institut
der Universität München
Luisenstrasse 14
80333 Munich
Germany

Rodolfo Bova
Dipartimento di Medicina
Sperimentale e Scienze
Biochimiche
II Università di Roma
Via Tor Vergata
00137 Rome
Italy

Gustavo Cactano-Anollés
Plant Molecular Genetics
The University of Tennessee
269 Ellington, Plant Science
Bldg
PO Box 1071
Knoxville, TN 37901–1071
USA

Lorraine Chalifour
Lady Davis Institute for Medical
Research
Sir Mortimer B. Davis – Jewish
General Hospital
3755 Chemin Cote Ste.
Catherine
Montreal, Quebec
Canada H3T 1E2

Andrew G. Clark
Institute of Molecular
Evolutionary Genetics
Department of Biology
Pennsylvania State University
University Park, PA 16802
USA

Benjamin Cobb
Microbial Pathogenicity Group
Bath University
Bath, BA2 7AY
UK

Ettore D'Ambrosio
Istituto di Medicina
Sperimentale
Consiglio Nazionale delle
Ricerche
Viale Marx 15/43
00137 Rome
Italy

Ismail Dweikat
Department of Agronomy
Lilly Hall of Life Sciences
Purdue University
West Lafayette, IN 47907
USA

Silvia Fancelli
Dipartimento di Biologia
Animale e Genetica
Università di Firenze
Via Romana 17
50125 Florence
Italy

Renato Fani
Dipartimento di Biologia
Animale e Genetica
Università di Firenze
Via Romana 17
50125 Florence
Italy

Kimber L. Fisher
Department of Medical Micro-
biology and Immunology, 420
SMI
University of Wisconsin Medical
School
1300 University Avenue
Madison
WI 53706 – 1532, USA

Timothy P. Fleming
Deptartment of Ophthalmology,
Box 8096
Washington University School of
Medicine
660 S. Euclid Ave.
St. Louis
MO 63110, USA

Fumihiro Fujimori
Nippon Roche Research Center
200 Kajiwara
Kamakura 247
Japan

Glenn C. Graham
Queensland Agricultural
Biotechnology Centre
4th Floor Gehrmann
Laboratories
University of Queensland
Research Road
St. Lucia
4072 Australia

Dario Grattapaglia
Plant Genetics Laboratory
CENARGEN-EMBRAPA
S.A.I.N. Parque Rural C.P. 02372
70879 – 970 Brasília
D.F. Brazil

Mette Grønborg
Department of Clinical
Chemistry
K.A.S. Glostrup
2600 Glostrup
Denmark

Claus Hansen
John F. Kennedy Institutet
Gl. Landevej 7
2600 Glostrup
Denmark

Robert J. Henry
Queensland Agricultural
Biotechnology Centre
4th Floor Gehrmann
Laboratories
University of Queensland
Research Road
St. Lucia
4072 Australia

René Hummel
Department of Growth and
Reproduction
Rigshospitalet
Blegdamsvej 9
2100 Copenhagen
Denmark.

Greg J. Hunt
Department of Entomology
Purdue University
Entomology Hall
West Lafayette, IN 47907 – 1158
USA

Takashi Ito
Human Genome Center
Institute of Medical Science
University of Tokyo
4 – 6-1 Shirokane-dai
Minato-ku, Tokyo 108
Japan

Marianne Jørgensen
Department of Growth and
Reproduction
Rigshospitalet
Blegdamsvej 9
2100 Copenhagen
Denmark

Katalyn Karikò
Rm 508 Johnson Pavilion
University of Pennsylvania
Medical School
3610 Hamilton Walk
Philadelphia, PA 19104
USA

Henrik Leffers
Department of Growth and
Reproduction
Section GR-5064, Rigshospitalet
Blegdamsvej 9
2100 Copenhagen
Denmark

Marlene Løfgren
Department of Growth and
Reproduction
Section GR-5064,
Rigshospitalet
Blegdamsvej 9
2100 Copenhagen
Denmark

Jan Lohmann
Zoologisches Institut
der Universität München
Luisenstrasse 14
80333 Munich
Germany

Sally Mackenzie
Department of Agronomy
Lilly Hall of Life Sciences
Purdue University
West Lafayette, IN 47907
USA

Michael McClelland
Sidney Kimmel Cancer Center,
room 290
11099 N. Torrey Pines Rd
La Jolla, CA 92037
USA

Maria Rita Micheli
Dipartimento di Biologia
Cellulare e Molecolare
Università di Perugia
Via Pascoli
06123 Perugia
Italy

Elena Mori
Dipartimento di Biologia
Animale e Genetica
Università di Firenze
Via Romana 17
50125 Florence
Italy

Noel B. Murphy
International Livestock Research
Institute
P.O. Box 30709
Nairobi
Kenya

Annette H. Nielsen
Laboratory for Gene Expression
Institute of Molecular Biology
Gustav Wieds Vej 10
8000 Aarhus
Denmark

Toru Okuda
Tanabe Seiyaku Co., Ltd
Lead Generation Research
Laboratory at Toda
2–2-50 Kawagishi
Toda, Saitama 335
Japan

Niels Pallisgaard
Institute of Molecular Biology
C. F. Møllers allé
8000 Aarhus
Denmark

Esterina Pascale
Dipartimento di Medicina
Sperimentale
Università de L'Aquila
67100 L'Aquila
Italy

Miguel A. Peinado
Institut de Recerca Oncologica
Autovia Castelldefels km 2,7
L'Hospitalet
08907 Barcelona
Spain

Roger Pellé
International Livestock Research
Institute
P.O. Box 30709
Nairobi
Kenya

Greg A. Penner
Agriculture and Agri-Food
Canada
Cereal Research Centre
195 Dafoe Road
Winnipeg, Manitoba
Canada, R3T 2M9

Avutu 'Sam' Reddy
Southern Crop Improvement
Facility
Texas A & M University
College Station, TX 77843–2123
USA

Steven A. Reeves
Molecular Neuro-Oncology
Laboratories
Neurosurgical Service Unit
Massachusetts General Hospital
and Harvard Medical School
149 13th Street
Charlestown, MA 02129–9142
USA

Mercedes Ricote
Cellular and Molecular
Medicine West
Department of Medicine
University of CA, San Diego
9500 Gilman Dr.
Department 0651
La Jolla, CA 92093-0651
USA

Mikkel Rohde
Departement of Cancer Biology
Danish Cancer Society
Strandboulevarden 49
2200 Copenhagen
Denmark

Yoshiyuki Sakaki
Human Genome Center
Institute of Medical Science
University of Tokyo
4–6–1 Shirokane-dai
Minato-ku, Tokyo 108
Japan

Richard M. Schultz
Department of Biology
University of Pennsylvania
415 South University Avenue
Philadelphia, PA 19104–6018
USA

C. Neal Stewart Jr.
Department of Biology
The University of North
Carolina
Greensboro, NC 27412
USA

Bradley Stone
Department of Urology, HSB
room I-340
University of Washington
mail stop 356510
Seattle, WA 98195
USA

Michael Strauss
Departement of Cancer Biology
Danish Cancer Society
Strandboulevarden 49
2200 Copenhagen
Denmark

Hiroki Takaya
Nippon Roche Research Center
200 Kajiwara
Kamakura 247
Japan

Silvia Tòrtola
Institut de Recerca Oncologica
Autovia Castelldefels km 2,7
L'Hospitalet
08907 Barcelona
Spain

Jennifer Walsh Weller
Perkin Elmer / Applied Biosystems Division
700 Building
850 Lincoln Centre Drive
Foster City, CA 94404-1128
USA

Mark A. Watson
Department of Ophthalmology,
Box 8096
Washington University School of
Medicine
660 S. Euclid Ave.
St. Louis, MO 63110
USA

John Welsh
Sidney Kimmel Cancer Center
11099 N. Torrey Pines Rd
La Jolla, CA 92037
USA

Jon P. Woods
Department of Medical Micro-
biology
and Immunology, 420 SMI
University of Wisconsin Medical
School
1300 University Avenue
Madison, WI 53706-1532
USA

James W. Zimmermann
Department of Biology
University of Pennsylvania
415 South University Avenue
Philadelphia, PA 19104-6018
USA

Part 1

DNA Fingerprinting

Overview

1
Introduction to DNA Fingerprinting Through Random PCR Amplification

A significant advance in DNA fingerprinting technology was made in the early 1990s through the development of a novel polymerase chain reaction (PCR)-based strategy which involves the use of single oligonucleotide primers of arbitrary sequence to amplify random genomic DNA fragments (Welsh and McClelland 1990; Williams et al. 1990; Caetano-Anollés et al. 1991). This strategy, often referred to as arbitrarily primed PCR, allows detection of polymorphisms between individuals (or strains) as differences between the patterns of DNA fragments amplified from the different DNAs using a given primer.

The increasing popularity of random amplification technology is due to the substantial advantages it provides over previously available DNA fingerprinting techniques, mainly the ability to easily and rapidly generate polymorphic markers using very small amounts of starting DNA, independently of any prior knowledge of the target DNA sequence. This feature makes random amplification technology a tool potentially useful in many areas of genetic research and particularly well suited to high throughput applications.

Random Amplification of Polymorphic DNA

While standard PCR reactions are based on the use of two specific oligonucleotides which selectively prime the amplification of the target DNA sequence, arbitrarily primed PCR reactions involve the use of a single oligonucleotide of arbitrary sequence which primes the amplification of several discrete DNA products. Each of these anonymous but reproducible fragments is derived from a region of the genome that contains, on opposite DNA strands, two primer binding sites located within an amplifiable distance of each other.

Principles   It is generally assumed that, thanks to the low stringency condition ensured by an appropriate annealing temperature, the arbitrary primer binds to a number of sites randomly distributed in the genomic DNA template and primes DNA synthesis even in those cases in which the match between primer and template is imperfect. The amplification process is initiated on those genomic regions in which two priming events occur on opposite DNA strands within a distance not exceeding a few thousand nucleotides. The outcome of the amplification reaction is determined by a competition in which those products representing the most efficient pairs of priming sites separated by the most easily amplifiable sequences (i.e., shorter sequences with little secondary structure) will prevail (McClelland et al. 1995). The assumption that not all amplifications are the result of perfect matching between primer and template is supported by the results of experiments performed on small size templates such as bacterial genomes. The number of fragments amplified from those genomes is higher than expected by statistics, if only perfect matching between primer and template is supposed to occur and, therefore, can only be explained through the assumption that imperfect matching is also involved (Williams et al. 1990).

A model to explain the amplification of DNA with arbitrary primers was proposed by Caetano-Anollés et al. (1992a). The model is based on the competitive effects of primer–template and template–template interactions established predominantly during the first few cycles of the amplification process. In the first few temperature cycles a group of cognate amplicons is selected for amplification during a "template screening" phase driven by primer–template–enzyme interactions that can accommodate primer–template mismatching events. First-round amplification products are initially single-stranded, but have palindromic termini that can establish template–template interactions, forming hairpin-loops and duplexes. The primer will have to recognize and displace these structures to allow enzyme anchoring and primer extension. In subsequent rounds of amplification, the different species of the reaction tend to establish at equilibrium, while the rare primer–template duplexes are enzymatically transformed into accumulating amplification products.

Reaction products are resolved by gel electrophoresis, and patterns generated from different genomic DNAs (e.g., DNAs from different individuals or strains) using the same primer are easily compared to identify possible differences (i.e., DNA fragments that are amplified from one genomic DNA but not from another). Any difference between the patterns of amplified fragments reveals a polymorphism in that it arises from a difference in template sequence which inhibits primer binding or otherwise interferes with amplification of the corresponding fragment.

All classes of mutations (substitutions, insertions, deletions, inversions) can be potentially detected by means of random amplification fingerprinting even though most of the detected polymorphisms are supposed to be caused by substitutions. Single base substitutions within the primer binding site may prevent amplification by introducing a mismatch; this is supported by the observation that most single base changes in a primer sequence result in a complete change in the amplification pattern (Williams et al. 1990). Substitutions outside the primer binding site can also be detected if they stabilize secondary structure thus preventing amplification through the inhibition of the primer binding or the elongation by *Taq* DNA polymerase. The other classes of mutations can be detected if they remove a primer binding site or change the distance between two priming sites, therefore preventing amplification or changing the size of the amplification product.

Basic methods

The random amplification strategy is shared by three methods called (arbitrarily primed PCR) (Welsh and McClelland 1990), random amplified polymorphic DNA (RAPD) (Williams et al. 1990) and DNA amplification fingerprinting (DAF) (Caetano-Anollés et al. 1991). They differ from one another in the length of primers used, amplification conditions, separation and visualization of amplified DNA fragments; consequently they generate markedly different fingerprint patterns varying from quite simple to highly complex. The acronym MAAP (multiple arbitrary amplicon profiling) was proposed by Caetano-Anollés et al. (1992b) to collectively define these techniques on the basis of the peculiar characteristics of the common underlying strategy: the multiple, arbitrary nature of targeted sites and the amplification of a range of characteristic DNA products. According to the Authors, this acronym would also be appropriate in view of the fact that this methodology is often used to place markers in a genetic map.

Although DNA amplification using arbitrary primers, as originally developed, involves a single primer, it has been shown that pairwise combinations of primers can be used as well: amplifications performed using two arbitrary primers generate reproducible patterns that are different from those obtained with each single primer (Welsh and McClelland 1991; Micheli et al. 1993). Welsh and McClelland (1991) have also shown that more than 50 % of the products generated by pairwise combinations of primers are different from those generated by either primer alone, that is the amount of products generated by both primers is higher than expected on a statistical basis.

An explanation for this observation could be the possibility that ssDNA products containing a primer at one end and its complement at the other end form a panhandle structure and are, therefore, at a disadvantage in competing for amplification compared to ssDNA products that do not form panhandle structures (McClelland and Welsh 1994).

The use of pairwise combinations of primers provides an advantage over the amplification by single primers: it allows generation of a much higher number of fingerprints when a given number of primers is used both individually and in all possible pairwise combinations. Furthermore, it is worth noting that amplification products generated by two different primers can be directly analyzed by cycle sequencing.

Further developments   Since the establishment of the basic methodology of DNA amplification by arbitrary primers, several variations aimed at further developing its potential have been proposed. These alternative approaches, by improving either primer design or amplification strategy, allow the researcher to tailor fingerprints in such a way that peculiar requirements of specific areas of application can be met, such as those pertaining to the complexity of amplification pattern, the level of polymorphisms detected and the nature of amplification products.

Strategies based on the improvement of primer design involve the use of the so-called mini-hairpin primers (Caetano-Anollés and Gresshoff 1994; Caetano-Anollés and Gresshoff 1996), which are primers containing a hairpin-turn structure at their 5' end, or primers with some degenerate bases (Sakallah et al. 1995). Furthermore, primers whose sequence has been biased to recognize particular sequence motifs, for example microsatellite repeats (Welsh et al. 1991), tRNA gene motifs (McClelland et al. 1992; Welsh and McClelland 1992), promoter consensus sequences or gene family motifs (Birkenmeier et al. 1992), have also been used. Useful comments on the application of primers encoding particular sequence motifs can be found in McClelland and Welsh (1994).

The improvement of amplification strategy is the basis of both ASAP analysis (Caetano-Anollés and Gresshoff 1996) and tecMAAP (Caetano-Anollés et al. 1993): the former is a dual-step amplification strategy in which fingerprints are generated by reamplification of a previous fingerprinting profile, while the latter involves endonuclease cleavage of the template prior to its amplification with arbitrary primers.

More recently, further alternative strategies (AFLP, RAHM, RAMPO and DS-PCR) have been proposed.

Vos et al. (1995) have described a technique called AFLP which produces highly complex profiles by arbitrary amplification of subsets of restriction fragments ligated to adaptor cassettes. Amplification is primed by tripartite oligonucleotides which consist of a sequence complementary to the adapter (at their 5' end), followed by another which is complementary to the restriction site and a very short arbitrary sequence (at their 3' end) which provides amplification selectivity: only restriction fragments in which the nucleotides flanking the restriction site match the 3' end of the primer will be amplified. The authors emphasize the high reliability of this technique resulting from the stringent reaction conditions used for primer annealing.

A method which combines random DNA amplification with hybridization to microsatellite-complementary oligonucleotide probes has been proposed by Cifarelli et al. (1995) and Richardson et al. (1995), who called the technique random amplified hybridization microsatellites (RAHM) and random amplified microsatellite polymorphisms (RAMPO), respectively. The procedure initially involves genomic DNA amplification with an arbitrary primer; after electrophoretic separation and staining, the amplification products are transferred to a nylon membrane and hybridized to oligonucleotide probes carrying simple sequence repeats (SSR). Fingerprint profiles not corresponding to the staining patterns and completely different for each probe are thus obtained. By enhancing the level of polymorphism detection, this approach can be usefully applied in genetic analyses of species where little or no intraspecific variation is detected by random amplification alone.

Double stringency PCR (DS-PCR) (Matioli and de Brito 1995) is a technique that combines microsatellite-specific priming with arbitrary priming: during the first amplification cycle, a single microsatellite oligonucleotide primes the amplification of DNA between two microsatellite regions; in subsequent cycles, the annealing temperature (and the consequent stringency) is lowered in order to allow the arbitrary oligonucleotide to prime amplification within some of the previously amplified fragments. In this approach, therefore, random amplification is targeted to highly polymorphic genomic regions.

Applications of Random Amplified Polymorphic DNA

Random amplification technology is an efficient tool to quickly and easily screen a very large number of loci for possible DNA polymorphisms which are usually referred to as random amplified polymorphic DNAs (RAPDs). These polymorphisms have proved able to discriminate between closely related individuals; this feature combined with their easy identification, makes RAPDs a valid type of marker, potentially useful in many areas of genetic research such as gene mapping, marker-assisted selection in breeding tasks and individual or strain identification. On the other hand, the dominant nature of RAPDs may be a drawback in some applications: fingerprint patterns are scored for the presence or absence of a specific band where the presence of the band does not allow discrimination between homozygosis and heterozygosis at the corresponding locus. Codominant RAPDs resulting from insertion/deletion events of moderate size, observed as amplification products of different size, are rarely found. The dominant nature of RAPDs must be taken into account when designing mapping experiments; a detailed discussion on the use of RAPDs in genetic mapping can be found in Williams et al. (1993).

Furthermore, RAPDs provide a powerful tool in population genetics as well as in phylogenetic analysis, though in the latter their utility is limited to closely related organisms.

Very interesting results have been obtained by applying random amplification to the comparison between genomic DNAs from tumor and normal tissues of the same individual (Peinado et al. 1992; Ionov et al. 1993; Kohno et al. 1994). These results demonstrate the ability of random amplification assay to detect tumor specific genetic alterations and indicate that this technique has a particularly promising future in experimental oncology.

2
Guide to Experimental Strategies

Fingerprinting Methods

A few significant differences in the experimental procedure distinguish the three basic random amplification techniques; such differences result in fingerprint patterns of varying complexity with a number of visible products ranging from about 5 to over 100. Lowest complexity patterns are generated by RAPD (Williams et al. 1990) which involves 10-base primers, low stringency amplification conditions, resolution of amplification products on agarose gels and detection by ethidium bromide staining. AP-PCR generates more complex patterns. As originally developed (Welsh and McClelland 1990), it involves longer primers (about 20 bases), low stringency conditions in the initial amplification cycles and high stringency conditions in the subsequent ones, resolution of amplification products on polyacrylamide gels and detection by autoradiography. DAF (Caetano-Anollés et al. 1991), which generates extremely complex fingerprint profiles, is typically performed by using a large excess of very short primers (8 nucleotides or even less) and involves polyacrylamide gels for the resolution and silver staining for the detection of amplified products.

When designing a random amplification experiment, the level of profile complexity best suited for the specific aim must first be singled out (for example, simple profiles are the first choice for gene mapping while highly complex ones are more suitable for fingerprinting). The most appropriate experimental strategy should consequently be chosen also keeping in mind that, besides the three basic techniques, further variations for modulating pattern complexity (such as 10-base primers in combination with the AmpliTaq Stoffel fragment; McClelland and Welsh 1994) are available.

Whatever approach is adopted, it is important to be aware that all these methods are rather sensitive to template DNA quality, reaction conditions and PCR temperature profiles. The issue of reproducibility of random amplification techniques has received particular attention and several factors potentially affecting fingerprint reliability have been discussed. Actually, quality and quantity of template DNA have emerged as the main factors affecting reproducibility (Williams et al. 1993; Micheli et al. 1994). Other factors, such as enzyme quality or buffer conditions, which are difficult to keep under control between different experiments, may affect the relative intensity of the bands within a single pattern, but seem unable to cause a band to appear or disappear in different experiments (McClelland and Welsh 1994).

Therefore, the first step of the experimental procedure, i.e., the preparation of the DNA to be used as template, is crucial. Hence, an extraction protocol providing high quality DNA must be chosen and the DNA precisely quantified. Control of both quality and quantity of template DNA in each single fingerprinting experiment is mandatory in order to ensure optimal fingerprint reliability.

Experimental Steps

The section on DNA extraction provides some procedures which have been tested for their ability to produce DNA suitable for random amplification. It spans the most frequently used experimental models from bacteria to mammals. Descriptions of a few other methods which have been successfully combined with these fingerprinting techniques are available in the literature (Zhu et al. 1993; Guidet 1994; Steiner et al. 1995; Gang and Weber 1995).

The first three chapters of the section on fingerprint production describe the three basic techniques (AP-PCR, RAPD, and DAF) and provide some insights into primer design and useful suggestions for a correct planning of the experiment. Resolution and detection steps are also included in AP-PCR and RAPD. The protocol for silver staining of polyacrylamide gels, typically associated to DAF, is given in a separate chapter (see section *Further Protocols for Amplicon Visualization*) due to its applicability also to products generated by different amplification protocols (see, for example, Schlegel et al. 1996).

The chapter by Weller and Reddy decribes two fluorescence-based procedures particularly well-suited for studies which involve the processing of large numbers of samples.

Chapter XI reports an interesting procedure aimed at optimizing the basic methodology: this approach is focused on the investigation of the effects of the interactions between variables, thus greatly reducing the number of experiments required compared to optimization strategies based on the sequential investigation of each reaction variable.

"Fingerprinting Tailoring" by Caetano-Anollés provides useful tools for all those applications in which the performance of the amplification reactions needs to be tailored. Included in this chapter are some fingerprinting alternatives which allow the modification of fingerprint patterns through the improvement of either primer design or amplification strategy.

The issue of amplicon visualization is further dealt with in other chapters. Besides the previously mentioned polyacrylamide gel electrophoresis and silver staining, two interesting techniques, denaturating gradient gel electrophoresis (DGGE) and temperature sweep gel electrophoresis (TSGE), both which are able to enhance the detection of DNA polymorphisms, are also reported. It is worth noting that the successful resolution of arbitrary amplification products on single strand conformational polymorphisms (SSCP) gels, reported by McClelland et al. (1994), has provided a further choice for amplicon visualization. The use of these gels increases throughput and reliability of scoring and may prove very useful in genetic mapping applications even though it is probably not well suited for phylogenetics and population genetics (McClelland and Welsh 1994).

Detection of specific amplification products ("diagnostic fragments") can easily be achieved through the dot blot hybridization technique described in Chap. XVI. This approach is extremely useful in those applications, such as marker-assisted selection, where thousands of samples have to be analyzed for the presence or absence of a given amplification product.

In the section *"Marker Isolation and Characterization"*, besides the procedures for amplification product recovery (both from polyacrylamide and agarose gels) and reamplification, two efficient strategies for RAPD marker cloning and a protocol for their direct sequencing (i.e., not requiring cloning) are also described.

Confirmation of the cloned product identity can be achieved by Southern blot hybridization analysis as described in Chap. XX. A gel containing the original fingerprint pattern is transferred to a blotting membrane and hybridized with a probe derived from the cloned product: the probe should hybridize to the expected band. The same procedure can be applied to confirm the equivalence of bands in patterns obtained from different templates and to detect codominant markers (differently sized products amplified from the same locus); in these cases fragments recovered from the original gel and reamplified can be used as probes (Williams et al. 1993).

Although not here reported, further developments allowed by RAPD marker sequencing are worth mentioning: on the basis of the nucleotide sequence of a given marker, specific oligonucleotides can be designed to be used in assays such as allele-specific PCR (AS-PCR) (Wu et al. 1989), allele-specific ligation (Landegren et al. 1988) or sequence-characterized amplified region (SCAR) assay (Paran and Michelmore 1993).

Research Applications

A few research applications of particular interest are reported in the last section, thus providing further experimental procedures. Chapter XX reports an interesting example of application of random amplification fingerprinting in human pathology. Examples of applications in genetic research are given in Chaps. XXI and XXII, both which provide strategies for linkage map construction, and Chap. XXIII, in which a method for estimating nucleotide divergence in population samples is described. The last two chapters are devoted to microorganisms. They report procedures for the elimination of duplicate strains in microbial screening programs (Chap. XXIV) and for the identification of microbial strains (Chap. XXV).

The wide range of issues dealt with in this section, as different as the identification of microbial strains and the detection of somatic mutations in cancer cells, well demostrates the high versatility of random amplification fingerprinting techniques.

References

Birkenmeier EH, Schneider U, Thurston SJ (1992) Fingerprinting genomes by use of PCR primers that encode protein motifs or contain sequences that regulate gene expression. Mamm Genome 2:537–545

Caetano-Anollés G, Gresshoff PM (1994) DNA amplification fingerprinting using arbitrary mini-hairpin oligonucleotide primers. Bio/Technology 12:619–623

Caetano-Anollés G, Gresshoff PM (1996) Generation of sequence signatures from DNA amplification fingerprints with mini-hairpin and microsatellite primers. BioTechniques 20, in press

Caetano-Anollés G, Bassam BJ, Gresshoff PM (1991) DNA amplification fingerprinting using very short arbitrary oligonucleotide primers. Bio/Technology 9:553–557

Caetano-Anollés G, Bassam BJ, Gresshoff PM (1992a) Primer-template interactions during DNA amplification fingerprinting with single arbitrary oligonucleotides. Mol Gen Genet 235:157–165

Caetano-Anollés G, Bassam BJ, Gresshoff PM (1992b) DNA fingerprinting: MAAPing out a RAPD redefinition? Biotechnology 10:937

Caetano-Anollés G, Bassam BJ, Gresshoff PM (1993) Enhanced detection of polymorphic DNA by multiple arbitrary amplicon profiling of endonuclease digested DNA: identification of markers linked to the supernodulation locus in soybean. Mol Gen Genet 241:57–64

Cifarelli RA, Gallitelli M, Cellini F (1995) Random amplified hybridization mycrosatellites (RAHM): isolation of a new class of microsatellite-containing DNA clones. Nucleic Acids Res 23:3802–3803

Gang DR, Weber DJ (1995) Preparation of genomic DNA for RAPD analysis from thick-walled dormant teliospores of *Tilletia* species. BioTechnique 19:92–97

Guidet F (1994) A powerful new techique to quickly prepare hundreds of plant extracts for PCR and RAPD analyses. Nucleic Acids Res 22:1772–1773

Ionov Y, Peinado MA, Malkhosyan S, Shibata D, Perucho M (1993) Ubiquitous somatic mutations in simple repeated sequences reveal a new mechanism for colonic carcinogenesis. Nature 363:558–561

Kohno T, Moroshita K, Takano H, Shapiro DN, Yokota J (1994) Homozygous deletion at chromosome 2q33 in human small-cell lung carcinoma identified by arbitrarily primed PCR genomic fingerprinting. Oncogene 9:103–108

Landegren U, Kaiser R, Sanders J, Hood L (1988) A ligase-mediated gene detection technique. Science 241:1077–1080

Matioli SR, de Brito R (1995) Obtaining genetic markers by using double-stringency PCR with microsatellites and arbitrary primers. BioTechnique 19:752–756

McClelland M, Welsh J (1994) DNA fingerprinting by arbitrarily primed PCR. PCR Methods Applic 4:s59-s65

McClelland M, Petersen C, Welsh J (1992) Length polymorphisms in tRNA intergenic spacers detected using the polymerase chain reaction can distinguish streptococcal strains and species. J Clin Microbiol 30:1499–1504

McClelland M, Arensdorf H, Cheng R, Welsh J (1994) Arbitrarily primed PCR fingerprints resolved on SSCP gels. Nucleic Acids Res 22:1770–1771

McClelland M, Mathieu-Daude F, Welsh J (1995) RNA fingerprinting and differential display using arbitrarily primed PCR. Trends Genet 11:242–246

Micheli MR, Bova R, Calissano P, D'Ambrosio E (1993) Randomly amplified polymorphic DNA fingerprinting using combinations of oligonucleotide primers. BioTechniques 15:388–390

Micheli MR, Bova R, Pascale E, D'Ambrosio E (1994) Reproducible DNA fingerprinting with the random amplified polymorphic DNA (RAPD) method. Nucleic Acids Res 22:1921–1922

Paran I, Michelmore RW (1993) Development of reliable PCR-based markers linked to downy mildew resistance genes in lettuce. Theor Appl Genet 85:985–993

Peinado MA, Malkhosyan S, Velazquez A, Perucho M (1992) Isolation and characterization of allelic losses and gains in colorectal tumors by arbitrarily primed polymerase chain reaction. Proc Natl Acad Sci USA 89:10065–10069

Richardson T, Cato S, Ramser J, Kahl G, Weising K (1995) Hybridization of microsatellites to RAPD: a new source of polymorphic markers. Nucleic Acids Res 23:3798–3799

Sakallah SA, Lanning RW, Cooper DL (1995) DNA fingerprinting of crude bacterial lysates using degenerate RAPD primers. PCR Methods Applic 4:265–268

Schlegel J, Vogt T, Münkel K, Rüschoff J (1996) DNA fingerprinting of mammalian cell lines using nonradioactive arbitrarily primed PCR (AP-PCR). BioTechniques 20:178–180

Steiner JJ, Poklemba CJ, Fjellstrom RG, Elliott LF (1995) A rapid one-tube genomic DNA extraction process for PCR and RAPD analysis. Nucleic Acids Res 23:2569–2570

Vos P, Hogers R, Bleeker M, Reijans M, van de Lee T, Hornes M, Frijters A, Pot J, Peleman J, Kuiper M, Zabeau M (1995) AFLP: a new technique for DNA fingerprinting. Nucleic Acids Res 23:4407–4414

Welsh J, McClelland M (1990) Fingerprinting genomes using PCR with arbitrary primers. Nucleic Acids Res 18:7213–7218

Welsh J, McClelland M (1991) Genomic fingerprinting using arbitrarily primed PCR and a matrix of pairwise combinations of primers. Nucleic Acids Res 19:5275–5279

Welsh J, McClelland M (1992) PCR-amplified length polymorphisms in tRNA intergenic spacers for categorizing staphylococci. Mol Microbiol 6:1673–1680

Welsh J, Petersen C, McClelland M (1991) Polymorphisms generated by arbitrarily primed PCR in the mouse: application to strain identification and genetic mapping. Nucleic Acids Res 19:303–306

Williams JGK, Kubelik AR, Livak KJ, Rafalski JA, Tingey SV (1990) DNA polymorphisms amplified by arbitrary primers are useful as genetic markers. Nucleic Acids Res 18:6531–6535

Williams JGK, Hanafey MK, Rafalski JA, Tingey SV (1993) Genetic analysis using random amplified polymorphic DNA markers. Methods Enzymol 218:704–740

Wu DY, Ugozzoli L, Pal BK, Wallace RB (1989) Allele-specific enzymatic amplification of β-globin genomic DNA for diagnosis of sikle cell anemia. Proc Natl Acad Sci USA 86:2757–2760

Zhu H, Qu F, Zhu L (1993) Isolation of genomic DNAs from plants, fungi and bacteria using benzyl chloride. Nucleic Acids Res 21:5279–5280

Chapter I

DNA Extraction from Mammals

E. D'Ambrosio and E. Pascale

1
Introduction

DNA fingerprinting methods based on PCR amplification using arbitrary primers such as RAPD (Williams et al. 1990), AP-PCR (Welsh and McClelland 1990) and DAF (Caetano-Anollés et al. 1991) require minimal amounts of DNA (generally a few ng per reaction); however, it has been shown that these methods may be significantly influenced by the quality of the template. The choice of an extraction procedure yielding high molecular weight DNA (free of contaminants which can interfere with the amplification process) is, therefore, an essential prerequisite in obtaining reliable DNA fingerprints. Furthermore, it is important that the same extraction procedure is used for all DNAs to be compared.

Several DNA extraction kits (which may be rather expensive) are commercially available. Those involving DNA purification by affinity column usually yield DNA suitable as a template for random amplification methods, while rapid extraction kits that include microspin columns are not recommended since the average DNA molecular weight is considerably reduced.

In this chapter, we will detail three protocols for mammalian DNA extraction which are quite simple and inexpensive, keep the use of hazardous solutions to a minimum, ensure efficient deproteination and yield abundant and high quality DNA.

2
Principles

All DNA extraction methods rely on extensive deproteination and subsequent precipitation of nucleic acids in alcohol in the presence of cations. Deproteination is typically obtained by phenol/chloroform extraction following protein digestion by proteinase K with SDS. The salting out method (Miller et al. 1988) is an alternative procedure which does not require the use of any hazardous solvent as the proteins are removed by centifugation in high salt concentration. A possible drawback of this method may result from the high protein content; in this case the protein pellet may not tightly stick to the bottom of the tube.

In the subsequent precipitation step contaminants may coprecipitate with DNA in alcohol solution. Ethanol precipitable contaminants have been shown to be a major source of variability in RAPD reactions (Micheli et al. 1994) and it is, therefore, crucial that they be removed if the DNA is to be used as a template in random amplifications. Purification can be easily carried out by affinity chromatography (e.g., column, silica powder, or glassmilk). Winding the DNA on a glass rod is an old, simple and inexpensive method that produces good quality DNA.

Although three different protocols are provided, each with its most appropriate application, several steps may well be interchangeable. DNA extraction from tissue (Sect. 4.1) requires thorough homogenization and deproteination by organic extraction which, in this case, ensures better results than the salting out method. Several contaminants that may impair PCR are contained in mammalian blood. The protocol detailed in Sect 4.2 outlines steps for lysing red blood cells and washing out hemoglobin with plasma from the white blood cell pellet.

Finally, if simultaneous preparation of DNA and RNA is required, the protocol described in Sect. 4.3 is appropriate since it involves a nuclei isolation step, thus leaving behind the cytosol from which RNA can be purified.

3
Materials

Equipment
- benchtop centrifuge
- variable speed microcentrifuge
- spectrophotometer with UV lamp

Supplies
- 15 ml centrifuge tube (e.g., Falcon)
- 1.5 ml centrifuge tube (e.g., Eppendorf)
- 50 µl capillary pipette (e.g., intraMARK, Brand)

Reagents and solutions
- PBS: (for 1 l) 8 g NaCl, 0.2 g KCl, 1.44 g Na_2HPO_4, 0.24 g KH_2PO_4 (pH 7.4)
- lysis buffer A: 10 mM Tris-HCl (pH 8.0), 75 mM NaCl, 25 mM EDTA

- 10 % SDS
- 10 mg/ml proteinase K
- freshly distilled phenol saturated with 0.1 M Tris-HCl (pH 8.0)
- chloroform:isoamyl alcohol (29:1)
- 3 M sodium acetate (pH 5.2)
- 100 % ethanol
- TE: 10 mM Tris-HCl (pH 8.0), 1 mM EDTA
- erythrocytes lysis solution: 0.85 % NH_4Cl
- lysis buffer B: 10 mM Tris-HCl (pH 7.6), 400 mM NaCl, 2 mM EDTA
- NaCl saturated solution (>6 M)
- nuclei lysis buffer: 0.14 M NaCl, 1.5 mM $MgCl_2$, 10 mM Tris-HCl (pH 8.6), 0.5 % Nonidet P-40 (NP40), 1 mM DTT (plus 1000 U/ml RNasin if RNA isolation is also planned)

4
Experimental Procedure

DNA Extraction from Tissue (Organic Extraction Procedure)

1. Mince tissue (usually 200–400 mg) in small pieces (about 2 mm³) cutting it with the tips of a scissor in a suitable Petri dish.

   Tissue preparation and homogenization

2. Transfer the pieces to a centrifuge tube containing a few ml of PBS or physiological solution.

3. Centrifuge for 4 min at 1600 rpm in a benchtop centrifuge. Discard the supernatant.

4. Collect the pellet with 1 ml of lysis buffer A and transfer to a 1.5 ml tube.

5. Centrifuge for 3 min at 3000 rpm.

6. Collect the pellet with 200 µl lysis buffer A and pour it into a tissue grinder or a 1 ml Dounce.

7. Add 40 µl of 10 % SDS and homogenize thoroughly using pestle A.

8. Transfer the homogenate to a 1.5 ml tube, rinse the Dounce with 200 µl of lysis buffer A, then add the buffer used to rinse the Dounce to the homogenate.

9. Add 10 µl of 10 mg/ml proteinase K.

   Protein digestion and removal

10. Incubate 1 h at 60 °C or overnight at 37 °C.

11. Extract proteins by adding 200 µl of saturated phenol. Shake for 20 s, then centrifuge for 3 min at full speed in a microcentrifuge.

12. Collect aqueous phase with a blue tip without disturbing the interphase.

13. Repeat extraction with 200 µl chloroform.

DNA precipitation and collection

14. Add 40 µl of 3 M sodium acetate to the aqueous phase collected after the last centrifugation.

15. Precipitate DNA by adding 1 ml of 100 % ethanol.

16. Remove the precipitated DNA, spooling it out of the solution by means of a flame sealed capillary pipette.

Note: Mix ethanol into aqueous phase by directly stirring the solution with the capillary pipette rather than by inverting the tube. In this way the DNA will stick to the pipette before it completely comes out of the solution. This is particularly important when dealing with small amounts of DNA since little "knots" of DNA may result which are difficult to reach.

17. Rinse the DNA briefly by shaking the pipette in 70 % ethanol and let it air dry for a few minutes.

18. Detach the DNA from the pipette shaking it in 200–400 µl of TE. If necessary, use a blue tip mounted pipetman to detach the DNA from the capillary.

19. Let the DNA rehydrate for at least 2 h before quantification.

Note: An accurate DNA quantification is crucial in order to obtain reproducible fingerprints.

20. Transfer a small aliquot to an appropriate volume of sterile water and check the O.D. If necessary adjust the volume so as to obtain a working solution not higher than 10 ng/µl.

21. Store DNA samples at 4 °C (repeated freezing and thawing must be avoided).

DNA Extraction from Blood (Salting Out Procedure)

This protocol works well when dealing with amounts of uncoagulated blood ranging from 0.5 to 2 ml. For larger volumes scale up solutions.

Red blood cell lysis

1. Place the uncoagulated whole blood in a 15 ml centrifuge tube.

2. Add 3 volumes of erythrocytes lysis solution and mix thoroughly by inverting the tube several times.

3. Centrifuge at 1600 rpm for 5 min and remove the supernatant by pouring off without disturbing the buffy coat pellet.

4. Repeat steps 2–3 one or more times with 5 ml of erythrocytes lysis solution until the pellet appears white or has very little red covering.

5. Suspend the cell pellet in 500 µl of erythrocytes lysis solution and transfer to 1.5 ml tube.

Protein digestion and removal

6. Centrifuge for 5 min at 3000 rpm and discard the supernatant.

7. Add 400 µl of lysis buffer B and thoroughly resuspend the pellet by pipetting.

8. Add 30 µl of 10 % SDS and 10 µl of 10 mg/ml proteinase K, then shake vigorously.

9. Incubate for 1 h at 60 °C or overnight at 37 °C.

10. Add 100 µl of NaCl saturated solution and shake vigorously for about 15 s.

11. Centrifuge for 15 min at 3000 rpm or 3 min at full speed.

12. Transfer the supernatant to a clean Eppendorf tube without disturbing the protein pellet.

DNA precipitation and collection

13. Precipitate DNA by adding 0.8 ml of 100 % ethanol, then proceed according to steps 16–21 of Sect. 4.1.

DNA Extraction from Cell Cultures

1. Remove culture medium from Petri dish ($\sim 5 \times 10^7$ cells) and wash with a few ml of PBS (Ca and Mg free).

Cell collection

2. Detach cells by means of a cell scraper.

3. Collect cells with a few ml of PBS and transfer to a 15 ml centrifuge tube.

4. Centrifuge for 3 min at 1600 rpm and remove the supernatant.

5. Suspend the cell pellet in 200 µl of nuclei lysis buffer and vortex.

Cell lysis and DNA extraction

6. Centrifuge for 2 min at 12000 rpm or at full speed and remove the supernatant.

Note: Do not discard the supernatant if total cytoplasmic RNA is to be purified.

7. Suspend the nuclei pellet in either lysis buffer A (for organic extraction) or lysis buffer B (for the salting out procedure) and proceed according to steps 4–21 of Sect. 4.1 (if using lysis buffer A) or steps 7–13 of Sect. 4.2 (if using lysis buffer B).

5
Results

DNAs isolated according to these protocols routinely have $A_{260/280}$ ratios of from 1.8 to 2.0 (demonstrating efficient deproteination) and have proven to be suitable templates for random amplifications.

6
Troubleshooting

Some problems may arise if the DNA solution is very diluted or highly concentrated.

- Very dilute DNA: Scale down volumes keeping in mind that, in order to have an efficient DNA precipitation, it is safe to have >0.1 µg/ml of DNA in a 0.1 M Na^+/70 % ethanol solution. Do not collect DNA with the capillary pipette but spin it down after lowering the SDS concentration to <0.5 % (0.2 % SDS being ideal);

- Highly concentrated DNA: Use a cut blue pipette tip (with an opening of 2–3 mm) to collect viscous DNA solution, being careful not to draw up proteins in the interphase (organic extraction procedure) or in the pellet (salting out procedure). If proteins come along with the DNA solution, perform an additional proteinase K digestion followed by a phenol/chloroform extraction.

References

Caetano-Anollés G, Bassam BJ, Gresshoff PM (1991) DNA amplification fingerprinting using very short arbitrary oligonucleotide primers. Bio/Technology 9:553–557
Micheli MR, Bova R, Pascale E, D'Ambrosio E (1994) Reproducible DNA fingerprinting with the random amplified polymorphic DNA (RAPD) method. Nucleic Acids Res 22:1921–1922
Miller SA, Dykes DD, Polesky HF (1988) A simple salting out procedure for extracting DNA from human nucleate cells. Nucleic Acids Res 16:1215
Welsh J, McClelland M (1990) Fingerprinting genomes using PCR with arbitrary primers. Nucleic Acids Res 18:7213–7218
Williams JGK, Kubelik AR, Livak KJ, Rafalski JA, Tingey SV (1990) DNA polymorphisms amplified by arbitrary primers are useful as genetic markers. Nucleic Acids Res 18:6531–6535

Chapter II

Insect DNA Extraction Protocol

G. J. HUNT

1
Introduction

The following procedure for obtaining insect DNA is similar to a plant DNA extraction procedure (Saghai-Maroof et al. 1984) and relies on a nonionic detergent, hexadecyltrimethylammonium bromide (CTAB) to lyse cells. The resulting DNA is suitable for use as a template in polymerase chain reactions to generate RAPD markers or for Southern blotting. This general protocol has been successfully applied for RAPD analyses of bees (*Apis*, *Bombus*, *Frieseomelitta*, *Trigona*, *Scaptotrigona*, *Melipona*), wasps (*Vespid* species) ants (*Leptothorax*, *Pogonomyrmex*), beetles (*Tribolium*) and mites (*Varroa*). The method used to extract honey bee DNA is presented here. Modifications in grinding techniques may be necessary for different species of arthropods. Grinding in a mortar and pestle with liquid nitrogen will usually improve yields and produce higher molecular weight DNA for Southern blotting, but the simple grinding method described here is adequate to provide template for PCR.

2
Materials

- electric hand-held drill   **Equipment**
- fluorometer (TKO100, Hoefer Scientific Instruments)
- microcentrifuge (e.g., Eppendorf 5415)

- plastic pestles (Kimble #95050–99)   **Supplies**

- phenol/chloroform: pre-equilibrate phenol to pH 7.4 and mix 1:1   **Reagents**
  with chloroform; store at –20°C in 50 ml aliquots   **and solutions**
- chloroform
- proteinase K: 20 mg/ml in water (store at –20°C)
- 3 M sodium acetate, pH 5.2

- ethanol (100 % and 70 %, keep at –20°C)
- high salt solution: 1.5 M NaCl, 50 mM Tris (pH 8)
- lysis solution: 1 % CTAB, 50 mM Tris (pH 8), 10 mM EDTA, 0.75 M NaCl (boil these last two solutions for 10 min and store at room temperature)
- TE: 10 mM Tris buffer (pH 7.6), 1 mM EDTA

3
Experimental Procedure

Honey bee samples

Samples of larvae or adult honey bees may be used for DNA extraction. Freeze the bees quickly on dry ice and store at –70°C prior to DNA extraction. Alcohol-preserved specimens can also be used, if necessary, by keeping the alcohol concentration high (about 95 %). This is done by adding fresh alcohol after bees have soaked for several hours.

DNA extraction

1. Thaw the phenol/chloroform for the day's extractions.

2. Put the frozen bee in a 1.5 ml microcentrifuge tube containing CTAB lysis solution. Use 200 μl for adult workers and 350 μl for drones and queens.

3. Add proteinase K to the lysis solution (2 μl for adult worker bees and 5 μl for the larger males) for a final concentration of about 100 μg/ml).

4. Grind the bee quickly and thoroughly with the electric drill (or by hand) as it thaws in the lysis solution. Use plastic pestles that fit the microcentrifuge tubes. About four individuals can be processed at one time with a drill. Use a clean pestle for each bee. Grind until the solution becomes viscous. Then immediately place samples in a 60°C water bath for 1–5 h.

5. Add one third volume of high salt solution after incubation to insure that CTAB and polysaccharide contaminants stay in solution during subsequent steps.

   Note: CTAB and polysaccharides will come out of solution if salt concentration is too low (see Fang et al. 1992). CTAB can also form a complex with DNA and can be used to precipitate it (Murray and Thompson 1980).

6. Extract the samples once by adding an equal volume of 1:1 phenol/chloroform to each tube. Invert all of the tubes about 20 times (very gently if relatively high molecular weight DNA is

needed for Southern blotting). Then, centrifuge at 12000–17000 g for 10 min (full speed in a microcentrifuge).

Caution: Phenol is both a strong oxidant and carcinogen and can cause burns. Gloves should be worn and extractions should be done in a fume hood.

7. Carefully draw off the aqueous (upper) phase and transfer this supernatant to a new, labeled tube. Avoid drawing up any of the interface. Discard the bottom phase with the old tube.

Note: If the phenol is not properly equilibrated to a high pH (>7.0), DNA will be lost in the phenol/chloroform extraction. Proteinase K that is no longer active or insufficient grinding in step 2 will also result in low yields.

8. The DNA solution is then extracted in the same way with chloroform and centrifuged for 2–3 min at high speed. An additional phenol extraction will help remove more protein and polysaccharide, but is not necessary for good PCR results. Always finish with the chloroform extraction to remove excess phenol.

9. DNA is precipitated by adding one tenth volume of 3 M sodium acetate and two volumes of cold ethanol and incubating the samples for 10 min at –70°C.

10. Centrifuge the samples for 20 min at approximately 4000–6000 g, then remove all of the ethanol with a pipetter. The pellet is then washed with cold 70 % ethanol and allowed to dry (tube inverted) for about 1–3 min to remove traces of ethanol. Do not let the pellet dry for too long or the DNA pellet might not dissolve (especially if it is high molecular weight DNA).

11. Each DNA pellet is dissolved in 50–100 µl of TE. After adding TE, heat the samples to 65°C in the water bath for 10 min and "flick" the tubes with your finger to help dissolve the DNA. Keep the tubes on ice or in a freezer.

12. Quantify the DNA with a fluorometer, following the manufacturers instructions. Always spin DNA samples at full speed for 2–3 min in a microcentrifuge prior to pipetting a sample for quantification or dilution. Centrifugation insures that any undissolved DNA is pelleted.

Caution: The dye used with the fluorometer is a potential mutagen because it binds to DNA.

13. It is convenient to dilute all of the concentrated DNA stocks to 100 ng/µl as they are quantified by adding cold TE. In addition to the concentrated stocks, prepare dilute aliquots for PCR (3 ng/µl) by adding modified TE that contains only 0.3 mM EDTA. The molecular weight of the DNA may be checked on an agarose gel, if desired. If the genomic DNA is somewhat degraded, it may be desirable to prepare more concentrated stocks for PCR (e.g., 20 ng/µl).

References

Fang G, Hammar S, Grumet R (1992) A quick and inexpensive technique for removing polysaccharides from plant genomic DNA. BioTechniques 13:52–56

Murray MG, Thompson WF (1980) Rapid isolation of high molecular weight DNA. Nucleic Acids Res 8:4321–4325

Saghai-Maroof MA, Jorgensen RA, Allard RW (1984) Ribosomal DNA spacer-length polymorphisms in barley: Mendelian inheritance, chromosomal location and population dynamics. Proc Natl Acad Sci USA 81:8014–8018

Chapter III

Rapid DNA Extraction from Plants

C. Neal Stewart, Jr.

1
Introduction

The first step of RAPD fingerprinting is the preparation of the target DNA template. Intuitively, minimal DNA template preparation should be necessary for RAPDs since, theoretically, PCR may amplify a single DNA molecule. It seems one would simply homogenize tissue and allow the PCR to "find" and amplify the target DNA. Indeed, many rapid DNA isolation methods designed for use with PCR actually involve little isolation of DNA. Rather, they employ a "grind and use" process (e.g.,Wang et al. 1993) or minimal purification (e.g., Edwards et al. 1991). These methods are very fast, require little tissue and amplify well with plants that are amenable to DNA extraction and that contain few interfering secondary metabolites. However, the resulting DNA templates are not very pure and may not be stable for long periods of time. Recently, DNA purity has been implicated as one of the most important factors in RAPD reproducibility (McClelland and Welsh 1994). McClelland and Welsh (1994) suggest that researchers use only high quality templates to assure reproducible RAPDs. The assay they suggest consists of replicate RAPD PCRs in which the DNA concentration is titrated over two orders of magnitude. If the DNA is of adequate quality the replicates should yield identical RAPD fingerprints. In addition, Heinze (1994) found that the addition of RNAse A improved RAPD reproducibility with gymnosperm embryos. However, the basal method was a crude prep similar to that of Edwards et al. (1991). Thus, it seems likely that there could be several interactive factors involving DNA purity or lack thereof in RAPD reproducibility. It seems prudent to researchers performing RAPD PCR to assure themselves that the isolated DNA is of sufficient quality to ensure reproducible RAPD fingerprints.

The objective of this chapter is to describe a rapid CTAB miniprep that yields relatively pure DNA from plants and that can also be used with fungi and fish. The procedure features the cationic detergent CTAB (hexadecyltrimethylammonium bromide) in the homogenization buffer, one chloroform extraction, and an alcohol precipitation. This procedure was published by Stewart and Via (1993) and has its origins in previously published procedures (Murray and Thompson 1980; Doyle and Doyle 1987, 1990). The procedure described in this chapter further streamlines the Stewart and Via (1993) miniprep procedure.

2
Materials

Equipment
- cordless electric drill capable of speeds between 500 and 1000 rpm (e.g., Black and Decker CD 2000 "Ranger")

Supplies
- disposable pellet pestles (e.g., Bio-Ventures catalog# P-50, Murfreesboro, Tennessee, USA). Alternatively, one may use a 1000 μl plastic pipette tip that has been pushed onto a deburring tool as a homogenizing pestle (Stewart and Via 1993). The end of the pipette tip is crimped upward when pressed on the tool (the tip is pressed against the bottom of the pipette tip box), thereby creating a "blade" for homogenization.

Reagents and solutions
- homogenization buffer (amounts stated are for 1 l; all reagents may be purchased through Sigma Chemicals):100 mM Tris-HCl (15.76 g in 800 ml of water, pH adjusted to 8.0 with NaOH), 20 mM EDTA (40 ml of 0.5 M EDTA, pH 8.0), 2 % w/v CTAB (20 g), 1.42 M NaCl (81.8 g), 2 % w/v PVP-40 (polyvinylpyrrolidone, average molecular weight 40000) (20 g), 5 mM ascorbic acid (0.88 g), 4.0 mM DIECA (diethyldithiocarbamic acid) (0.69 g). Bring volume to 1 l. This solution may be kept indefinitely at room temperature. Just prior to use, add 6 μl 2-mercaptoethanol per ml of homogenization buffer.
- extraction solution: chloroform:isoamyl alcohol (24:1). This may be stored indefinitely at room temperature in a sealed container.
- ethanol or isopropanol

3
Experimental Procedure

1. Harvest between 0.02 and 0.2 g fresh leaf, callus, embryo, stipe, (or appropriate animal tissue) into a microfuge tube. Place on ice. These may be sufficiently stable for hours to days, depending upon species.

Note: For plants, best quality and yield is obtained when newly flushing leaves are used. However, the procedure has been used with a plethora of tissues, including herbarium specimens.

Caution: All subsequent steps should be performed in a fume hood.

2. Aliquot 500 µl of homogenization buffer (including the 2-mercaptoethanol) into each microfuge tube. For tough leaves from woody plants (exemplified by oaks or cranberry) heating the buffer to 65°C may increase yield.

3. Chuck pestle directly onto drill and homogenize tissue. The disposable pestles may be reused after decontamination with 10 % commercial bleach followed by subsequent washes with water.

4. Incubate tubes at 65°C for 5–60 min. The incubation time is not crucial for most species and can be determined by investigator convenience.

5. Add 500 µl extraction solution to tubes and mix. For most plant species this may consist of ten inversions. For recalcitrant species (such as woody plants) placing tubes horizontally on a shaker (500 rpm) for 5–10 min may increase yield.

6. Centrifuge at 3000 g for 5 min. Slightly higher g or longer centrifugation periods may be necessary to keep the interface compact and the top layer clear if many (>12) samples are centrifuged at a time.

7. Pipette top (aqueous) layer (about 400 µl) into a fresh tube.

8. Add 800 µl ethanol or 270 µl isopropanol and invert several times to precipitate DNA. The sample may be allowed to set at room temperature or 4°C for 5–60 min to increase DNA yield. Longer periods will not affect DNA quality, but may slightly increase yield.

9. Centrifuge at 13000–16000 g for 10–20 min at room temperature.

10. Decant supernatant, air-dry pellet, and resuspend in 50–500 µl water.

4
Results and Comments

This method has been used with success in several plants (*Arabidopsis*, banana, rapid cycling *Brassicas*, canola, chestnut, cranberry, and tobacco, to name a few), fungi (such as *Russula* spp.) and fish (such as *Rivulus marmoratus*). DNA yields, as quantified using Hoechst 33258 dye fluorometry, are typically between 1and 50 µg depending on species, starting material and extraction conditions mentioned earlier. This represents enough DNA for between 40 and 2000 RAPD PCRs. RAPD fingerprints using this method compare favorably to those using very pure DNA isolated by the method of Doyle and Doyle (1990) and subsequent purification through a CsCl ethidium bromide gradient (Stewart and Via 1993). This method has recently been used to isolate DNA for restriction digests and subsequent Southern blotting (e.g., our lab; May et al. 1995). Restrictability is another indication of high quality DNA. The prepared DNA is stable in time. For example, cranberry DNA templates that were extracted over 2 years ago and stored at −20°C yield identical RAPD fingerprints now as when the DNA was first extracted. In addition, they continue to pass the McClelland and Welsh (1994) test and produce reproducible RAPD fingerprints when template amounts are varied over two orders of magnitude (5–500 ng). In summary, the CTAB miniprep method is rapid (one person can process 100–200 samples in a typical workday) and yields adequately pure DNA for RAPD fingerprinting.

References

Doyle JJ, Doyle JL (1987) A rapid DNA isolation procedure for small quantities of fresh leaf tissue. Phytochem Bull 19:11–15

Doyle JJ, Doyle JL (1990) Isolation of plant DNA from fresh tissue. Focus 12:13–15

Edwards K, Johnstone C, Thompson C (1991) A simple and rapid method for the preparation of genomic plant DNA for PCR analysis. Nucleic Acids Res 19:1349

Heinze B (1994) RAPD reactions from crude plant DNA. Mol Biotechnol 1:307–310

May GD, Afza R, Mason HS, Wiecko A, Novak FJ, Arntzen CJ (1995) Generation of transgenic banana (*Musa acuminata*) plants via *Agrobacterium*-mediated transformation. Bio/Technology 13:486–492

McClelland M, Welsh J (1994) DNA fingerprinting by arbitrarily primed PCR. PCR Methods Applic 4:s59-s65

Murray HG, Thompson WF (1980) Rapid isolation of high molecular weight DNA. Nucleic Acids Res 8:4321–4325

Stewart CN Jr, Via LE (1993) A rapid CTAB DNA isolation technique useful for RAPD fingerprinting and other PCR applications. BioTechniques 14:748–751

Wang H, Qi M, Cutler AJ (1993) A simple method of preparing plant samples for PCR. Nucleic Acids Res 21:4153–4154

Chapter IV

Preparation of Fungal Genomic DNA for PCR and RAPD Analysis

G. C. GRAHAM and R. J. HENRY

1
Introduction

The successful extraction of DNA from fungal isolates is essential to the investigation of fungal genetics at a molecular level (Graham et al. 1994). The ever increasing array of molecular techniques available to scientist and clinician has provided a means for more detailed investigations into genetic diversity, reproduction, population biology and phylogenetics of fungi (Chihlar and Sypherd, 1980; Cuberta et al. 1991). This understanding of molecular based genetics has also produced an increasing series of diagnostic assays with higher specificities and increased sensitivity.

Random amplified polymorphic DNA (RAPD) is an arbitrarily primed, PCR based method for developing molecular markers in genome analysis (Welsh and McClelland 1990; Williams et al. 1990). The RAPD method provides a simple and relatively inexpensive approach to fungal genome analysis (Graham et al. 1994; McDermott et al. 1994).

2
Principles and Applications

Nucleic acid preparation for fungal genome analysis requires the extraction of high molecular weight and highly purified DNA. The extraction system should accomplish several important things:

- The fungal cell walls should be broken to release the cellular contents. This is usually done by freezing in liquid nitrogen and grinding in a mortar and pestle. After grinding of the fungal tissue, care must be taken to avoid thawing of the sample. This allows the nucleases to become active and degrade the nucleic acids.

- The fungal cellular organelle membranes must be disrupted to allow the release and exposure of nucleic acids in the extraction buffer. This is generally achieved by the addition of a detergent such as CTAB (cetyl-trimethyl ammonium bromide) or SDS (sodium dodecyl sulfate), to the extraction buffer.

- The nucleic acids must be protected from endogenous nucleases. The detergents will denature and disassociate proteins and so inactivate the nucleases. EDTA is added to the extraction buffer to chelate the metal ions in the DNA extraction buffer and in the fungi. These metal ions are required by the nucleases to catalyze the degradation of the nucleic acids.

- The extraction buffer containing the nucleic acid is emulsified with chloroform to denature and remove proteins from the nucleic acid preparation.

- Shearing of DNA should be kept to a minimum. The DNA can be broken by exposure to turbulence such as achieved by vigorous mixing. DNA of lengths up to 100 kb can be obtained with care particularly if pipetting is kept to a minimum.

- The addition of PVP 40 (polyvinylpyrrolidone, Sigma) to the extraction buffer assists in the absorbtion of phenolics liberated by disruption of fungal cell walls. Phenolics can be used by polyphenol oxidase enzymes to bind nucleic acids to carbohydrate molecules (Katterman and Shattuck 1983).

3
Materials

Equipment	– microfuge (capable of $14000 g$)
	– 55°C water bath
	– gel electrophoresis apparatus
	– spectrophotometer
Supplies	– mortar and pestle
	– liquid nitrogen
	– microfuge tubes
Media	– potato dextrose broth/agar
Reagents and solutions	– CTAB extraction buffer: 2 % (w/v) cetyl-trimethyl ammonium bromide (CTAB), 100 mM Tris-HCl (pH 8.0), 1.4 M NaCl, 20 mM EDTA, 2 % (w/v) PVP 40
	– chloroform:isoamyl alcohol (24:1)
	– 7.5 M ammonium acetate
	– absolute ethanol (ice cold)

- 70 % ethanol (ice cold)
- TE buffer (pH 8.0): 10 mM Tris, 1 mM EDTA
- RNase: 10 mg/ml
- 1x TBE buffer: 89 mM Tris-borate, 2 mM EDTA, pH 8.3
- agarose
- λ HindIII/EcoRI molecular weight markers (Boehringer Mannheim, Germany)
- ethidium bromide stock solution: 10 mg/ml (working solution: 0.5 μg/ml)

4
Experimental Procedure

1. Culture the fungi to be analyzed in an appropriate medium, i.e., potato dextrose broth/agar.

   Preparation of DNA

2. Using a spatula transfer approximately 100 mg of fungal mycelia or spores to a precooled mortar (–20°C).

3. Grind the material to a very fine powder after freezing in liquid nitrogen.

   Caution: Care should be taken here to avoid cold burns from the liquid nitrogen and be sure to keep the ground material frozen at all times. Transfer to a pre-cooled microcentrifuge tube.

4. Add as quickly as possible 1 ml of CTAB extraction buffer and mix thoroughly to protect the nucleic acids from degradation by nucleases. Incubate for 15–20 min at 55°C in a recirculating water bath or heating block.

5. After incubation, centrifuge the CTAB/fungal extract mixture at 14000 g for 5 min to remove the cell debris. Transfer the supernatant to a clean microfuge tube using a wide bore pipette tip.

6. Add approximately one half volume of chloroform:isoamyl alcohol (24:1) and mix to a homogenous liquid by inversion. After mixing, centrifuge at 14000 g for 1 min and transfer the upper aqueous phase to a clean microfuge tube.

7. Repeat the chloroform extraction, making sure the interface layer between the chloroform phase and the aqueous phase is not disturbed when transferring the aqueous phase.

8. Add 1/10 volume of 7.5 M ammonium acetate followed by 2 volumes of ice cold (–20°C) absolute ethanol. Slowly invert the tube several times to precipitate the DNA. Generally the DNA can be

seen to precipitate out of solution and will appear as a cotton wool-like material. Alternatively the tubes can be placed for 1 h at –20°C after the addition of ethanol if the DNA does not precipitate within 20 s.

9. Following the precipitation, the DNA can be pipetted off with a wide bore tip into two changes of ice cold (–20°C) 70 % ethanol. The DNA can be washed by slow inversion.

10. The DNA is pelleted by centrifugation at 14000 g for 1 min.

11. The DNA can then be dried and resuspended in either water or TE buffer, pH 8.0.

   Note: The TE buffer or water should contain RNase (10 μg/ml) to remove any RNA in the preparation.

12. After resuspension the DNA is incubated at 65°C for 20 min to destroy any nucleases that may be present and stored at 4°C until ready for use.

13. Agarose gel electrophoresis (1.2 % agarose in 1x TBE) of the DNA can be used to show the integrity of the DNA, while UV spectrophotometry will give an indication of the concentration and purity.

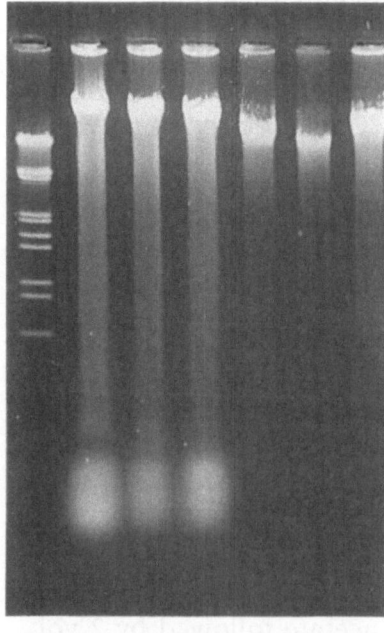

Fig. 1. Genomic DNA isolation from fungal sporangiospore. *Lane 1*, λ *HindIII/EcoRI* molecular weight marker (Boehringer Mannheim); *lanes 2–7*, genomic DNA from six isolates of *Pseudocercospora* (resolution was on 1.2 % agarose gel and the DNA visualized with ethidium bromide). *Lanes 2–4* have not been RNase treated while *lanes 5–7* have been treated with RNase according to the protocol

1   2   3   4   5   6   7

Fig. 2. RAPD profiles of fungal isolates. *Lane 1,* λ *Hind*III/*Eco* RI molecular weight marker (Boehringer Mannheim); *lanes 2–7,* PCR products from isolates of *Pseudocercospora* (see Fig. 1) using OPA-01. Amplification of RAPD profiles were cycled using the following reaction conditions: Obtain oligonucleotide primers (decamers) of arbitrary nucleotide sequence by synthesis from a commercial source, such as Operon Technologies (Alameda, CA), and use the following reaction mix to a total volume of 25 μl: reaction buffer (10 mM Tris-HCl, pH 8.3; 50 mM KCl), 1.5 mM MgCl2, 200 μM each of dATP, dTTP, dCTP and dGTP, 0.2 μM of a single 10-mer oligonucleotide, 1.0 U *Taq* polymerase (Boehringer Mannheim) and 20 ng of template DNA. Cycling was done by one initial cycle of denaturation at 95°C for 2 min, annealing at 40°C for 30 s and extension at 72°C for 60 s. The subsequent 40 cycles were: denaturation at 95°C for 5 s, annealing at 40°C for 30 s and extension at 72°C for 60 s. All RAPD profiles were electrophoresed in 1.5 % agarose gels with 1x TBE buffer at 15 V/cm

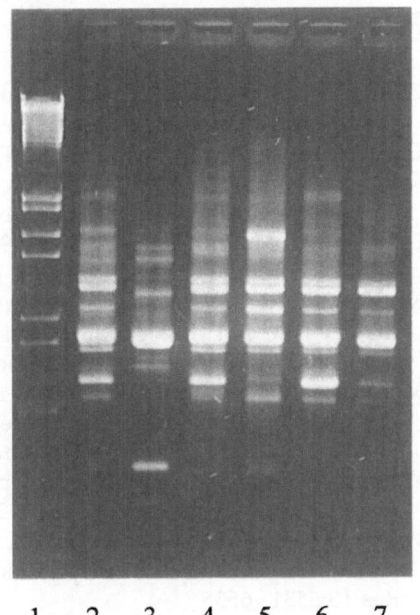

1   2   3   4   5   6   7

5
Results and Comments

A typical DNA extraction electrophoresed in agarose is shown in Fig. 1. The comparison between RNase treated and nontreated DNAs demonstrates the need for removal of RNA if quantitation of DNA template for RAPD-PCR is to be made by spectrophotometry. RNA will contribute to the absorption such that inaccurate values of DNA concentration will result. The amount of DNA template being added to a RAPD-PCR has to be very carefully controlled. It is important to make sure that the correct amounts of DNA template are added precisely to the amount determined from the optimization experiments.

Generally, standard PCR protocols only need approximate concentrations of DNA, which demonstrates an important and fundamental difference between the two techniques. Deviation from the precise amount of DNA in the RAPD-PCR will change the amplification profiles of the RAPD (nonreproducibility). These changes manifest themselves as DNA markers disappearing or appearing from the profiles. An example of RAPD-PCR profiles is shown in Fig. 2.

References

Chihlar, RL, Sypherd, PS (1980) The organisation of the ribosomal RNA genes in the fungus *Mucor racemosus*. Nucleic Acids Res 8:793–804

Cuberta MA, Echandi E, Abernethy T, Vilgalys R (1991) Characterisation of anastomosis groups of binucleate *Rhizoctonia* species using restriction analysis of an amplified ribosomal RNA gene. Phytopathology 81:1395–1400

Graham GC, Mayers P, Henry RJ (1994) A simplified method for the preparation of fungal genomic DNA for PCR and RAPD analysis. BioTechniques 16:48–50

Katterman FRH, Shattuck VI (1983) An effective method of DNA isolation from the mature leaves of *Gossypium* species that contain large amounts of phenolic terpenoids and tannins. Preparative Biochemistry 13:347–359

McDermott J M, Brandle U, Dutly F, Haemmerli UA, Keller S, Muller KE, Wolfe MS (1994) Genetic variation in powdery mildew of barley: Development of RAPD, SCAR, and VNTR markers. Phytopathology 84:1316–1321

Welsh J, McClelland M (1990) Fingerprinting genomes using PCR with arbitary primers. Nucleic Acids Res 18:7213–7218

Williams JGK, Kubelik AR, Livak KJ, Rafalski JA, Tingey SV (1990) DNA polymorphisms amplified by arbitrary primers are useful as genetic markers. Nucleic Acids Res 18:6531–6535

Williams JGK, Hanafey MK, Rafalski JA, Tingey SV (1993) Genetic analysis using random amplified polymorphic DNA markers. Methods Enzymol 218:704–740

Chapter V

Extraction of *Histoplasma capsulatum* DNA for PCR

J. P. WOODS, D. E. BERG, and K. L. FISHER

1
Introduction

Histoplasma capsulatum is a pathogenic fungus that is a significant cause of respiratory and systemic infection worldwide. The organism displays temperature-regulated dimorphism, with the infectious mold morphological type (mycelia and conidia) growing in the soil or at room temperature in laboratory media, and the pathogenic yeast morphological type growing in the mammalian host or at 37 °C in the laboratory. Human infection is common following inhalation of conidia or mycelial fragments, which convert to yeasts that survive and replicate in macrophage phagolysosomes. In many immunocompetent individuals, primary infection may be subclinical or associated with relatively mild disease. However, the fungus causes latent infection and can be carried for years or decades without significant symptoms. Severe (disseminated, fulminant, perhaps fatal) disease can occur either on primary exposure or from reactivation of latent infection and is particularly a problem in immunocompromised hosts, including HIV-infected individuals (Wheat 1988; Eissenberg and Goldman 1991).

Molecular epidemiological techniques have been used to distinguish different isolates, to examine outbreaks of histoplasmosis, and to study virulence. DNA-based typing schemes have included pulsed field gel electrophoresis (PFGE) (Steele et al.1989), restriction fragment length polymorphism (RFLP) analyses using mitochondrial DNA or Southern blotting with ribosomal and *yps* nuclear gene probes (Vincent et al.1986, Spitzer et al.1989, Keath et al.1992), and analyses using the arbitrary primer polymerase chain reaction (AP-PCR), or the random amplified polymorphic DNA (RAPD) technique (Kersulyte et al.1992; Woods et al.1993). Rigorous purification of large amounts of DNA as intact chromosomes for PFGE or suitable for restriction enzyme digestion for RFLP analysis is arduous. Published protocols to obtain high molecular weight DNA for restriction enzyme digestion require at least 2 days and involve grinding mycelia in liquid nitrogen or enzymatic spheroplasting of yeast cells and then extensive protease and RNase digestions, organic extractions, alcohol precipitations, and optional cesium chloride density gradient centrifugation (Spitzer et al. 1989; Woods and Goldman 1992). An additional complication is that this organism, especially in its airborne mold form, presents a significant potential biohazard in the laboratory.

PCR protocols have the advantages of requiring only tiny amounts of DNA, which does not need to be intact or double-stranded. We have designed a method for preparing *H. capsulatum* DNA for PCR that requires only about 2 h, readily enables processing of multiple samples, and markedly reduces the biohazard risk. This technique includes simple boiling of a fungal suspension to kill the organism and lyse the cells. We have found that supernatants of boiled *H. capsulatum*, unlike some other organisms, are not suitable for PCR, possibly due to the presence of inhibitors of DNA polymerases, such as carbohydrates that are present in large quantities in fungal cell walls. This problem is resolved by including a single organic extraction and alcohol precipitation (Woods et al. 1993).

2
Materials

Equipment
- microcentrifuge
- boiling water bath
- agarose gel electrophoresis chamber and power supply

Media
- HMM (*Histoplasma*-macrophage medium) (Worsham and Goldman 1988). Broth is F-12 medium with L-glutamine and phenol red without sodium bicarbonate (Gibco BRL) containing (per liter) 18.2 g glucose, 1.0 g glutamic acid, 84 mg cystine, and 5.96 g HEPES (pH 7.5). Solid medium also contains 0.8 % (w/v) agarose and 10 μM $FeSO_4$.

Reagents and solutions
- distilled, deionized water (dd H_2O)
- 2x lysing solution: 0.5 M Tris (pH 8.0), 3 % (w/v) SDS
- phenol/chloroform:isoamyl alcohol (volume ratio 25:24:1)
- 3 M sodium acetate (pH 5.2)
- absolute ethanol
- agarose
- 10x TAE: 0.4 M Tris-acetate, 10 mM EDTA

- 6x loading buffer: 0.25 % bromphenol blue, 0.25 % xylene cyanol FF, 30 % glycerol in water
- ethidium bromide stock solution: 10 mg/ml

3
Experimental Procedure

1. Scrape cells from an HMM agarose plate and suspend in 1 ml dd H_2O. Use at least a heavy inoculation loopful of cells (approximately 10^8 cells). Alternatively, cells growing in liquid HMM may be used after pelleting by centrifugation.

2. Pellet cells in a microcentrifuge at full speed for 10 s. Discard supernatant.

3. Wash cells with 1 ml dd H_2O and repeat spin.

4. Suspend cell pellet in 0.25 ml dd H_2O.

5. Add 0.25 ml 2x lysing solution and mix.

6. Incubate in a boiling water bath for 30 min.

7. Vortex for 2 min.

8. Extract once with phenol/chloroform:isoamyl alcohol, vortexing for 2 min and spinning for 5 min in a microcentrifuge.

9. Precipitate DNA from the aqueous phase using 1/10 volume 3 M sodium acetate (pH 5.2) and 2 volumes absolute ethanol.

10. Pellet DNA in a microcentrifuge at full speed for 30 min.

11. Optionally, wash DNA with 70 % ethanol.

12. Dry DNA and resuspend in 75 µl dd H2O.

13. Add 1 µl 6x loading buffer to a 5 µl aliquot of DNA, electrophorese in a 1 % agarose gel using 1x TAE in the gel and running buffer; stain with ethidium bromide (0.5 µg/ml), and estimate DNA quantity.

14. Use approximately 20 ng DNA (typically 1–15 µl of preparation) for a single RAPD reaction. Much less DNA may be used for stringent PCR. Thus, each DNA preparation yields enough template for 5–75 RAPD reactions or many more stringent PCR reactions.

Extraction of DNA

4
Results and Comments

Shown in Fig. 1A are 5 µl aliquots of *H. capsulatum* DNAs prepared using this protocol. The DNA is extensively degraded, but provides a suitable template for PCR.

Shown in Fig. 1B are stringent PCR reactions resulting in an amplified fragment of the *Histoplasma URA5* gene (Woods and Goldman, in preparation). Identical results were obtained using extensively purified, high molecular weight DNA (Woods and Goldman 1992) or DNAs prepared using this protocol as templates.

Shown in Fig. 1C are RAPD reactions using extensively purified, high molecular weight DNAs or DNAs prepared using this protocol as templates. These patterns were obtained using a random oligonucleotide primer available in our laboratory and demonstrate that DNAs prepared using this protocol compare favorably with extensively purified, high molecular weight DNA as templates for RAPD reactions and that different strains of *H. capsulatum* can be differentiated using the RAPD technique. More reliable RAPD patterns displaying arrays of 5–15 prominent bands for different *H. capsulatum* strains were obtained with other previously reported primers (Kersulyte et al.1992; Woods et al.1993).

This protocol is suitable for both the yeast and the mold morphological types of the organism. *H. capsulatum* grown in broth or scraped from agarose plates may be used. Initially we used relatively large quantities of cells (approximately 3×10^9, corresponding to 5 ml late-log or stationary phase broth cultures) (Woods et al.1993). Subsequently, as shown here, we obtained satisfactory results using inoculation loopfuls of cells scraped from agarose plates (approximately 10^8 cells). The technique has been used successfully with multiple isolates and strains of *H. capsulatum*. The method provides DNA suitable for highly stringent PCR or RAPD analysis, with the former requiring much less DNA.

5
Troubleshooting

- Omission of any of the steps included in the protocol can reduce the consistency and reliability of RAPD patterns obtained subsequently, especially for amplification products more than 1.5 kb in size. Small differences in intensities of amplified products may be observed when limiting DNA concentrations are used. Therefore,

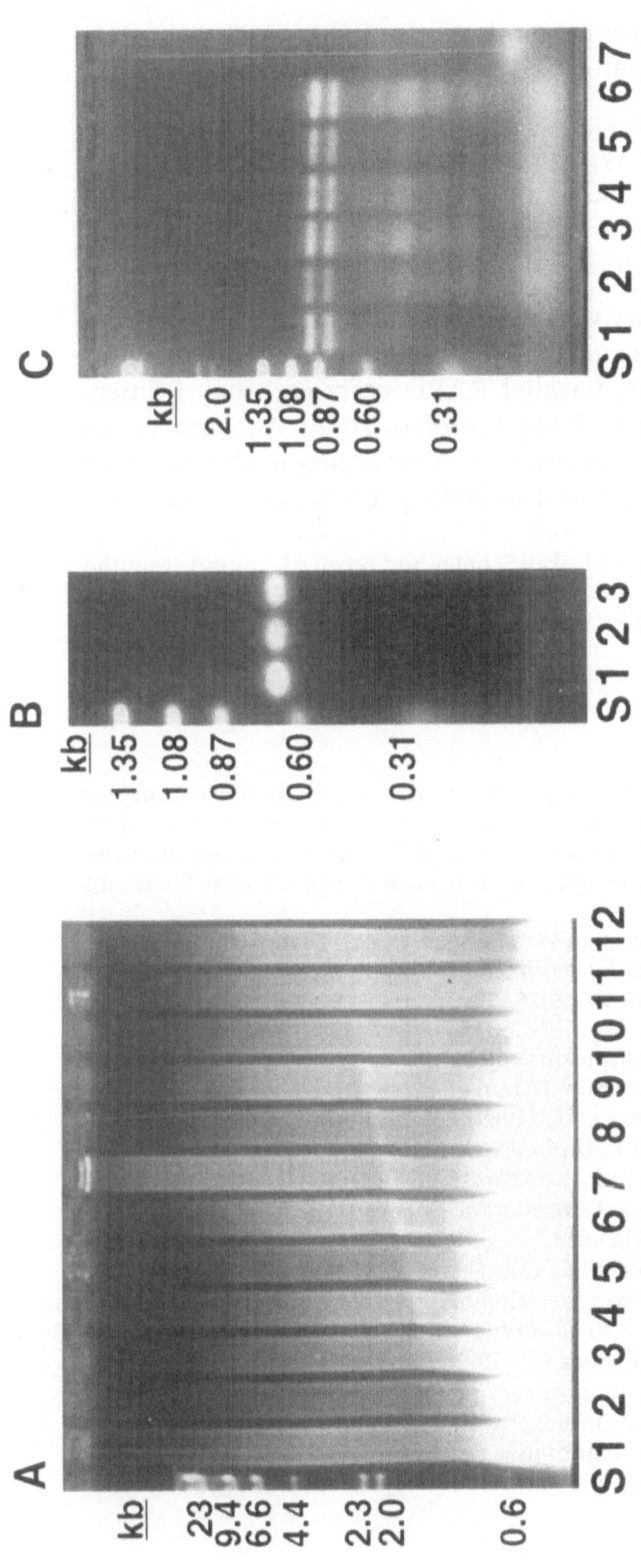

Fig. 1A–C. A DNA preparations from isolates of *H. capsulatum* strains G184A (*lanes 1–3 and 7–9*) and G186A (*lanes 4–6, 10–12*) were electrophoresed in a 1% agarose gel and stained with ethidium bromide. DNAs were prepared by the method described here from loopfuls of cells grown on HMM agarose plates, and 1/15 of each preparation is shown. *Lane S* contains λ DNA digested with *Hind*III, with fragment sizes indicated to the *left*. **B** Stringent PCR reactions using *Histoplasma URA5* gene oligonucleotide primers designed to generate a 0.65 kb amplification product (Woods and Goldman, in preparation) were electrophoresed in a 1.5% agarose gel and stained with ethidium bromide. Template DNAs were prepared from *H. capsulatum* strain G217B DNA. For *lane 1*, extensively purified, high molecular weight DNA, prepared as described previously (Woods and Goldman 1992), was used. For *lanes 2 and 3*, DNAs were prepared by the method described here from loopfuls of cells grown on HMM agarose plates, and 1/150 of each of two preparations was used for PCR under specific amplification conditions designed for the two primers. *Lane S* contains ΦX174 DNA digested with *Hae*III, with fragment sizes indicated to the *left*. **C** RAPD reactions using the 22-mer oligonucleotide 5′ ATTGTAATACGACTCACTATAG 3′ were electrophoresed in a 2% agarose gel and stained with ethidium bromide. Template DNAs were prepared from *H. capsulatum* strains G184A (*lanes 1–6*) or G217B (*lane 7*). For *lanes 1 and 7*, extensively purified, high molecular weight DNAs, prepared as described previously (Woods and Goldman 1992), were used. For *lanes 2–6*, DNAs were prepared by the method described here from loopfuls of cells grown on HMM-agarose plates, and 2/15 of each of five preparations was used for RAPD testing under amplification conditions described previously for long primers (Kersulyte et al.1992, Woods et al.1993). *Lane S* contains λ DNA digested with *Hind*III and ΦX174 DNA digested with *Hae*III, with fragment sizes indicated to the *left*.

we recommend using similar and nonlimiting DNA amounts when comparing RAPD patterns.

- Comparison of RAPD patterns should be made using cultures of the same morphological type, i.e., yeast or mold. Small differences may appear in intensities of amplified products using different morphological types of the same strain, especially with limiting DNA concentrations and for products larger than 2 kb.

- Differences in band intensity in RAPD patterns could also reflect the presence of an inhibitor of DNA polymerases. When needed to control for this possibility, parallel RAPD reactions may be performed with 5 ng as well as 20 ng template DNA. An increase in intensity of bands or the appearance of new bands in the presence of less template DNA is consistent with the presence of an inhibitor.

Acknowledgements. JPW is a Lucille P. Markey Scholar and received support from the Lucille P. Markey Charitable Trust. DEB received support from the National Institutes of Health (AI38166 and DK48029). JPW and KLF also received support from the University of Wisconsin.

References

Eissenberg LG, Goldman WE (1991) *Histoplasma* variation and adaptive strategies for parasitism: new perspectives on histoplasmosis. Clin Microbiol Rev 4:411–421
Keath EJ, Kobayashi GS, Medoff G (1992) Typing of *Histoplasma capsulatum* by restriction fragment length polymorphisms in a nuclear gene. J Clin Microbiol 30:2104–2107
Kersulyte D, Woods JP, Keath EJ, Goldman WE, Berg DE (1992) Diversity among clinical isolates of *Histoplasma capsulatum* detected by polymerase chain reaction with arbitrary primers. J Bacteriol 174:7075–7079
Spitzer ED, Lasker BA, Travis SJ, Kobayashi GS, Medoff G (1989) Use of mitochondrial and ribosomal DNA polymorphisms to classify clinical and soil isolates of *Histoplasma capsulatum.* Infect Immun 57:1409–1412
Steele PE, Carle GF, Kobayashi GS, Medoff G (1989) Electrophoretic analysis of *Histoplasma capsulatum* chromosomal DNA. Mol Cell Biol 9:983–987
Vincent RD, Goewert R, Goldman WE, Kobayashi GS, Lambowitz AM, Medoff G (1986) Classification of *Histoplasma capsulatum* isolates by restriction fragment polymorphisms. J Bacteriol 165:813–818
Wheat LJ (1988) Histoplasmosis. Infect Dis Clin N Am 2:841–859
Woods JP, Goldman WE (1992) In vivo generation of linear plasmids with addition of telomeric sequences by *Histoplasma capsulatum.* Mol Microbiol 6:3603–3610
Woods JP, Kersulyte D, Goldman WE, Berg DE (1993) Fast DNA isolation from *Histoplasma capsulatum*: methodology for arbitrary primer polymerase chain reaction-based epidemiological and clinical studies. J Clin Microbiol 31:463–464
Worsham PL, Goldman WE (1988) Quantitative plating of *Histoplasma capsulatum* without addition of conditioned medium or siderophores. J Med Vet Mycol 26:137–143

Chapter VI

DNA Extraction from Bacterial Cultures

M. Bazzicalupo and S. Fancelli

1
Introduction

There are several different protocols available for the extraction of DNA from bacteria. These have been developed over the past 30 years, starting with the first and best-known method described in the early 1960s by Marmur (1961). Of course, the choice of a particular method ultimately depends on the bacterial species from which the DNA must be extracted. However there are a few recent protocols, including the following one, that are suitable for use with almost any species, at least any gram-negative bacteria. Gram-positive bacteria, because of their thick cell wall, usually require more severe treatments during the first steps of the extraction, i.e., those that are devoted to lysis of the cells. Recently, many kits for the extraction of DNA from biological samples have become commercially available. These procedures are usually very simple, fast, and inexpensive; however they are mostly designed for extraction of DNA from tissues or body fluids of higher organisms and can fail with bacterial cultures.

Two considerable advantages of the RAPD amplification are the small amount of DNA required and the possibility to work with partially broken molecules. Nonetheless, in order to obtain consistent results with RAPD amplification it is essential that the template DNA is prepared following exactly the same procedure for each sample. Finally, when the RAPD amplification is carried out for fingerprinting purposes, it is recommended that the DNA extraction procedure is suitable for treating large number of samples simultaneously.

The DNA extraction procedure described below is intended to fulfil the above mentioned requirements and, in our hands, has given satisfactory results with different bacterial species, not only for RAPD amplification but also for specific PCR, restriction digestion and Southern blot, RFLP, etc. It is based on the following steps:

- Growth of the culture
- Harvest of the cells
- Lysis
- Extraction
- RNAse treatment
- Precipitation of DNA with alcohol

2
Materials

Equipment
- incubator for plates
- shaker bath for bacterial growth
- microcentrifuge for 1.5 ml Eppendorf type tubes. All microcentrifuges have approximately the same rotor radius, centrifuge speed is given as rpm rather than g.
- water or dry bath for microcentrifuge tubes
- vacuum pump

Media
- The media for bacterial growth should be chosen according to the characteristics of the strain to grow. Whenever possible it is better to choose a rich medium, as bacteria reach higher density than in minimal media. A general purpose medium in which many bacterial species grow very well is LB, composed of: tryptone 10 g; yeast extract 5 g; NaCl 10 g; water to 1000 ml. To prepare solid medium for plates add 15 g/l of bacteriological agar.

Reagents and solutions
- TE: 10 mM Tris-HCl (pH 8.0), 1 mM EDTA (pH 8.0) (sterilize in autoclave)
- CTAB: dissolve 10 g CTAB (cetyl-trimethyl ammonium bromide) in 100 ml H_2O. **Note:** The resulting solution is very viscous and must be kept at 65°C for 10 min before use.
- SDS: make 10 % solution with sterile water, store at room temperature
- 5 M NaCl
- phenol:chloroform: phenol saturated with Tris-HCl (pH 8.0) and containing 0.1 % hydroxyquinoline; chloroform; isoamyl alcohol (25:24:1)
- chloroform:isoamyl alcohol (24:1)
- isopropanol
- ethanol: 70 % in H_2O

- proteinase K: dissolve 20 mg/ml in sterile water and store frozen at –20°C
- RNAse: dissolve RNAse A at 10 mg/ml in sterile water boiling for 5 min and store at –20°C

3
Experimental Procedure

Standard Protocol

1. Isolate the bacterial culture by streaking on agar plates with the appropriate medium.

2. Incubate the plate for the time required to produce visible colonies, usually 2 days, but avoid using colonies that are too old.

3. Inoculate a single colony in 1–3 ml of the appropriate medium and allow to grow in a shaker bath up to mid-log phase.

4. Collect the cells by centrifugation of 1.5 ml of the culture for 2 min at 8000 rpm in the microcentrifuge. If growth of the culture was poor, after centrifugation, a second 1.5 ml aliquot of the culture can be centrifuged in the same tube.

5. Suspend the cell pellet in 500 µl of TE and add 30 µl of SDS together with 5 µl of proteinase K. Mix the tube by inversion and incubate at 37°C for 1 h to allow cell lysis.

6. Add 100 µl of 5 M NaCl and vortex for few seconds. Add 80 µl of CTAB; mix and heat 10 min at 65°C.

7. Add an equal volume (about 800 µl) of chloroform:isoamyl alcohol, vortex for a few seconds and centrifuge 5 min at 11000 rpm. Collect the aqueous upper phase in a new tube, add an equal volume of phenol:chloroform:isoamyl alcohol, mix by vortex and centrifuge 5 min at 11000 rpm.

8. Recover the upper aqueous phase in a fresh tube, add 2 µl of RNAse and incubate 30 min at 37°C.

9. Add approximately an equal volume of isopropanol and precipitate DNA for 5 min at room temperature; centrifuge 5 min at 11000 rpm. Discard the supernatant, wash the pellet with 70% ethanol and centrifuge again 5 min at 11000 rpm. Dry the pellet under vacuum and solubilize in 10–20 µl of sterile TE.

Bacterial growth *(margin note, steps 1–4)*

Lysis *(margin note, steps 5–6)*

Extraction *(margin note, steps 7–9)*

Note: A suitable amount of template to be used for each amplification is usually included in 1 µl of the DNA extracted by this protocol.

3.2
Simplified Procedure

For some bacterial species we have found a very useful procedure that can be applied to the preparation of template DNA for PCR or RAPD amplifications.

1. Starting from young isolated colonies, pick the whole colony, or two or three small colonies, with a microbiology loop or with a sterile toothpick.

2. Dissolve the colony in 20 µl of sterile water in a 1.5 ml tube.

3. Heat the tube at 95°C for 10 min.

4. Use the cell lysate as such in the amplification mixture: 1 µl is usually sufficient for RAPD.

4
Comments

The simplified procedure works well with many gram-negative strains such as *Rhizobium*, *E. coli*, several *Pseudomonas* strains, and *Azospirillum*, but should be tested with any other strains. For application to RAPD amplification, it is essential that growth of the colonies is exactly standardized, as we found that with this procedure the age of the cell can dramatically influence the results. For example, a 3 day old colony often gives a pattern of RAPD bands markedly different from that of a 2 day old colony. However, after a few tests to determine the right conditions, this protocol gives very consistent and reliable results and is highly recommended when routine extraction from many strains is required.

5
Troubleshooting.

● Failure to extract DNA with the standard protocol can be ascribed to poor or absent cell lysis. This problem is encountered with some particular bacterial species that have a hard cell wall. It is possible to overcome this problem by the addition of lysozyme

(5 mg/ml final concentration dissolved in TE) and incubation for 30–60 min at 37°C before addition of SDS. Another method consists of freezing and thawing the bacteria, suspended in TE, two or three times before starting the lysis.

- Another possible reason for failure could be, in some species and under some growth conditions, a high concentration of nucleases. This problem can be overcome by collecting the cells very early in the log phase (nucleases tend to accumulate in the late log and stationary phases) and by increasing the concentration of EDTA in TE to up to 3 mM.

References

Marmur J (1961) A procedure for the isolation of deoxyribonucleic acid from microorganisms. J Mol Biol 3:208–217

Chapter VII

Arbitrarily Primed PCR and RAPDs

R. Arribas, S. Tòrtola, J. Welsh, M. McClelland, and M. A. Peinado

1
Introduction

The arbitrarily primed polymerase chain reaction (AP-PCR) is a PCR-based DNA fingerprinting technique using primers whose nucleotide sequence is arbitrarily chosen (Welsh and McClelland 1990; Williams et al. 1990). This method has also been called random amplified polymorphic DNA (RAPD). Initial cycles of the reaction are performed under low stringency conditions (usually achieved with relatively low temperatures during the annealing step and/or high magnesium concentration in the reaction buffer). Under these circumstances the arbitrary primer anneals to the best matches in the template. Competition between these annealing events results in reproducible and quantitative amplification of many discrete bands. Further amplification of these sequences under high stringency conditions produces a fingerprint when they are resolved by gel electrophoresis. The band pattern obtained in a simple experiment is characteristic and representative of the genome used as template.

Although the parameters governing the actual priming and amplification events are quite complex and not totally understood, the empirical evidence shows that for a determined set of experimental conditions the pattern of DNA bands that is amplified is unique and reproducible. In addition, and in contrast with other DNA fingerprinting approaches, the AP-PCR method permits direct cloning of the in vitro amplified DNA sequences.

Two aspects of this method are of special interest. First, the amplified bands usually, though not always, originate from single copy sequences. Second, the amplification is quantitative, that is, the intensities of the amplified bands are almost proportional to the concentration of the corresponding target sequences in the DNA preparation. It can be concluded that, in this situation, information about the overall allelic composition of the genome can be generated by a simple experiment which is done in a single microtube.

The large number of bands amplified with a single arbitrary primer generates a complex fingerprint that can be utilized to detect relative differences in the arbitrarily amplified DNA sequences from two different but closely related genomes. Such differences correspond to polymorphisms or somatic genetic alterations. The priming events during the initial low stringency cycles are arbitrary since they depend on the nucleotide sequence of the PCR primer, which is arbitrarily chosen. Thus, the amplified sequences are in principle a representative, arbitrary small sample of the donor DNA cell genome. As a consequence, this type of analysis presents multiple applications, including genetic mapping of polymorphisms, taxonomy and detection of somatic mutations in cancer cells. Its usefulness has been demonstrated in various prokaryotic and eukaryotic systems, including plants, rodents and humans (reviewed by McClelland and Welsh 1994).

2
Materials

Equipment
- thermal cycler (e.g., MJ Research model PTC100, Perkin Elmer model 9600)
- sequencing gel electrophoresis apparatus (40 cm long/30 cm wide/0.4 mm thick)
- gel dryer

Supplies
- PCR tubes
- mineral oil
- autoradiogram markers
- exposure cassettes and X-ray film

Reagents and solutions
- 10x PCR buffer: 100 mM Tris (pH 8.0), 500 mM KCl, 15 mM $MgCl_2$, 0.01 % gelatin
- 100 mM $MgCl_2$
- 10x dNTPs: 1.25 mM of each dNTP
- arbitrary primer 62R1 (5'-TTTCTCGTTAACTTATTTCATCTTG-3'): 100 pmol/μl
- $[\alpha^{33}P]dATP$: >2000 Ci/mmol (2 μCi/reaction tube) or $[\alpha^{32}P]dCTP$: >3000 Ci/mmol (1–2 μCi/reaction tube)
- *Taq* polymerase: 5 U/μl (Perkin Elmer Cetus or Boehringer Mannheim)
- TE: 10 mM Tris-HCl (pH 7.5), 1 mM EDTA
- genomic DNAs diluted to 20 and 10 ng/μl
- 1x TBE buffer: 0.089 M Tris-borate, 0.025 M disodium EDTA, pH 8.3

- 6 % polyacrylamide/8 M urea gel in 1x TBE buffer
- denaturing loading buffer: 95 % formamide, 0.1 % bromophenol blue, 0.1 % xylene cyanol, 10 mM EDTA

3
Experimental Procedure

As an example, an exact protocol for the analysis of eight different genomic DNAs is described. As in every PCR experiment, a blank control (without genomic DNA) is absolutely necessary to confirm the absence of contamination by DNA in the reaction mixture.

DNA amplification

1. Genomic DNAs are prepared in two different concentrations (20 and 10 ng/µl), a reaction mixture for 17 tubes is prepared. Add components sufficient for 18 reactions in the following order:

Twice distilled water	16.3 x 18 =	293.4
10x PCR buffer	2.5	45.0
100 mM MgCl$_2$	0.25	4.5
10x dNTP mix	2.5	45.0
arbitrary primer	0.5	9.0
5 µCi [α^{33}P]dATP	0.5	9.0
1 U *Taq* polymerase	0.2	3.6
Total volume	22.5 µl	

2. Distribute 22.5 µl to each reaction tube. Add 2.5 µl of genomic DNA (50 and 25 ng, respectively) to each tube. Add one drop of mineral oil to each tube.

3. The reaction is carried out in a thermal cycler for five low stringency cycles (94 °C, 45 s; 40 °C, 30 s; 72 °C, 1 min 15 s) and 35 high stringency cycles (94 °C, 45 s; 60 °C, 30 s; 72 °C, 1 min). A final extension step at 72 °C for 5 min is performed.

4. Dilute 10 µl of the reaction mix with 10 µl of denaturing loading buffer and incubate at 90°–95 °C for 3 min. Immediately chill the tubes on ice and load 2 µl on to an 8 M urea/6 % polyacrylamide sequencing gel and electrophorese for 5 h 30 min at 55 W.

Gel electrophoresis and autoradiography

5. Dry the gel under vacuum at 80 °C and stick two or more luminescent labels (autoradiogram markers) to the dried gel in case isolation of one of the bands is desired. Expose the gel directly to an X-ray film at room temperature without an intensifier screen. Exposure time will depend on the isotope used. ^{32}P-labeled gels usually require exposure times of 1–2 days. It is strongly recom-

mended not to use an intensifier screen, because although a longer exposure is required, bands are considerably sharper. ^{33}P-labeled gels require longer exposures (2–4 days), but band resolution is much better.

Tips
- Optimal AP-PCR conditions vary considerably. Thus it is often necessary to make a few modifications to the standard protocols. In general, changing the annealing temperature (especially in the low stringency cycles) is sufficient. Although intra-assay reproducibility is usually good, significant differences may occur from assay to assay. Such variations are not important if experiments are designed that only require intra-assay reproducibility. In our hands the most critical parameters that affect the band pattern are PCR machine brand (probably due to differences in the temperature profiles of the cycles), primer synthesis lot and *Taq* polymerase brand.

- When using a primer for the first time, reaction conditions as described in the protocol are appropriate. If there are too many bands or a high background, a higher annealing temperature should be used. Although it is possible to obtain a fingerprint with almost any primer, not all of them work well, resulting in poorly reproducible patterns or diffuse bands and high background.

- Although primers as short as ten nucleotides may work fine, we prefer the use of longer primers (about 20 bases) because, in general, the fingerprints are more reproducible and the complexity of the band pattern may be easily adjusted by modifying the reaction conditions. In most cases, any single primer designed for a different purpose (i.e, specific PCR amplification or sequencing) can be used succesfully for AP-PCR if reaction conditions are set properly.

- If a sufficient number of cycles is performed, the fingerprint is relatively independent of the genomic DNA concentration (as demonstrated in Fig. 1), nevertheless, two different DNA concentrations that differ by twofold per sample should be used. Another important parameter to be considered is the quality of DNA: degraded template may produce altered patterns and should be discarded. We always check genomic DNA quality and concentration by electrophoresis analysis in a 0.75% agarose gel with ethidium bromide staining.

Fig. 1. Intra-assay reproducibility. Arbitrarily primed PCR of different DNA dilutions of the same genomic DNA. Experimental conditions are as explained in the text , but the experiment was done in triplicate. *Numbers on top* indicate the DNA concentration in ng/assay tube

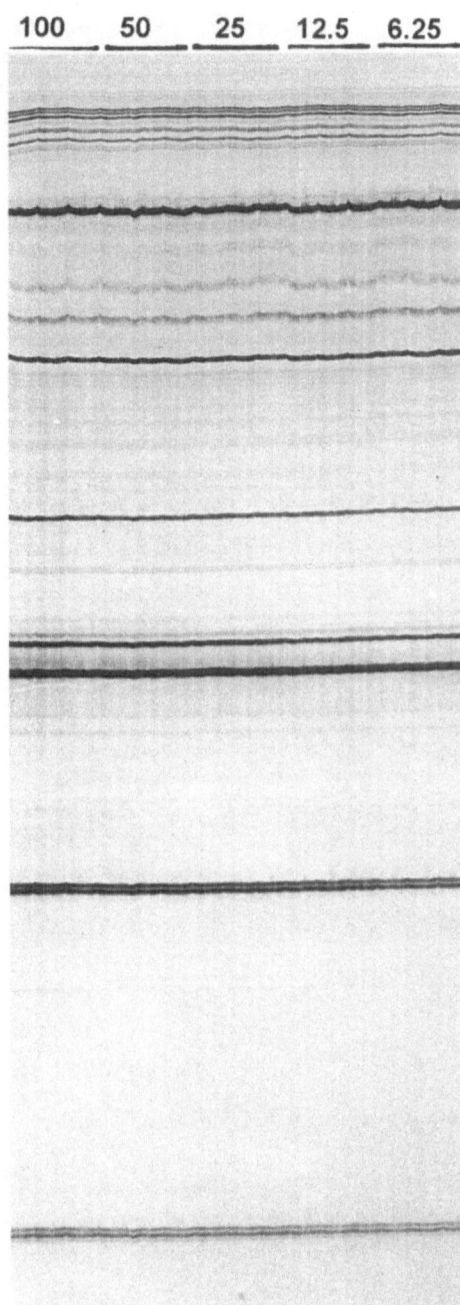

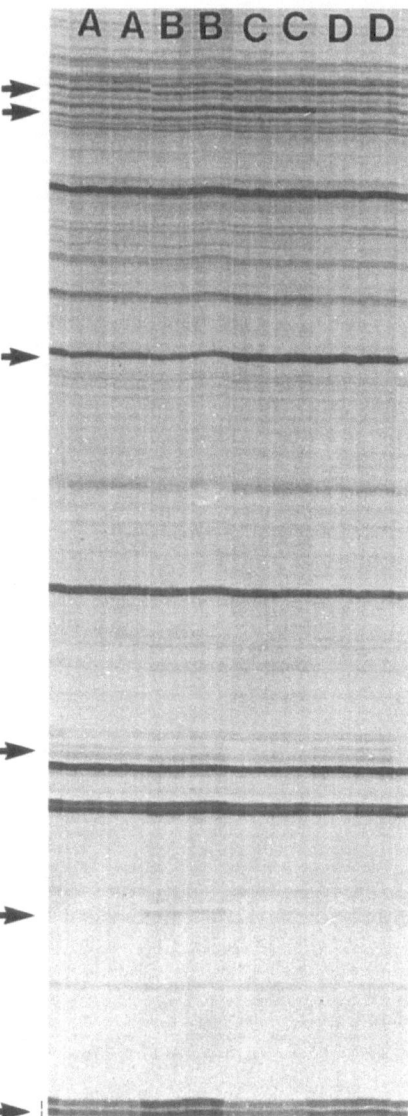

Fig. 2. Arbitrarily primed PCR analysis of four human cell lines. Reaction and electrophoresis analysis were performed in duplicate and as described in the text. *Arrows at left* indicate polymorphic bands

4
Results and Comments

As shown in Fig. 2, it is possible to identify multiple polymorphisms in a single experiment.

In many cases it is necessary to isolate one or several of the amplified bands, either to confirm the differences by an alternative technique or to identify the DNA sequence of interest. Although direct reamplification may be sufficient, cloning of the product is highly recommended in order to obtain a unique fragment (see Chap. XVIII).

Confirmation that the cloned band corresponds to the band visualized in the AP-PCR gel is performed by Southern blot hybridization analysis (see Chap. XX). Once the identity of the cloned fragment with the AP-PCR band has been confirmed, it can be sequenced directly from a plasmid miniprep or rescued phagemid using standard DNA sequencing methods.

Acknowledgements. This work was supported in part by U.S. National Institutes of Health grants CA 68822, NS33377, AI32644 and AI34829 (to JW and MMcC) and Spanish CICYT SAF93–511 and FIS 94/37 (to MAP). RA and ST are fellows of the Spanish Ministry of Education.

References

McClelland M, Welsh J (1994) DNA fingerprinting by arbitrarily primed PCR. PCR Methods Applic 4:s59-s65

Welsh J, McClelland M (1990) Fingerprinting genomes using PCR with arbitrary primers. Nucleic Acids Res 18:7213–7218

Williams JGK, Kubelik AR, Livak KJ, Rafalski JA, Tingey SV (1990) DNA polymorphisms amplified by arbitrary primers are useful as genetic markers. Nucleic Acids Res 18:6531–6535

Chapter VIII

Random Amplified Polymorphic DNA Assay

M. R. MICHELI, R. BOVA, and E. D'AMBROSIO

1
Introduction

A common strategy underlies three PCR-based methods for DNA fingerprinting developed in the early 1990s. Arbitrarily primed PCR (AP-PCR) (Welsh and McClelland 1990), random amplified polymorphic DNA (RAPD) assay (Williams et al. 1990) and DNA amplification fingerprinting (DAF) (Caetano-Anollés et al. 1991) are all based on the use of arbitrary primers to perform the PCR amplification of random genomic DNA fragments. Each primer (or combination of primers) generates a characteristic pattern of amplification products which is visualised by either radionuclide incorporation or ethidium bromide staining or silver staining. Polymorphisms between individuals (or strains) are detected as differences between the patterns of DNA fragments amplified from the different DNAs using a given primer(s).

This PCR-based strategy provides a number of advantages over classic DNA fingerprinting through restriction fragment length polymorphism (RFLP) analysis since it allows easy and rapid generation of polymorphic markers by using very small amounts of starting DNA. Furthermore, unlike other PCR-based fingerprinting methods, it does not require any knowledge of target DNA sequence.

The above cited methods, as originally developed, differ from one another in the length of primers used, amplification conditions, separation and, as already mentioned, visualisation of amplified DNA fragments. Fingerprint patterns produced are also markedly different, varying from quite simple (RAPD) to highly complex (DAF).

RAPD assay involves PCR amplification of genomic DNA using short primers (9 or 10 bases), separation of the amplification products on agarose gels and their detection by ethidium bromide staining. Compared with AP-PCR and DAF, this method is easier, faster and less expensive (agarose gels vs polyacrylamide gels and ethidium bromide staining vs radionuclide incorporation or silver staining). However, the fingerprint patterns it generates contain relatively few bands which could, in some cases, be disadvantageous.

The RAPD technique allows detection of polymorphisms in closely related organisms (e.g., different populations of single species or individuals within a population) and, therefore, provides a powerful tool for such tasks as gene mapping, marker-assisted selection in breeding programs, population and pedigree analysis, phylogenetic studies and individual and strain identification.

2
Principles

The RAPD assay is a PCR amplification performed on genomic DNA templates using a short, arbitrary oligonucleotide primer and a low annealing temperature, conditions that ensure the generation of several discrete DNA products. Each of these anonymous but reproducible fragments is derived from a region of the genome that contains, on opposite DNA strands, two primer binding sites located within an amplifiable distance of each other (e.g., within a few thousand nucleotides). Polymorphisms between individuals (or strains) result from sequence differences which inhibit primer binding or otherwise interfere with amplification; therefore, they can be simply detected as DNA fragments that are amplified from one individual (or strain) but not from another.

Although in AP-PCR and RAPD protocols, as first described (Welsh and McClelland 1990; Williams et al. 1990), a single arbitrary primer is involved, it has been shown that performing amplifications using two arbitrary primers generates reproducible patterns that are different from those obtained with each single primer (Welsh and McClelland 1991; Micheli et al. 1993). It has also been shown that more than 50 % of the products generated by pairwise combination of primers are different from those generated by either primer alone (Welsh and McClelland 1991). This modification increases the analytical power of the RAPD technique since a higher number of polymorphisms can be detected when the number of primers used are equal; for instance, 15 different patterns are obtained with only 5 available primers (used either individually or in pairwise combinations), thus increasing the chance to detect polymorphic bands. The use of pairwise combinations of primers provides a further advantage: sequencing of RAPD products generated by two different primers can be directly performed by cycle sequencing (see Chap. XIX)

Although the sequence of RAPD primers is arbitrarily chosen, two basic criteria, indicated by Williams et al. (1990, 1993), must be met: a minimum of 40 % G+C content (50 %–80 % G+C content is generally used) and the absence of palindromic sequences. Furthermore, when using pairwise combinations of primers, sequences must be designed so as to avoid complementarity between primers to be used together (Micheli et al. 1993).

Primer design

The problem of RAPD fingerprint reproducibility has received particular attention since this technique has been developed (Ellsworth et al. 1993; Meunier and Grimont 1993; Muralidharan and Wakeland 1993; Penner et al. 1993). Two parameters, quality and quantity of template DNA, have emerged as the factors mainly affecting reproducibility. Hence, controlling both these factors is absolutely required in order to ensure reproducible fingerprints.

Reproducibility

Therefore, the choice of an appropriate DNA extraction protocol is highly recommended and, in general, methods yielding undegraded DNA are to be preferred. Presence of contaminants capable of interfering with the amplification process must be avoided. In particular, it has been shown (Micheli et al. 1994) that ethanol precipitable contaminants in DNA are a major source of variability in RAPD reactions (Fig. 1). Template DNA quality can be easily checked by performing RAPD reactions on serial dilutions of the template over a sufficiently wide range (McClelland and Welsh 1994). DNA is assumed to be a good quality template if identical fingerprints are obtained within a significant concentration range.

DNA to be used as template has to be accurately quantified since template concentration is also a crucial parameter for obtaining reproducible fingerprint patterns. Once an optimal concentration range has been determined for a given template DNA, concentrations within that range must always be used in further RAPD assays. Either too high or too low template concentrations may lead to unreliable patterns. Optimal concentration for mammalian DNA ranges from 5 to 25 ng per 25 μl reactions (Micheli et al. 1993). To check the reliability of differences observed between RAPD patterns of different template DNAs, the amplification of each template DNA must always be performed using two different concentrations: only differences occurring at both DNA concentrations are reliable.

Some primer/template combinations appear unable to give reproducible patterns when amplification is performed under standard conditions, thus requiring the optimization of reaction parameters (e.g., magnesium concentration). However, a better strategy to be adopted is to screen a large number of primers and select for further

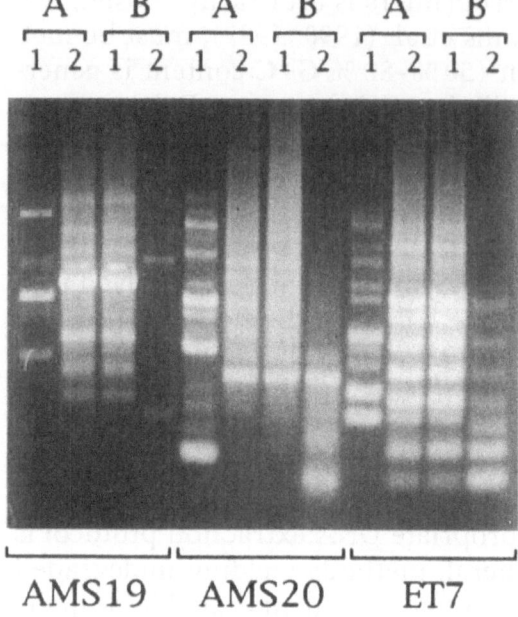

Fig. 1. RAPD patterns from different preparations of the same genomic DNA. *A* and *B* indicate two different preparations of the same DNA: ethanol precipitated DNA was collected either by centrifugation (*A*) or by winding on a glass rod (*B*). *Lane A1*, centrifuged DNA; *lane A2*, centrifuged DNA after a further ethanol precipitation and subsequent collection by glass rod; *lane B1*, wound DNA; *lane B2*, wound DNA to which its supernatant material has been added back. A total of 10 ng of template was used in each reaction; AMS19, AMS20 and ET7 indicate the 10mers used as primers. (Reproduced from Micheli et al. 1994 by permission of Oxford University Press)

use only those that give reproducible patterns with the templates of interest.

3
Materials

Equipment
- laminar flow hood (e.g., Biological Safety Cabinet, Forma Scientific)
- adjustable micropipettes (e.g., Gilson P-2, P-10, P-100)
- oven thermal cycler (e.g., Gene Machine Junior, Scientific Plastics) or block-based thermal cycler (e.g., PTC100, MJ Research)
- gel electrophoresis apparatus (e.g., Minnie Submarine Agarose Gel Unit HE33, Hoefer Scientific Instruments)
- power supply (e.g., GPS 200/400, Pharmacia LKB)
- UV transilluminator

– photographic apparatus (e.g., MP-4+ Camera System or DS-34 Camera System, Polaroid) with an orange UV filter

– sterile gloves
– sterile micropipette tips
– sterile tubes (e.g., 0.5 ml microcentrifuge tubes, Eppendorf)
– mineral oil (Sigma) (if using a thermal cycler without a heated lid)
– films (e.g., Type 57 or Type 667, Polaroid)

– deionized distilled water sterilized by autoclaving
– TE: 10 mM Tris-HCl (pH 8.0), 1 mM EDTA
– 25 mM $MgCl_2$ (Promega, supplied with enzyme)
– 10x magnesium-free *Taq* polymerase buffer (Promega, supplied with enzyme)
– dNTPs (Pharmacia) stock solution: 10 mM of each dNTP in TE; dilute in sterile water when preparing working solution (1 mM)
– 10-base primers (Genosys) stock solution: 200 μM in TE; dilute in sterile water when preparing working solution (2 μM)
– genomic DNA: 5 ng/μl and 25 ng/μl in sterile water
– *Taq* DNA polymerase (Promega)

Note: Store all PCR reagents at –20°C except template DNA to be stored at 4°C.

– 5x TBE: 0.45 M Tris-borate, 10 mM EDTA (pH 8.0); (for 1 l: 54 g Tris base, 27.5 g boric acid, 20 ml 0.5 M EDTA)
– agarose (e.g., Bio-Rad)
– 10x loading buffer: 70 % (w/v) glycerol, 0.1 M EDTA, 0.5 % (w/v) bromophenol blue; store at 4°C (for longer storage, keep at –20°C)
– DNA size marker: ΦX174 DNA *Hae*III digest
– ethidium bromide stock solution: 10 mg/ml; store in a dark bottle

Caution: Ethidium bromide is a powerful mutagen: wear gloves when handling this dye; also wear a mask when weighing it.

4
Experimental Procedure

Note: In order to minimize risk of contaminations (e.g., amplified DNA carry-over) the following measures should be taken when preparing RAPD reactions:
– prepare reactions in a laminar flow hood
– dedicate a set of pipettes to the preparation hood
– wear sterile gloves during the reaction set-up

Each experiment should be designed so as to amplify two concentrations of each involved template and to include a template-free control.

1. Prepare a master mix of those reagents common to all the programmed reactions. Prepare a quantity sufficient for the total number of samples and controls to be run, plus one extra aliquot to ensure adequate volume for pipetting. In a single tube combine the following amounts per sample: 2 µl of 25 mM $MgCl_2$, 2.5 µl of 10x *Taq* polymerase buffer and 2.5 µl of 1 mM dNTPs. Add distilled water to a final volume of 21.5 µl per sample (if using a single primer) or 19 µl per sample (if using pairs of primers) and mix well.

   Note: Making a reaction mix greatly reduces the chances of pipetting errors for individual samples.

2. Dispense 21.5 µl (or 19 µl) of the master mix to each reaction tube.

3. Add 2.5 µl of 2 µM primer(s) and 1 µl of template DNA (5 or 25 ng, respectively).

   Note: When using a single template DNA with several different primers or a single primer with several different templates, the single template DNA or primer, respectively, can be included in the master mix.

4. Gently mix the reaction mixtures by tipping the tubes, briefly spin them to remove air bubbles, then overlay the mixtures with one drop of mineral oil (if required).

5. Place the tubes in the thermocycler and perform an initial denaturation step at 96°C for 5 min

6. Chill the tubes on ice, then, making sure to place the pipette tip under the oil overlay, add 1 U of *Taq* DNA polymerase.

7. Place the tubes in the thermocycler programmed for 45 cycles of denaturation at 94°C for 1 min, annealing at 36°C for 1 min and extension at 72°C for 2 min.

   Note: An optimization of the reaction times may be required depending on the brand of the thermocycler.

8. Store samples at 4°C.

Gel electrophoresis analysis

9. Prepare a 1.5 % agarose gel in 1x TBE buffer. It is advisable to use a comb with teeth as thin as possible: the thinner the teeth, the sharper the bands will appear.

10. Add 1 μl of 10x loading buffer to a 9 μl aliquot of each sample. Be careful in removing the aliquot from the reaction tube: put the pipette tip on the bottom of the tube in order to not draw up the oil along with the sample.

11. Load the samples and 10 μl of the size marker (10 ng/μl in 1x loading buffer).

12. Run the gel in 1x TBE buffer at a voltage not higher than 80 V (better resolution of the RAPD patterns is obtained at lower voltages) until the dye front has covered a distance of at least 6 cm from the wells.

13. Stain the gel in ethidium bromide 0.5 μg/ml for 30 min and then destain in deionized water for 30 min with several changes.

14. Photograph the gel on UV transilluminator with a UV radiation wavelength not shorter than 254 nm if you plan to recover the bands from the gel: shorter wavelengths are more harmful to DNA molecules.

5
Results and Comments

RAPD assay, according to the described protocol, generates reproducible fingerprints provided that optimal concentrations of good quality DNAs are used. Our results show that mammalian DNA concentrations ranging from 5 to 25 ng per 25 μl reaction ensure good reliability for amplifications both with single and coupled primers.

Pairwise combinations of primers generate patterns different from those obtained with each single primer and increase the chance of detecting polymorphic bands. Figure 2 shows the amplification profiles of genomic DNAs from different rat strains produced by pairwise combinations of arbitrary primers: two of the three possible combinations of three primers allowed us to detect three polymorphic bands (indicated by arrows), while amplifications with each single primer generated identical patterns (not shown).

Usually the size of the RAPD products ranges from a few hundreds to about 2000 bp. As the examples in Figs. 1 and 2 show, a limited number of visible bands (generally not more than ten) are obtained from a typical RAPD reaction. A much larger number of bands can be visualised by resolving RAPD products according to the protocols described in Chaps. XIII–XV.

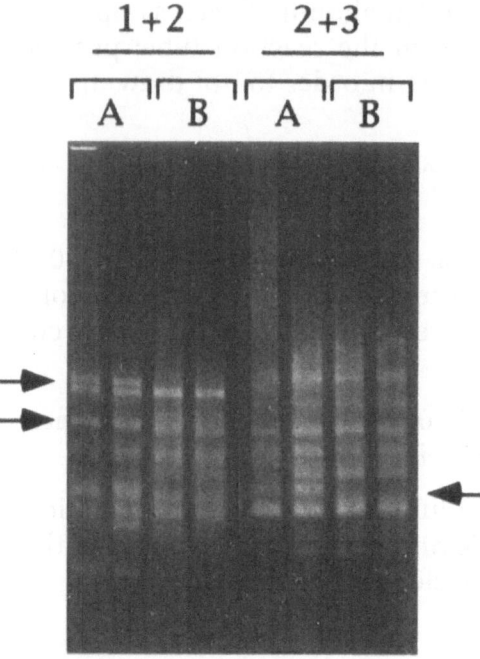

Fig. 2. RAPD patterns generated from two different rat strains by pairwise combinations of primers. DNAs have been prepared, respectively, from liver of Wistar rat (*A*) and Fisher rat (*B*). Two concentrations (25 ng and 5 ng per 25 µl reaction) were used for each template. Oligomers used to prime amplifications are indicated by *numbers*. Polymorphic bands are indicated by *arrows*. (Reproduced from Micheli et al. 1993, by permission of Eaton Publishing Co., 154 E. Central St., Natick, MA 01760–3644, USA)

6
Troubleshooting

- Unreproducible patterns: poor quality or insufficient amount of template DNA are most likely involved. Check DNA quality by performing RAPD reactions on serial dilutions of the template over a sufficiently wide range: if identical fingerprints are not obtained within a significant concentration range, discard the DNA and perform a more accurate DNA extraction. In case of good quality template this test will also allow one to ascertain whether an insufficient amount of DNA has been used (i.e., below the concentration range giving reliable fingerprints in the test). In some cases an optimization of the reaction conditions may be required depending on the brand of the thermocycler or the particular primer used.

- Smearing: an excessive amount of template DNA or primer or *Taq* DNA/polymerase has most likely been used. Perform test reactions with a reduced amount of each one of these components at a time.

- Low intensity of the bands: insufficient amount of primer or dNTPs. Try increasing the amount of each one of these components at a time.

Acknowledgements. The work of the authors reported in the two papers cited in this chapter was supported by Progetto Finalizzato "Applicazioni Cliniche della Ricerca Oncologica" (Consiglio Nazionale delle Ricerche).

References

Caetano-Anollés G, Bassam BJ, Gresshoff PM (1991) DNA amplification fingerprinting using very short arbitrary oligonucleotide primers. Bio/Technology 9:553–557

Ellsworth DL, Rittenhouse KD, Honeycutt EL (1993) Artifactual variation in randomly amplified polymorphic DNA bending patterns. BioTechniques 14:214–217

McClelland M, Welsh J (1994) DNA fingerprinting by arbitrarily primed PCR. PCR Methods Applic 4:s59-s65

Meunier JR, Grimont PAD (1993) Factors affecting reproducibility of amplified polymorphic DNA fingerprint. Res Microbiol 144:373–379

Micheli MR, Bova R, Calissano P, D'Ambrosio E (1993) Randomly amplified polymorphic DNA fingerprinting using combinations of oligonucleotide primers. BioTechniques 15:388–390

Micheli MR, Bova R, Pascale E, D'Ambrosio E (1994) Reproducible DNA fingerprinting with the random amplified polymorphic DNA (RAPD) method. Nucleic Acids Res 22:1921–1922

Muralidharan K, Wakeland EK (1993) Concentration of primer and template qualitatively affects products in random-amplified polymorphic DNA PCR. BioTechniques 14:362–364

Penner GA, Bush A, Wise R, Kim W, Domier L, Kasha K, Laroche A, Scoles G, Molnar SJ, Fedak G (1993) Reproducibility of random amplified polymorphic DNA (RAPD) analysis among laboratories. PCR Methods Applic 2:341–345

Welsh J, McClelland M (1990) Fingerprinting genomes using PCR with arbitrary primers. Nucleic Acids Res 18:7213–7218

Welsh J, McClelland M (1991) Genomic fingerprinting using arbitrarily primed PCR and a matrix of pairwise combinations of primers. Nucleic Acids Res 19:5275–5279

Williams JGK, Kubelik AR, Livak KJ, Rafalski JA, Tingey SV (1990) DNA polymorphisms amplified by arbitrary primers are useful as genetic markers. Nucleic Acids Res 18:6531–6535

Williams JGK, Hanafey MK, Rafalski JA, Tingey SV (1993) Genetic analysis using random amplified polymorphic DNA markers. Methods Enzymol 218:704–740

Chapter IX

DNA Amplification Fingerprinting

G. Caetano-Anollés

1
Introduction

The generation of characteristic signatures from virtually any nucleic acid, even of anonymous nature, has been made possible by the invention of nucleic acid scanning (Livak et al. 1992; Bassam et al. 1995). These signatures are composed of arbitrary collections of amplification products that result from the targeting of a multiplicity of anonymous sites (amplicons) in template DNA or RNA molecules. The targeted sites are amplified with one or more oligodeoxynucleotides to produce arbitrary but entirely characteristic "fingerprint" patterns. Nucleic acid scanning allows detection of polymorphic DNA or RNA without prior knowledge of sequence or cloned and characterized probes.

Three techniques were originally described: randomly amplified polymorphic DNA (RAPD) (Williams et al. 1990), arbitrarily primed PCR (AP-PCR) (Welsh and McClelland 1990), and DNA amplification fingerprinting (DAF) (Caetano-Anollés et al. 1991). These techniques produce markedly different amplification profiles, varying from quite simple (RAPD) to highly complex (DAF). Several variations of the original strategies provide higher multiplex ratios and detection of polymorphic DNA (e.g., Caetano-Anollés et al. 1993; Caetano-Anollés and Gresshoff 1994; Caetano-Anollés and Gresshoff 1996).

DAF uses at least one primer of at least 5 nucleotides (nt) in length to produce characteristic and highly informative DNA patterns (Caetano-Anollés et al. 1991). These patterns are adequately resolved by polyacrylamide gel electrophoresis and silver staining, but can be produced using agarose gel electrophoresis, denaturing gradient gel electrophoresis (DGGE), temperature sweep gel electrophoresis (TSGE) or by automated analysis using DNA sequencers or capillary electrophoresis (CE). DAF can be distinguished from other genome scanning techniques by the high primer to template ratios, excellent reproducibility and high multiplex ratios.

The DAF technique, in its different forms, can be used to finger-print a wide variety of genomes including those of high (plants and animals) and low (fungi, bacteria, mitochondrial and plastid DNA) complexity, as well as subgenomic fragments like PCR amplified products, cloned DNA and cDNA populations. Examples of these applications can be found in Chap. XII. A DAF protocol usually invol-ves two major steps: DNA amplification, and the separation and visualization of amplification products. The sequence of typical experimental manipulations is shown in Fig. 1. Here, I will describe the DNA amplification step in detail, highlighting the role of primer design and appropriate optimization of experimental variables. The separation and visualization step is described in Chap. XIII.

2
Principles

The amplification of nucleic acids with arbitrary primers is mainly driven by the interaction between primer, template annealing sites and enzyme, and determined by complex kinetic and thermody-namic processes. The outcome is the generation of a population of polynucleotide products usually representing those genomic regions (amplicons) that have been predominantly amplified. Polymorphisms result from changes in DNA sequence, initially within primer-defined sites in the genome. However, they can also arise from the deletion, insertion or inversion of a priming site or segments between priming sites and from conformational changes in DNA that would alter the efficiency of amplification or priming.

A typical DAF reaction involves: (1) the use of a single primer usu-ally composed of 8–12 nt, a length that approaches the minimum configuration for DNA amplification (Caetano-Anollés et al. 1992; Vincent et al. 1994); (2) amplification conditions that discriminate bonafide amplicons from those of artifactual origin; and (3) a non-stringent reaction environment that ensures the reproducible target-ing of multiple sites (Bassam et al. 1992). The annealing of a single arbitrary primer to short and complementary inverted repeats that are in near proximity but scattered throughout the template and the successful strand extension of the annealed oligonucleotides are the basis of the amplification reaction. Targeting of template sites occurs under a nonstringent reaction environment that is both adequate for annealing of short primers and specific enough to provide dis-crimination of legitimate and illegitimate amplicons. How can the

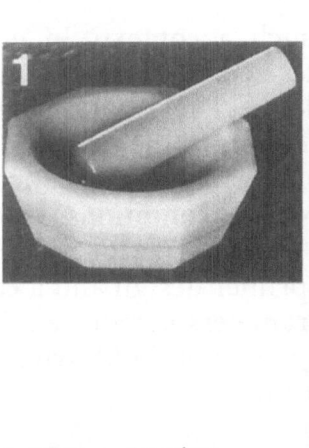 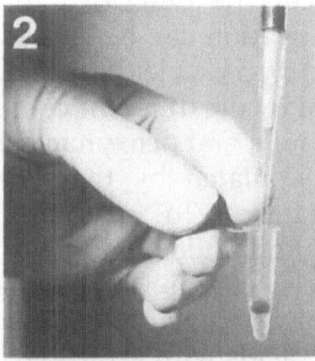

Buffer
Nucleotides
Magnesium
Primer
Polymerase
Template

1. DNA extraction
2. Reaction assemblage
3. Thermal cycling
4. Electrophoresis
5. Sample storage
6. Silver staining and
   gel preservation
7. Image processing,
   analysis and storage

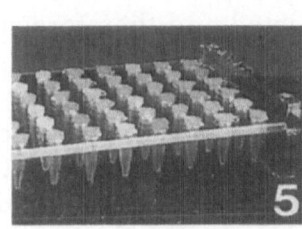

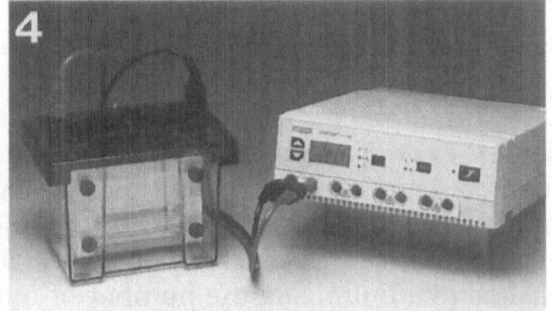

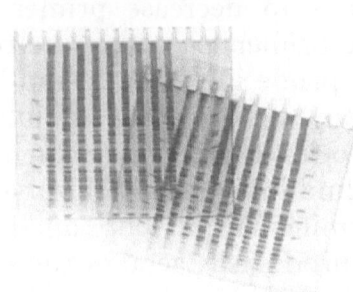 

Fig. 1. Steps in DNA amplification fingerprinting

amplification process be described in such a context? In general terms, DNA amplification can be characterized by three parameters: **specificity, efficiency** (i.e., yield) and **fidelity**. These parameters are strongly influenced by the different components of the reaction (such as primer, magnesium, and deoxynucleoside triphosphate concentrations) and are modulated by thermocycling conditions (such as annealing temperature). To guarantee optimal performance and reproducibility, a good understanding of primer design and a careful optimization exercise of amplification parameters is required. Please note that assumptions made for PCR may not hold true when amplifying nucleic acids with arbitrary primers.

3
Experimental Design

Primer Design

Fingerprint complexity and detection of DNA polymorphisms are inherently dependent on primer length, primer sequence and the number of primers in the reaction.

Primer length   Specificity in the PCR is related to the ability to amplify only the targeted site, the predicted product being either present or absent. In DAF, a multiplicity of arbitrary sites is targeted and therefore specificity is expressed as the ability to produce a "consensus" fingerprint characteristic of such a multiplex reaction. In the search for this "most parsimonious" fingerprint, the nucleic acid scanning reaction must be defined as much as possible. This can be generally accomplished by minimizing the number of interactions established within and between the different primer and template species during amplification. A simple way to do this is to decrease primer length (Caetano-Anollés et al. 1992). While primer length can be reduced down to 5 nt, these oligonucleotides prime inefficiently and have to be used at high concentrations, sometimes leading to inhibitory effects from concentrated primer stocks. Furthermore, their use is compromised by the existence of palindromic termini in first-round amplification products capable of forming hairpin loops and interfering with the amplification of certain products (Caetano-Anollés et al. 1992). In turn, increasing primer length favors mismatching during primer annealing allowing generation of amplification products in those cases in which none are to be expected on theoretical grounds (Caetano-Anollés 1994). The existence of "multiple mismatch annealing" events (see Venugopal et al. 1993) should therefore be decreased to a minimum. Our studies indicate that while amplification requires a primer of at least 5 nt and annealing sites with perfect homology to the first 5 or 6 nt from the 3' terminus (Caetano-Anollés et al. 1992; Caetano-Anollés 1994; Caetano-Anollés and Gresshoff 1994), a length of 8 nt provides a good compromise between efficiency and specificity.

For PCR, ideal primers should have 40%–60% G+C content and a 3' terminal clamp of 2–3 nt, should be free of palindromes, repetitive motifs, excessive degeneracy and long stretches of purines or pyrimidines, and primer pairs should have similar size (18–25 nt), melting temperatures (T_m) and nucleotide ratios (Roux 1995). Furthermore, annealing temperatures used in the PCR reaction should straddle 2°–10°C below the calculated T_m values for primer-template sequences. None of these requirements are necessary or fulfilled when using arbitrary primers. To illustrate this point, Table 1 shows a series of typical DAF oligonucleotides known as successful DAF primers. Within this arbitrary primer set, fractional G+C contents ranged from 62% to 100%, T_m values were in the range of –14°–31°C, and

Primer sequence

Table 1. Properties of a selected group of arbitrary primers[a]

Primer[b]	T_m	T_d	ΔG	% GC	M_r	E	
	(°C)	(°C)	(Kcal/mol)			nmol/A_{260}	µg/A_{260}
CCTGTGAG	–14.2	–7.5	–13.0	62.5	2506	13.19	32.6
GACGTAGG	– 9.4	–3.5	–13.5	62.5	2555	11.83	30.2
GTATCGCC	0.1	5.6	–15.2	62.5	2466	13.61	33.5
GTAACGCC	1.6	6.9	–15.3	62.5	2475	13.09	32.4
CGAGGTGG	3.5	9.8	–16.0	75	2571	12.50	32.1
GCAGGTGC	3.9	10.1	–16.0	75	2531	13.16	33.3
GCTGGTCG	4.5	10.7	–16.2	75	2522	13.81	34.8
GAAACGCC	7.2	12.6	–16.5	62.5	2448	12.67	31.5
GGACCCGC	17.2	22.8	–18.9	87.5	2476	13.85	34.3
GCCCGCCC	30.7	35.6	–22.2	100	2412	15.85	38.2

[a] Parameters were calculated using the program Oligo (v. 4.0; National Biosciences, Plymouth, MN). T_m, melting temperature calculated using nearest-neighbor thermodynamic values (1 M salt, 0.6 pM primer); T_d, dimer dissociation temperature; ΔG, free energy (25°C); %GC, fractional GC content; M_r, molecular weight; E, extinction coefficient.
[b] Primers were from a group selected for distribution within the framework of a joint FAO/IAEA Division Coordinated Program. These primers generate DAF profiles from a wide variety of templates.

some primers lacked G+C clamps or contained repetitive sequences. However, no obvious variation in amplification parameters were detected in the analysis of bacterial, fungal and plant DNA even at annealing temperatures of 50 °C. In particular, several studies also demonstrated insensitivity of DAF to primer G+C content (e.g., Caetano-Anollés 1994; Prabhu and Gresshoff 1994). Overall, results suggest that primer annealing is governed by the kinetic component of the reaction rather than by thermodynamic parameters.

While palindromic sequences should always be avoided, it should be noted that amplification failure of some primers or production of few amplification products can be related to the many known compositional inhomogeneities characteristic of DNA sequence. Using optimized conditions, DAF renders scorable patterns in more than 90 % of primers studied.

Multiplex DAF   Arbitrary primers can be used in pairwise combinations in what has been termed multiplex DAF. This strategy (Caetano-Anollés et al. 1991) has reportedly increased detection of polymorphic DNA (Callahan et al. 1993; Micheli et al. 1993). It also permits the combinatorial use of a limited set of primers. For example, a set of ten oligonucleotides can be used in 90 different pairwise combinations (excluding those in which they are used alone). However, a small fraction of amplification products (about 20 %) is actually generated by each contributing primer. These fingerprint overlaps only partially compromise the potential of this multiplex DAF approach.

3.2
Optimization of DAF Reactions

Each investigator should set up his or her own amplification protocol by choosing appropriate reaction components, concentrations and thermocycling conditions. To do so one must determine the widest range of values for a particular reaction component or thermocycling condition within which amplification parameters exhibit little or no variation. This range of values defines a "reproducibility window" (Bassam and Bentley 1994) and provides a measure of central tendency with which to define an experimental concentration, temperature or cycle number and avoid borderline conditions.

The optimization of the amplification reaction is a laborious process as a large group of interacting factors have profound effects on fingerprint, product number and efficiency of amplification. It relies on the sequential investigation of each variable and the design of large experiments. It should be noted that in reality optimum conditions are seldom identified.

Table 2. DAF reagents[a]

| Reagents | DNA amplification: | | | |
| | Plants and animals | | Fungi and bacteria | |
	Optimal range	Recommended	Optimal range	Recommended
Primer	2–10 μM	3–6 μM	3–9 μM	3–6 μM
Template	0.01–2 ng/μ	0.1 ng/μl	0.1–10 ng/μl	1 ng/μl
MgCl$_2$	1–8 mM	1.5 mM	4–8 mM	6 mM
dNTPs	50–300 μM	200 μM	50–300 μM	200 μM
Enzyme[b]	0.2–0.8 U/μl	0.3 U/μl	0.2–2 U/μl	0.3 U/μl

[a] Using buffer containing 10 mM Tris-HCL and 10 mM KCl (pH 8.3).
[b] AmpliTaq Stoffel DNA polymerase.

Table 2 shows recommended conditions for amplification of genomes of low and high complexity. These conditions should serve as
a start for a DAF optimization exercise.

Optimization can be achieved in different ways. The influence of different reaction components on the DAF reaction was determined for
bacterial (Bassam et al. 1992) and turfgrass DNA (Weaver et al. 1995)
by using an iterative process of analysis. This optimization strategy is
based on a simple matrix analysis in which several values for those
experimental variables determined a priori to be most important are
tested in combination with each of the other variables. This approach
simplifies the otherwise overwhelming full matrix analysis. Similarly,
Wolff et al. (1993) used a fractional factorial design to study the significance of several reaction components in RAPD analysis of *Chrysanthemum*. Finally, the Taguchi method (Taguchi 1986) has also been
applied to the study of interactions between specific reaction components in PCR and RAPD (see Chap. XI).

Reaction Components

The most important variable in the amplification reaction is the ratio
between concentrations of primer and template. Operationally, both
primer concentration and length define the different nucleic acid
scanning techniques (RAPD, AP-PCR and DAF). DAF uses over ten
times more primer than RAPD (at least 3 μM), and uses high primer-
template mass ratios in the range of 5–50000.

Primer-template ratio

DAF can reproducibly amplify very low DNA template levels and tole-
rates template concentrations that span over a thousand-fold range

Template

(Table 2). DNA concentrations of 0.1 ng/μl produce consistent DAF fingerprints from most plant and animal genomes. However, too little template causes amplicon stoichiometric misrepresentation and lack of reproducibility. This is particularly important in the analysis of low complexity genomes (viruses, bacteria and most fungi) for which a template concentration of at least 1 ng/μl should be used (Bassam et al. 1992; Caetano-Anollés and Gresshoff 1994).

A reasonable effort to use relatively good quality DNA should also be invested. Avoid DNA isolation methods that produce severely degraded DNA or do not eliminate contaminants that can inhibit the activity of the DNA polymerase. It should be noted that for certain genomes purity and integrity of DNA have minimal effect on the amplification reaction. Therefore, invest some time in designing a suitable DNA extraction protocol that provides high throughput and consistent amplification of the template material. Recently, some rapid DNA extraction protocols have been described (Williams and Ronald 1994; Guidet 1994).

Enzyme   The activity of thermostable DNA polymerases is highly variable (Bej and Mahbubani 1994). This variability is particularly evident in nucleic acid scanning applications in which even subtle differences in specificities manifest as changes in efficiency, multiplex ratio and fingerprint composition (Bassam et al. 1992; Schierwater and Ender 1993; Aldrich and Cullis 1993; G. Caetano-Anollés and B.J. Bassam, unpublished). Different eubacterial DNA polymerases exhibit widely different optima and result in variant DNA patterns. In general, truncated DNA polymerases such as *Thermus aquaticus* Stoffel fragment produce better defined fingerprints with stronger products of low molecular weight (≤500 bp), are more thermostable, and have a broader tolerance for wide ranging magnesium concentrations. Overall, these thermostable equivalents of Klenow fragment DNA polymerase are more tolerant of experimental variables (Bassam et al. 1992) and are here recommended. Choosing a polymerase is usually accompanied by the selection of appropriate buffer components. Recommended buffers are generally tailored for PCR and should only serve as the starting point for an optimization exercise to establish the influence of ionic components and concentrations.

Ionic components   Ionic components are crucial determinants of amplification. Magnesium is one important example. Consistent fingerprints can be obtained with relatively low levels of magnesium (1.5–4 mM) for plant and animal DNA and with high levels (4–8 mM) for bacteria and fungi (Table 2). However, magnesium requirements are dependent on the counterion and other buffer components. Activity is also modulated by the concentrations of primer, template and deoxyribo-nucleoside triphosphates (Weaver et al. 1995). An excess of any of these components can inhibit the amplification reaction due to the sequestration of free magnesium cation. In turn, an excess of magnesium levels decreases amplification stringency and increases primer-template mismatching. Potassium, ammonium and detergents like Triton X-100 alter amplification efficiency and specificity with Tris or Tricine buffers (Caetano-Anollés et al. 1994). In contrast, pH had little effect. A study of the effect of these components defined reaction buffers that use an uniform $MgSO_4$ concentration (4 mM) to amplify templates of low and high complexity (Caetano-Anollés et al. 1994; see Sect. 5). These buffers were formulated for use with the Stoffel enzyme but had deleterious effects on the activity of other enzymes such as $Vent_R$ DNA polymerase (New England Biolabs, Beverly, MA). In general, our results argue against the possible formulation of an universal buffer for use with enzymes from different sources.

Thermal Cycling Parameters

A number of thermal cycling parameters should be considered, including annealing and template denaturation temperatures, cycle number, temperature effects on enzyme and nucleic acids, and times of annealing, denaturation and strand extension. Temperature affects the interaction between enzyme and nucleic acid species and the kinetics of the amplification reaction. For example, annealing temperature causes changes in yield, number and distribution of amplification products (Caetano-Anollés et al. 1992). When using octamer primers, annealing temperatures as high as 65 °C can be used. Figure 2 shows the effect of annealing temperature on the amplification of bacterial DNA with a pentamer primer. In this case, yield and number of amplification products was maximal at 55 °C.

Identical DAF profiles can be produced using different thermocyclers and many thermal cycling parameters have minimal effect in the DAF reaction once fixed within a range. Examples include denaturing temperature, cycle number, and times of annealing, denaturation and strand extension. However, it is important to minimize denaturation time so as to extend the life of the enzyme and to provide enough time of annealing (especially during the first cycles) as to maximize stochastic processes.

DAF amplification products are usually up to 2000 bp in size. When using truncated DNA polymerases their average length decreases to about less than 500 bp. We found that a two step cycling between annealing (30°–60 °C) and denaturation (90°–96 °C) temperatures

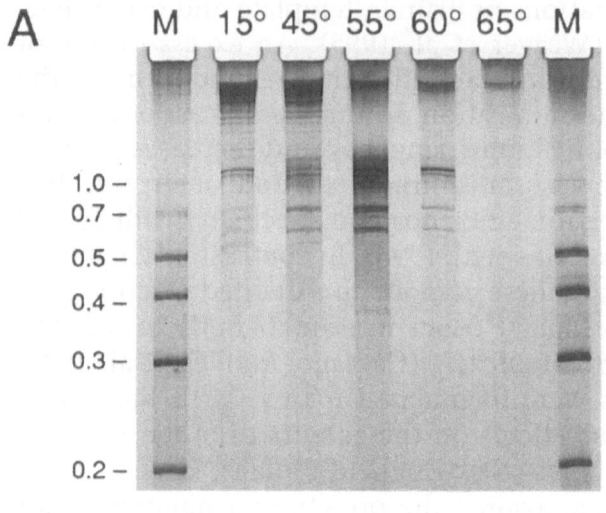

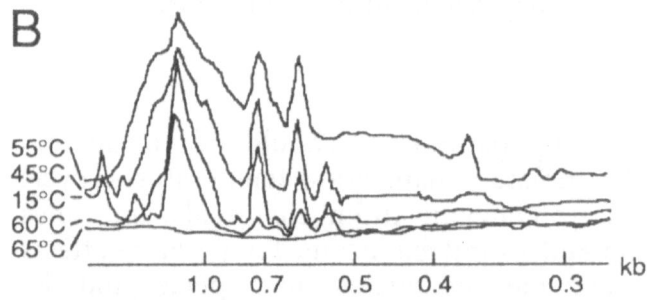

Fig. 2A,B. Effect of annealing temperature on DNA amplification fingerprinting of *Escherichia coli* strain Smith 92 DNA with the pentamer AGCTG. Note that short primers produce very few amplified products, fewer than predicted from the total number of annealing sites (Caetano-Anollés et al. 1992; Caetano-Anollés 1994), and the absence of amplification products of low molecular weight. These effects result because shorter primers have greater difficulty in displacing hairpin loop structures formed by the amplification products. The figure shows PAGE fingerprints (**A**) and scanned profiles analyzed with the Image program (**B**). Molecular weights are given in kb

provided optimal yield and reproducibility. The customary extension step was not needed if heating and cooling rates (10°–20 °C/min) provided adequate time for primer extension. If not, a short (30 s) extension step at 68°–72 °C can be performed. Other cycle regimens that keep enzyme exposure to high temperatures to a minimum can also be used with little if no variation in the DAF profile. For example, Bassam and Bentley (1994) use a "touch-down" alternative resembling that of Yu and Paul (1992): 94 °C for 2 min, followed by 35 cycles of 90 °C for 30 s and 50 °C for 1 min. Finally, cycle number should be kept to a minimum (usually 35 cycles) to avoid "plateau" amplification effects.

4
Materials

- block-based thermocycler (Ericomp, San Diego, CA) or oven thermocycler (Bios, New Haven, CT)

- high quality, deionized, double distilled water (>10 MΩ cm), preferably HPLC grade
- Stoffel (STF) buffer: 100 mM Tris-HCl, 100 mM KCl (pH 8.3). The following buffers stocks (all adjusted to pH 8.6) can also be used: TTNK10 (200 mM Tris-HCl, 1 % Triton X-100, 40 mM $(NH_4)_2SO_4$, 100 mM KCl); TTK10 (200 mM Tris-HCl, 1 % Triton X-100, 100 mM KCl); TTK30 (200 mM Tris-HCl, 1 % Triton X-100, 40 mM $(NH_4)_2SO_4$, 300 mM KCl); or TB (100 mM Tricine).
- deoxynucleoside triphosphate stock solution: 2 mM of each dNTP
- magnesium solution: 25 mM or 100 mM $MgCl_2$ solution
- oligodeoxinucleotide primer: 30 μM or 300 μM stock solution
- thermostable DNA polymerase enzyme. Truncated versions of *Thermus aquaticus* DNA polymerase, such as AmpliTaq Stoffel fragment (Perkin-Elmer/Cetus, Norwalk, CT), are preferred.
- template: 1–10 ng/μl stock solutions

Note: Store all reagents at –20 °C; however, keep the magnesium stock in use at 4 °C to ensure reagent uniformity.

5
Experimental Procedure

DNA amplification usually takes 3–8 h of experimentation time depending on the thermocycler apparatus used. This time can decrease with availability of faster thermocycling units. DNA separation in the typical laboratory is usually done by polyacrylamide gel electrophoresis and individually resolved fragments are generally identified by silver staining. These two procedures may take about 1.5–2.5 h. Alternatively, real-time analysis of amplification products is possible but limited to proper instrumentation, such as access to DNA sequencers or CE units.

1. Assemble a typical amplification mixture in 20 μl total volume by adding components in the following order: 10.2 μl water, 2 μl deoxynucleoside triphosphates (stock with 2 mM of each dNTP), 1.2 μl

magnesium (25 or 100 mM MgCl$_2$ stock), 2 μl STF reaction buffer, 0.6 μl DNA polymerase enzyme (10 U/μl AmpliTaq Stoffel fragment), 2 μl oligonucleotide primer (30–300 μM), and 2 μl template (usually diluted to 1–10 ng/μl). Shorter primers require higher oligonucleotide concentrations and low complexity genomes higher MgCl$_2$ levels. TTNK10, TTK10, TTK30 and TB buffers can also be used successfully (Caetano-Anollés et al. 1994). When possible, prepare a master mix with reagents that are common to avoid pipetting errors and aliquot the mix into 0.2 μl or 0.5 μl microcentrifuge tubes. The total volume of the reaction can be decreased to 10 μl by generally doubling concentrations of reagents in stocks, or alternatively by using more accurate pipettors. These precautions are necessary to ensure reproducibility.

2. If necessary, cover the amplification mix with 1–2 drops of heavy mineral oil. With certain thermal cyclers the mineral oil layer may not be required.

3. Amplify the DNA in the temperature cycler for the desired number of cycles (usually 35). Several protocols can be used depending on primer, template and thermal cycler selected. Generally, use two-step cycles of 20 s at 96 °C and 20 s at 30 °C or 50 °C in a block-based thermocycler, and three-step cycles of 30 s at 96 °C, 30 s at 30 °C or 50 °C and 30 s at 72 °C in an oven thermocycler.

4. Retrieve amplification mixtures by adding 100–200 μl of chloroform to each tube and pipetting out the aqueous droplet, or directly by using a long pipette tip or by rolling the amplification mixture over Parafilm.

5. Dilute the samples five- to tenfold prior to electrophoresis to avoid gel overloading. Only a small aliquot (3%–6%) of the amplification reaction is generally used for electrophoretic separation, leaving the rest of the sample for future reference or analysis. Samples can be stored at 4 °C and kept for years without visible degradation of amplification products. Evaporation can be prevented by maintaining the oil overlays.

Separation and visualization of amplification products

DNA amplification products are usually electrophoresed in 0.45 mm thick polyacrylamide slab gels (8×10 cm) backed on polyester film, and the DNA can be silver stained using the procedure of Bassam et al. (1991). Silver stained profiles can be scanned and the images stored in electronic files. A description of the separation and visualization of amplification products as well as suggestions for image processing and data analysis can be found in Chap. XIII.

- Maintain a nucleic acid-free clean environment in which the only **Tips** DNA that enters the reaction is the template added by the investigator (Dieffenbach and Dveksler 1993, Dragon 1993). Standard DAF can be considered a contamination-insensitive amplification technique as long as adequate levels of template DNA are used ($>0.1\,\text{ng}/\mu\text{l}$) and no isolation and cloning of amplification products are done in the area. While there is no need to have a separate sample preparation workplace for DNA extraction and dilution, preamplification and post-amplification areas should be separated either physically or by working in contained environments (Dieffenbach and Dveksler 1993). Sterilize bench areas routinely with UV light. UV light reduces contamination by several orders of magnitude but is less effective with DNA fragments shorter than 300 bp (Sarkar and Sommer 1990, 1991). Therefore expose bench and materials to prolonged UV doses.

- Sample preparation and preamplification reagents should be rendered free from nucleic acid contamination. Handle reagents and manage the laboratory bench much as described by Dragon (1993). Gloves should be worn during sample and reagent preparation, setting up of amplification reactions, and retrieving amplified samples for analysis. Use sterile double-distilled water in all operations. All reagents should be prepared in large volumes, aliquoted, and if possible maintained frozen at $-20\,°\text{C}$. This will guarantee reagent consistency and decrease contamination. If possible, use aerosol barrier pipette tips or positive-displacement pipettes for sample preparation reagents and try to maintain a set of dedicated pipettes for pre- and postamplification activities. Always use sterilized pipette tips and disposable sterile bottles or tubes.

- While standard DAF does not require the use of barrier or positive-displacement pipettes, minimize the production of aerosols by briefly centrifuging the tubes prior to opening. Do not pop-open tubes. Remember that airborn contamination in aerosols is the main cause of false positives at the bench. For maximum efficiency and minimum contamination prepare "ready-to-use" master mixes containing all reagents except for one or two missing components. Exert caution when preparing and using primer stock solutions, and include routine negative and positive amplification controls. As a final rule, establish an unidirectional traffic flow from the preamplification area to a contamination-contained postamplification workplace.

6
Troubleshooting

Troubleshooting can be a distressing experience, especially when the source of the problem is not readily apparent. Two scenarios are the most commonly observed in the laboratory: (1) amplification fails for no apparent reason, and (2) spurious bands appear in the amplification or control reactions. These problems usually arise because of poor optimization, contaminated reagents, or inadequate handling of reagents and assemblage of the reaction mix.

- Absence of amplification may indicate the presence of inhibitors in the DNA sample, such as ionic detergents, stains (e.g., bromphenol blue), phenol and heparin. Try diluting the template DNA and improving the DNA extraction procedure. It may also indicate the recalcitrant behavior of template DNA. Try increasing the denaturation temperature and the length of the denaturation step, at least during the first cycle.

- Pipetting can influence reproducibility by inadequate delivery of reagents. It is **strongly** recommended to assemble master mixes containing all but one crucial or variable reagent. When running reactions in multiple thermal cyclers, make sure that all units are calibrated in their performance. Also be aware of temperature inhomogeneities during thermocycling. Well-to-well variations can have profound effects on reproducibility.

- Intraexperiment variability has been commonly observed in RAPD analysis as the result of incorrectly prepared DNA containing ethanol-precipitable contaminants (Micheli et al. 1994). This potential problem can also affect DAF and AP-PCR techniques, but can be diagnosed by amplifying serially diluted DNA: reliable fingerprints should be obtained within a large reproducibility window in these experiments.

References

Aldrich J, Cullis CA (1993) RAPD analysis in flax. Optimization of yield and reproducibility using KlenTaq 1 DNA polymerase, Chelex 100, and gel purification of genomic DNA. Plant Mol Biol Rep 11:128–141
Bassam BJ, Bentley S (1994) DNA fingerprinting using arbitrary primer technology (APT): a tool or a torment. Australasian Biotechnol 4:232–236
Bassam BJ, Caetano-Anollés G, Gresshoff PM (1991) Fast and sensitive silver staining of DNA in polyacrylamide gels. Anal Biochem 196:81–84

Bassam BJ, Caetano-Anollés G, Gresshoff PM (1992) DNA amplification fingerprinting of bacteria. Appl Microbiol Biotech 38:70–76

Bassam BJ, Caetano-Anollés G, Gresshoff PM (1995) Method for profiling nucleic acids of unknown sequence using arbitrary oligonucleotide primers. US Patent 5,413,909

Bej AK, Mahbubani MH (1994) Thermostable DNA polymerases for *in vitro* DNA amplifications. In: PCR Technology: Current Innovations (Griffin HG and Griffin AM, eds), CRC Press, Boca Raton. pp. 219–237.

Caetano-Anollés G (1994) MAAP: a versatile and universal tool for genome analysis. Plant Mol Biol 25:1011–1026

Caetano-Anollés G, Gresshoff PM (1994) DNA amplification fingerprinting using arbitrary mini-hairpin oligonucleotide primers. Bio/Technology 12:619–623

Caetano-Anollés G, Gresshoff PM (1996) Generation of sequence signatures from DNA amplification fingerprints with mini-hairpin and microsatellite primers. BioTechniques 20:1044–1056

Caetano-Anollés G, Bassam BJ, Gresshoff PM (1991) DNA amplification fingerprinting using very short arbitrary oligonucleotide primers. Bio/Technology 9:553–557

Caetano-Anollés G, Bassam BJ, Gresshoff PM (1992) Primer-template interactions during DNA amplification fingerprinting with single arbitrary oligonucleotides. Mol Gen Genet 235:157–165

Caetano-Anollés G, Bassam BJ, Gresshoff PM (1993) Enhanced detection of polymorphic DNA by multiple arbitrary amplicon profiling of endonuclease digested DNA: identification of markers linked to the supernodulation locus in soybean. Mol Gen Genet 241:57–64

Caetano-Anollés G, Bassam BJ, Gresshoff PM (1994) Buffer components tailor DNA amplification with arbitrary primers. PCR Methods Applic 4:59–61

Callahan LM, Weaver KR, Caetano-Anollés G, Bassam BJ, and Gresshoff PM (1993) DNA fingerprinting of turfgrasses. Int Turfgrass Soc Res J 7:761–767

Dieffenbach CW, Dveksler GS (1993) Setting up a PCR laboratory. PCR Methods Applic 3:S2-S7

Dragon EA (1993) Handling reagents in the PCR laboratory. PCR Methods Applic 3:S8-S9

Guidet F (1994) A powerful new technique to quickly prepare hundreds of plant extracts for PCR and RAPD analysis. Nucleic Acids Res 22:1772–1773

Livak KJ, Rafalski JA, Tingey SV, Williams JG (1992) Process of detecting polymorphisms on the basis of nucleotide differences. US Patent 5,126,239

Micheli MR, Bova R, Calissano P, D'Ambrosio E (1993) Randomly amplified polymorphic DNA fingerprinting using combinations of oligonucleotide primers. BioTechniques 15:388–390

Micheli MR, Bova R, Pascale E, D'Ambrosio E (1994) Reproducible DNA fingerprinting with the random amplified polymorphic DNA (RAPD) method. Nucleic Acids Res 22:1921–1922

Prabhu R, Gresshoff PM (1994) Mendelian, maternal and paternal inheritance of polymorphic markers generated by short single arbitrary oligonucleotide primers in soybean. Plant Mol Biol 26:105–116

Roux KH (1995) Optimization and troubleshooting in PCR. PCR Methods Applic 4:S185-S194

Sarkar G, Sommer SS (1990) Sheding light on PCR contamination. Nature 343 27

Sarkar G, Sommer SS (1991) Parameters affecting susceptibility of PCR contamination to UV inactivation. BioTechniques 10:590–594

Schierwater B, Ender A (1993) Different thermostable DNA polymerases may amplify different RAPD products. Nucleic Acids Res 21:4647–4648

Taguchi G (1986) Introduction to quality engineering. In: Asian productivity organisation. UNIPUB, New York, New York

Vincent J, Gurling H, Melmer G (1994) Oligonucleotides as short as 7-mers can be used for PCR amplification. DNA & Cell Biol 13:75–82

Venugopal G, Mohapatra S, Salo D, Mohapatra A (1993) Multiple mismatch annealing: basis for random amplified polymorphic DNA fingerprinting. Biochem Biophys Res Commun 197:1382–1387

Weaver KR, Callahan LM, Caetano-Anollés G, Gresshoff PM (1995) DNA amplification fingerprinting and hybridization analysis of centipedegrass. Crop Sci 35:881–885

Welsh J, McClelland M (1990) Fingerprinting genomes using PCR with arbitrary primers. Nucleic Acids Res 18:7213–7218

Williams JGK, Kubelik AR, Livak KJ, Rafalski JA, Tingey SV (1990) DNA polymorphisms amplified by arbitrary primers are useful as genetic markers. Nucleic Acids Res 18:6531–6535

Williams CE, Ronald PC (1994) PCR template-DNA isolated quickly from monocot and dicot leaves without tissue homogenization. Nucleic Acids Res 22:1917–1918

Wolff K, Schien ED, Peters-van Rijn J (1993) Optimizing the generation of random amplified polymorphic DNAs in chrysanthemum. Theor Appl Genet 86:1033–1037

Yu K, Pauls KP (1992) Optimization of the PCR program for RAPD analysis. Nucleic Acids Res 20:2606

Fluorescent Detection and Analysis of RAPD Amplicons Using the ABI PRISM DNA Sequencers

J. Walsh Weller and A. Reddy

1
Introduction

DNA markers represent an important resource for creating genetic maps, distinguishing individuals and investigating genetic diversity, and as a practical application in streamlining breeding programs. Genotype identification through DNA analysis has undergone several changes in the past 8 years. The advent of the polymerase chain reaction (PCR) (Mullis and Faloona 1987; Saiki et al. 1988) coupled with random amplified polymorphic DNA technology (Williams et al. 1990; Welsh and McClelland 1990; Caetano-Anollés et al. 1991) allows DNA typing with minute amounts of target DNA. Since both breeding populations and genetic diversity may be large in plants, the development and routine use of molecular markers require that large numbers of samples be processed. Thus, automation becomes an important consideration.

2
Principles and Applications

PCR-based DNA typing with fluorescent primers and using automated analysis provides a sensitive assay with enhanced accuracy. The advantages of analyzing data using an automated DNA sequencer were first described by Skolnick and Wallace (1988). Ziegle et al. (1992) reported automated DNA sizing technology with four-color fluorescence-based techniques for genotyping microsatellite loci.

When investigating RAPD markers, it became clear that the more complex the fingerprint the more reproducible it tended to be (McClelland and Welsh 1994). A minimum of ten bands has therefore been recommended as the cut-off for keeping a primer for analysis of samples. Some primers will give 20–50 bands or more, at which point resolution becomes an issue, as does the difficulty of bookkeeping with so many markers. Thus not only is automation of analysis important for speed and statistically relevant numbers of data points, but linking a sample in a lane to the signal in that lane accurately can significantly decrease the errors associated with the transfer of data.

The recent availability of fluorescent random primers and dUTPs ([F]dUTPs; Perkin-Elmer/Applied Biosystems Division) has made this approach attractive for the analysis of RAPDs. Although RAPD markers are often scored on agarose gels, the comigrating DNA fragments can be difficult to resolve. Radiolabeling in conjunction with sequencing-grade polyacrylamide gels allowed excellent resolution of RAPD markers in sorghum (Pammi et al. 1994). However, handling radioactivity and the limited throughput achievable when only one reaction can be loaded per lane can limit fingerprinting using RAPDs. The advantage of a distinct dye for use as an in-lane size standard is considerable for accuracy and reproducibility of the results. Moreover, the ability to distinguish three other dyes, and thereby multiplex reaction products, in a single lane greatly enhances the throughput capabilities of the technology. The use of what are essentially sequencing gels gives a high resolution for the PCR products such that one can distinguish bands differing by only a few nucleotides with confidence. The analysis software allows lanes to be superimposed for quick, accurate pattern comparison. If one wishes to follow a particular subset of bands in a large population there is additional software (Genotyper, Perkin-Elmer/Applied Biosystems Division) that automates the scoring and provides a variety of output options. Because of the standardization provided by the in-lane sizing and the lane-overlaying capabilities of the software, it is possible to directly compare the results of many gels together, without the ambiguities that result from gel to gel variations. We describe here two fluorescence-based DNA fingerprinting techniques using either fluorescently labeled primers or incorporation of [F]dUTPs for internally labeling RAPD amplicons. The running conditions and analysis parameters for the ABI automated DNA sequencers are also described.

3
Materials

Equipment
  – Perkin-Elmer Lambda 2 spectrophotometer (Perkin-Elmer)
  – Perkin-Elmer LS-50B luminescence spectrometer (Perkin-Elmer)
  – PE 9600 or 480 thermocycler (catalogue #N801–0001 or N801–0100, Perkin-Elmer)

- ABI 373 or 377 DNA sequencer (catalogue #373–01 or 377–01, Perkin-Elmer/ABI)
- Thermolyne heating unit and block (catalogue #42–0010–01 and 42–0020–02, PGC Scientific)

Note: All catalogue numbers for equipment are for 120V instruments. Please contact the companies if other voltage requirements exist.

- GeneScan 2.0.1 analysis software (catalogue #604–205, Perkin-Elmer/ABD)
- Genotyper 1.0 software (catalogue #401–614, Perkin-Elmer/ABD)

Analysis programs

- GeneAmp reaction tubes (catalogue #N801–0533, Perkin-Elmer/ABD)
- GeneAmp caps for tubes (catalogue #N801–0534, Perkin-Elmer/ABD)
- deionizing resin 20–50 mesh (catalogue# AG 501-X8(D), BioRad)

Supplies

- dye-labeled primers (ABI) or [F]dUTPs (catalogue #401–894, Perkin-Elmer/ABD)
- standard RAPD primers (e.g., from Operon Tech, Alameda, CA)
- oligonucleotide buffer: 10 mM Tris-HCl (pH 8.0), 50 mM NaCl
- TE buffer: 10 mM Tris-HCl (pH 8.0), 1 mM EDTA
- AmpliTaq DNA polymerase (catalogue #N801–0055, Perkin-Elmer/ABD) or Stoffel fragment (catalogue #N808–0038, Perkin-Elmer/ABD)
- 10 mM dNTPs (GeneAmp dNTPs catalogue #N808–0007, Perkin-Elmer/ABD Inc.)
- 10x PCR buffer and $MgCl_2$ (catalogue#N808–0010, Perkin-Elmer/ABD)
- mineral oil
- 29:1 acrylamide stock (for ABI 373 DNA Sequencer): 29 g acrylamide (catalogue #161–0101, BioRad) and 1 g bis-acrylamide (catalogue #70102, International Biochemicals Inc.) dissolved in deionized water
- 50 % Long Ranger gel solution (for ABI 377 DNA Sequencer) (catalogue #4730–02, JT Baker)
- urea, molecular biology grade (catalogue #5505UX, Gibco BRL Life Technologies)
- TEMED (catalogue #5524UB, Gibco BRL Life Technologies)
- ammonium persulfate (catalogue #7727–54–0, Sigma)
- 10x TBE buffer (catalogue #15581–036, Gibco BRL Life Technologies)
- PicoGreen (Molecular Probes, catalogue #P-7581)

Reagents and solutions

- salmon sperm DNA (Gibco/BRL, catalogue #15632–011)
- molecular biology certified water (catalogue#W-4502. Sigma)
- formamide (catalogue #5515UA, Gibco BRL Life Technologies)
- GeneScan 2500-Rox (catalogue #401–100, Perkin-Elmer/ABD)

Note: This comes with a blue dextran/EDTA gel-loading solution that is used in conjuction with formamide.

- *Taq* terminator matrix standard (catalogue #401–071, Perkin-Elmer/ABD)
- dye primer matrix standard (catalogue #401–114, Perkin-Elmer/ABD)

4
Experimental Procedure

Oligonucleotide quantitation

Most oligonucleotides arrive in a dry state after HPLC or OPC purification.

1. Resuspend the oligonucleotide in 200 μl molecular biology grade water. Make a dilution (1:100 for 1–5 O.D.) using the oligonucleotide buffer.

2. Measure the absorbance at 260 nm in a spectrophotometer. The Perkin-Elmer Lambda 2 spectrophotometer has a program that allows entry of the nucleotide sequence and a nearest-neighbor-based calculation of the extinction coefficient and concentration. These must be modified for the presence of the fluor, as described in the next step.

3. Calculate the extinction coefficient of the primer. For oligonucleotides of length less than 20, a simple additive function works well for determining the extinction coefficient, if the spectrophotometer available does not provide this function. Multiply the number of each nucleotide by the extinction coefficient contribution given below, add them together, multiply by 0.9.

$\varepsilon(M)dG=12010$
$\varepsilon(M)dA=15200$
$\varepsilon(M)dC=7050$
$\varepsilon(M)dT=8400$
$\varepsilon(M)dD=26662$ (dD stands for fluoresceine- or rhodamine-based dyes)

4. Calculate the molar concentration of the primer using Bier's law. The molar concentration of the oligonucleotide: $A=\varepsilon cl$.

The DNA should be quantitated fluorescently with a dye that distinguishes between double and single-stranded nucleic acids. An example that has worked well in our hands follows:

DNA quantitation

1. Make a 1:400 dilution of PicoGreen in TE buffer.

2. Aliquot 100 μl per well of a 96-well microtiter dish for reading on an LS50B luminescence spectrometer from Perkin-Elmer, or an equivalent instrument.

3. Add 1–10 μl of sample or standard to the well.

4. Read on the LS-50B luminescence spectrometer using the WPR software. Use the following settings for detection and quantitation:
   - Excitation wavelength   480 nm
   - Excitation slit width   5 nm
     Emission wavelength   520 nm
   - Emission slit width   4 nm
   - Emission filter   515 nm

Use the generic WPR macro and template for generating a spreadsheet that displays the concentrations in μg/ml.

Note: A good standard is salmon sperm DNA(Gibco/BRL) which comes in a stock concentration of 10 mg/ml.

In this case, a preliminary screen has to be performed using nonlabeled primers (e.g., from Operon Tech, Alameda, California), mini-denaturing polyacrylamide gels (Idea Scientific, Minneapolis, MN, USA) and ethidium bromide staining for detection. The screening allows identification of primers yielding stable polymorphic amplicons. These informative primers will then be ordered with dye labels from the Custom Synthesis group at Perkin-Elmer/ABD. Three different NHS-ester dyes are available (called FAM, JOE and TAMRA) and the useful primers can be arbitrarily divided into three groups, allowing the products to be multiplexed for efficient gel use.

RAPD protocol for dye-labeled primers

Make a PCR master mix: For 50 reactions, the total volume will be 1.0 ml. Mix the following on ice:

1. 100 μl 10x PCR buffer (no MgCl$_2$)

2. 30 μl 100 mM MgCl$_2$ stock

3. 3 μl 100 μM dye-labeled primer

Note: While the product must be diluted for running on a gel, even five-fold dilution of the primer slows down the PCR reaction, because of concentration effects in the annealing kinetics, to the point where detectable product will not be formed using the indicated thermocycling conditions. It is possible to use much smaller amounts of primer if the PCR thermocycling conditions are adjusted, but we cannot provide detailed guidelines for carrying out this process.

4. 10 µl of each dNTP (as 10 mM stock)

5. 50 U *Taq* polymerase (e.g., 10 µl of Perkin-Elmer AmpliTaq at 5 U/µl) or 100 U of the Stoffel fragment of *Taq* polymerase

6. Add molecular biology grade water to a volume of 750 µl.

7. Aliquot 15 µl per reaction tube (thin-walled tubes are recommended)

8. Add 5 µl of template, from 0.4 ng/µl to 4 ng/µl.

   Note: The assayed templates are of high complexity. The amount of template will have to be adjusted upwards as the complexity of the genome decreases.

9. Layer with 20 µl mineral oil if the thermocycler does not have a heated lid.

10. Cycle as follows (do not hot-start):
    Initial denaturation: 94°C, 1 min
    Then, 25 cycles with a ramp time no faster than 1°C/s: 94°C, 30 s; 36°C, 1 min; 72°C, 2 min
    Final extension: 65°C for 10 min

Store at 4°C, but this is not recommended if amplicons are to be cloned.

RAPD protocol for dye-labeled nucleotides

In this case, the RAPD primers may be obtained from many sources, since the labeling is internal. This is an advantage in terms of time per assay, especially as the cost of internally labeling the amplicons is quite low. Three different fluorescently labeled nucleotides are available from Perkin-Elmer/ABD. The concentration (Con), excitation wavelength (Ex) and emission wavelength (Em) and colors using filter wheel A (on the 373) or virtual filter A (on the 377) are given below:

dUTP	Con	Ex(nm)	Em(nm)	color
[R110]dUTP	100 μM	502	530	blue
[R6G]dUTP	100 μM	528	555	green
[TAMRA]-dUTP	400 μM	552	580	yellow

Note: TAMRA is four times less fluorescent than the other two [F]dUTPs. The procedure described below for PCR conditions is as set forth in Pammi et al. (1994). For fluorescent labeling, one of the [F]dUTPs is added to the PCR amplification reaction without changing the concentration of dTTP.

1. Make a PCR master mix with each [F]dUTP, label three 0.5 ml tubes (as blue, green and yellow) and mix the following on ice:

Components	Volume(μl)	Final concentration
water	63.75	
10x PCR buffer	34.0	1x
25 mM MgCl$_2$	34.0	2.5 mM
1 mM dNTPs	136.0	100 μM
Primer	34.0	2 μM
Stoffel enzyme	2.55	0.38 U
[F]dUTP	1.7	0.05 μM
Total volume	306	

2. Label 96 0.2 ml thin-walled tubes as follows:
   - for [R110]dUTP reactions, label B1–B32
   - for [R6G]dUTP reactions, label G1–G32
   - for [TAMRA]dUTP reactions, label Y1–Y32

3. Add 1 μl of template at 10 ng/μl.

   Note: Amounts of template DNA to be added to the RAPD reactions depend on the genome complexity

4. Aliquot 9 μl of PCR master mix into each thin-walled tube and mix the contents.

5. Cycling parameters using the Perkin-Elmer PE9600 PCR machine are as follows:
   Initial denaturation: 94°C, 5 min
   35 cycles: 92°C, 1 min;48°C, 40 s;72°C, 1.15 min
   Final extension:72°C, 10 min

Detection and Analysis of RAPD Amplicons Using the ABI 373 DNA Sequencer

On the ABI 373 DNA sequencer a 24 cm gel with square wells (24 or 36) is recommended.

Gel casting polyacrylamide

For 50 ml of a 6% acrylamide gel solution:

1. Use 10 ml of the 29:1 acrylamide stock.

2. Add 22.5 g urea (this will be 7.5 M final concentration).

3. Add 5 ml 10x TBE buffer.

4. Add ddH$_2$O to 50 ml.

5. Stir to dissolve, then add 0.5 g AG-501X8 resin and stir an additional 5 min.

6. Filter through nitrocellulose and de-gas for 5 min.

7. Add 35 µl of TEMED.

8. Add 250 µl of freshly made 10% ammonium persulfate in water, swirling to mix.

9. Pour the gel using a square-tooth comb, allow to polymerize for at least 2 h before use.

Gel running parameters

When first running samples labeled with NHS-ester dyes or [F]dUTPS, it is necessary to set up the appropriate multicomponenting matrix. Standards can be purchased, and it is only necessary to use eight lanes (one dye in every other lane) of the first gel. Thereafter, the same matrix will be valid for quite some time. It is important not to use a Taq terminator matrix when interpreting NHS-ester-attached dyes or vice versa. The instrument manuals provide complete instructions and explanations.

Gel sample preparation

1. Samples are diluted, for dye-primer produced amplicons use a dilution of 1:100 using formamide/EDTA/blue dextran or for the [F]dUTP-labeled amplicons use a dilution of 1:5 with formamide/EDTA/blue dextran mixture (50 µl EDTA/blue dextran and 150 µl formamide).

2. Put 1 µl of the diluted sample in a tube. Note that three samples may be combined, as long as the amplicons were all labeled with different dyes (i.e., a FAM, a JOE and a TAMRA reaction can be coelectrophoresed).

3. Add GeneScan 2500-ROX molecular weight standards, 0.5 µl per tube.

4. The samples are heated to 95°C for 3 min and then held at 4°C prior to loading on the gel.

5. A sample volume of 2–5 µl can be loaded per well.

6. The samples are preelectrophoresed for 3–4 min, and the form-amide is rinsed from the wells.

7. The gel is run for 10 h at 2400 V (limiting parameter).

8. Analysis then proceeds using the GeneScan analysis 2.0.1 soft-ware.

Running conditions

Detection and analysis of RAPD amplicons using the ABI 377 DNA Sequencer

On the ABI 377 DNA sequencer a 36 cm gel with 34 square wells is recommended.

For 50 ml of a 5 % Long Ranger gel for the ABI 377:

1. Use 5 ml of the 50 % stock as it comes from the manufacturer.

2. Add 5 ml 10x TBE buffer.

3. Add 18 g urea (for 6 M final concentration).

4. Add deionized water to 50 ml.

5. Stir to dissolve, add 0.5 g AG-501X8 resin, stir 5 min.

6. Filter through nitrocellulose, de-gas for 5 min

7. Add 25 µl TEMED.

8. Add 250 µl freshly made 10 % ammonium persulfate, with swirl-ing to mix

9. Pour the gel using a square-tooth comb, and allow to set for at least 2 h before use.

Gel casting Long Ranger

Gel running parameters are as described for ABI 373.

Gel running parameters

1. Dye-primer labeled amplicons are diluted 1:200 using formam-ide/EDTA/blue dextran as above (conditions for running [F]dUTP-labeled samples have not been worked out).

2. Put 1 µl of the diluted sample in a tube. Again, three samples may be combined per tube, as long as the dyes on the primers are dis-tinct.

Gel sample preparation

3. Add GeneScan 2500-ROX molecular weight standards, 0.2 µl per lane to the tube.

4. The samples are heated to 95°C for 3 min and then held at 4°C prior to loading on the gel.

5. A sample volume of 1–3 µl can be loaded per well.

Running conditions

6. The samples are preelectrophoresed for 2–3 min, and the formamide is rinsed from the wells.

7. The gel is run for 5 h at 3000 V (limiting parameter).

8. Analysis then proceeds using the GeneScan analysis 2.0.1 software.

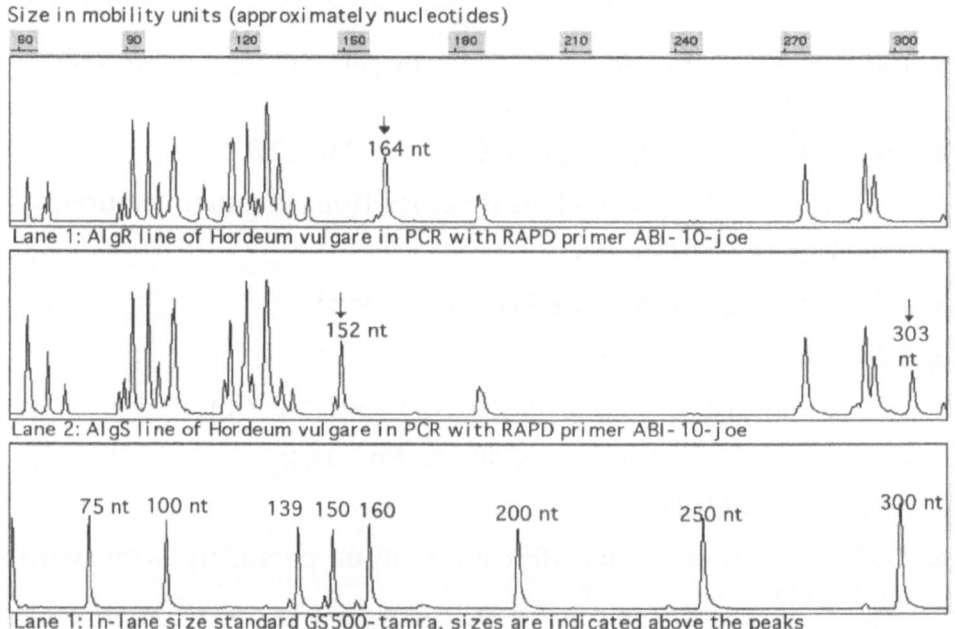

Fig. 1. Results from using dye-labeled primers. Shown are two electropherograms comparing the RAPD patterns and polymorphisms occurring between near-isogenic lines of barley. The in-lane size standard of the first sample is also shown. In a color figure the electropherograms can be superimposed using a different color for each, making it exceptionally easy to discriminate the polymorphisms, here demarked with *arrows*. Once the positions of these polymorphisms have been determined they may be set up as categories in the program Genotyper, which will automatically import the data from additional samples, scan them for these categories, and tabulate the results

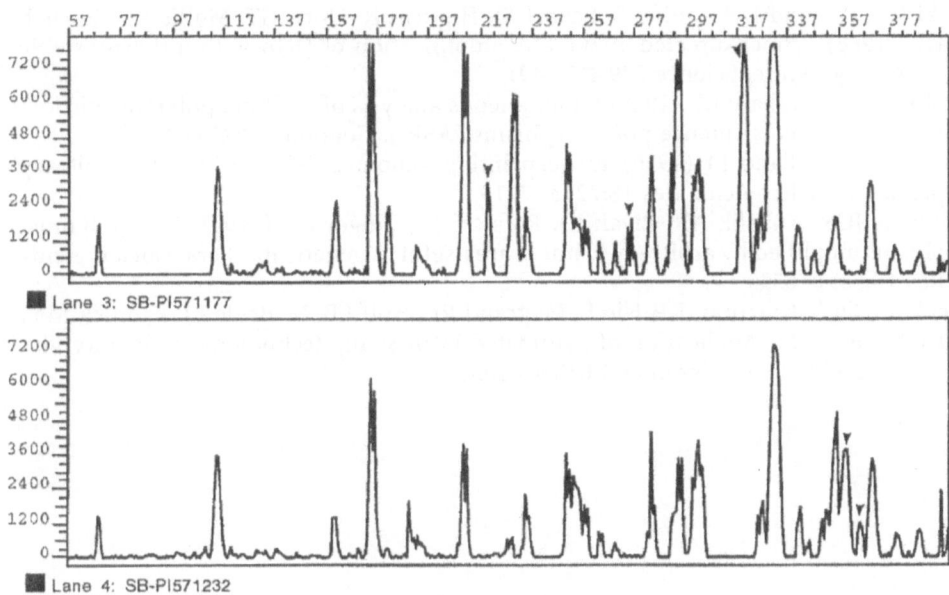

Fig. 2. Results from using dye-labeled nucleotides. Electropherograms comparing the RAPD products derived from two accessions of sorghum are shown. *Arrows* indicate polymorphic RAPDs between the two sorghum genotypes. Each panel depicts the fluorescence intensity (*Y-axis*) and the apparent molecular weight in nucleotides (*X-axis*)

5
Results

The analysis programs are easy to use and come with complete manuals that include extensive troubleshooting sections. The most commonly observed error has been that researchers do not sufficiently dilute the samples, with severe over-loading leading to smeary bands and poor resolution. In Figs. 1 and 2 are shown examples of the results obtained using either of the labeling methods.

References

Caetano-Anollés G, Bassam BJ, Gresshof PM (1991) DNA amplification fingerprinting using short arbitrary oligonucleotide primers. Bio/Technology 9:553–557

McClelland M, Welsh J (1994) DNA Fingerprinting by Arbitrarily Primed PCR. Manual Supplement. PCR Methods Applic 4:s59-s65

Mullis K.B., Faloona F.A. (1987) Specific synthesis of DNA in vitro via a polymerase-catalyzed chain reaction. Methods Enzymol 155:335–350

Pammi S, Schertz K, Xu G, Hart G, Mullet JE (1994) Random-amplified-polymorphic DNA markers in sorghum. Theor Appl Genet 89:80–88

Saiki RK, Gelfand DH, Stoffel S, Scharf SJ, Higuchi R, Horn GT, Mullis KB, Ehrlich HA (1988) Primer-directed enzymatic amplification of DNA with a thermostable DNA polymerase. Science 239:487–491

Skolnick MH, Wallace RB (1988) Simultaneous analysis of multiple polymorphic loci using amplified sequence polymorphisms (ASPs). Genomics 2:273–279

Welsh J, McClelland M (1990) Fingerprinting genomes using PCR with arbitrary primers. Nucleic Acids Res 18:7213–7218

Williams JGK, Kubelik AR, Livak KJ, Rafalski JA, Tingey SV (1990) DNA polymorphisms amplified by arbitrary primers are useful as genetic markers. Nucleic Acids Res 18:6531–6535

Ziegle JS, Su Y, Corcoran KP, Nie L, Mayrand PE, Hoff LB, McBride LJ, Kronick MN, Diehl SR (1992) Application of automated DNA sizing technology for genotyping microsatellite loci. Genomics 14:1026–1031

Chapter XI

Optimization of RAPD Fingerprinting

B. COBB

1
Introduction

RAPD fingerprinting has been extensively used to detect DNA sequence polymorphism in many plant, bacterial and fungal species (Williams et al. 1990; Kresovich et al. 1992; Cobb and Clarkson 1993). In addition, tightly linked RAPD markers have been used to detect breeding lines containing specific gene blocks facilitating marker-based selection and detection of cryptic genotypes (Crowhurst et al. 1991; Kresovich et al. 1992). Although the basic methodology is well documented (Welsh and McClelland 1990; Williams et al. 1990), optimization is often required when transferring protocols between thermocyclers and in generating RAPD fingerprints with modified size distribution. A sequential investigation of each reaction variable on the reaction yield generally results in prohibitively large experiments in order to include every possible component combination. Since the polymerase chain reaction (PCR) forms a complex series of interactions, an alternative approach is to investigate interaction effects (Cobb and Clarkson 1994). This chapter describes an optimization strategy for RAPD fingerprinting, based on modified methods from Taguchi and Wu (1980), which dramatically reduces the number of experiments required (Fig. 1).

2
Principles

Optimization of RAPD fingerprinting would normally require each variable to be tested independently. A trial investigating the effects and interactions of four reaction components, each at three concentration levels, would require an experiment with 81 (i.e., 3^4) separate reactions. However, using Taguchi methods only nine reactions are required to perform the same optimization. Here an estimate of the

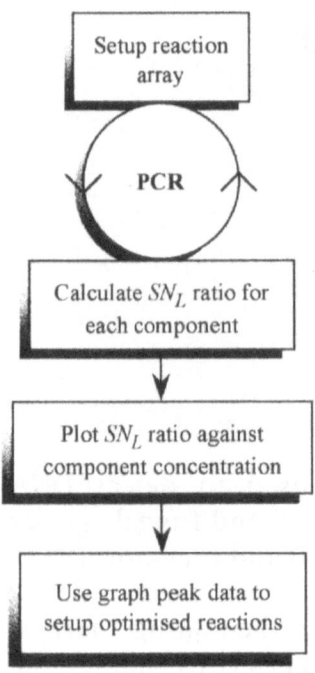

Fig. 1. Flow-sheet for the optimization of RAPD fingerprinting using orthogonal arrays

effect of individual components is achieved by looking at the effects component interactions have on the fingerprint.

These interactions are determined by arranging those components thought to effect the reaction into an orthogonal array (Table 1). With the RAPD PCR, each column represents the individual reaction components tested, and each row represents individual reaction vessels. Each component occurs at one of three predetermined levels (A, B and C), sufficiently separated so that their effects on the reaction can be determined. Some knowledge may be helpful in predicting "ballpark" values. The properties of an orthogonal array are such that, between each pair of columns, each combination of levels (A, B or C) occurs an equal number of times (Table 1).

The number of experiments required (E) is calculated from the equation $E=2k+1$, where k is the number of factors to be tested, providing that three concentrations are used for each of the reaction components tested. If the calculated number is not a multiple of three, then the required number is the next higher multiple. Hence, as the number of components to be tested increases, the reduction in the number of experiments necessary becomes more marked; e.g., to test nine factors would require 19683 experiments (3^9) to analyze fully, whereas using Taguchi methods this could be reduced to just 21 ($2 \times 9+1=19$, next integer divisible by three is 21).

Table 1. Orthogonal arrays for four variables testes at leves, A, B and C

Reaction	Component 1	2	3	4
1	A	A	A	A
2	A	B	B	B
3	A	C	C	C
4	B	A	B	C
5	B	B	C	A
6	B	C	A	B
7	C	A	C	B
8	C	B	A	C
9	C	C	B	A

Component 1: $MgCl_2$ (A, 1.5 mM; B, 2.5 mM; 3.5 mM).
Component 2: dNTP (A, 0.2 mM; B, 0.4 mM; C, 0.6 mM).
Component 3: primer (A, 3.0 pmoles; B, 6.0 pmoles; C, 9.0 pmoles).
Component 4: DNA (A, 50.0 ng; B, 100.0 ng; C, 150.0 ng).

The product yield for each reaction is used to estimate the effects of individual components on amplification. This is done using quadratic loss functions which Taguchi refers to as signal to noise ratios (Taguchi and Wu 1980; Taguchi 1986; Charteris 1992). These mathematically penalize small deviations from a theoretical target. Generally, the theoretical target of the RAPD fingerprinting is to increase the number of scorable products in the profile, although the size distribution may also be useful. The following quadratic loss function is used, which keeps the deviation between the number of amplification products as low as possible whilst keeping the mean number of amplification products as high as possible:

$$SN_L = -10 \log \left(\frac{1}{n} \sum_{i=1}^{n} \frac{1}{y_i^2} \right)$$

where SN_L is the signal to noise ratio, n is the number of levels and y is the yield. For each component the optimal conditions are those that give the largest SN_L. The reaction can be further refined by using polynomial regression from the SN_L values for each component to obtain curves whose maxima represent the reaction optima.

3
Materials

Equipment
- microcentrifuge
- thermocycler
- agarose gel electrophoresis apparatus
- UV transilluminator

Reagents and solutions
- deionized water
- 10x reaction buffer: 100 mM Tris-HCl (pH 8.4), 500 mM KCl, 1 mg/ml gelatin (Promega, Madison, WI)
- 20 mM $MgCl_2$
- 2 mM dNTP mix (dATP, dCTP, dGTP, dTTP; Promega, Madison, WI)
- 3 pmoles/ml oligonucleotide primer (OPA08 $^{5'}$gTgACgTAgg$^{3'}$; Operon Technologies, Alameda, CA)
- 50 ng/ml genomic DNA
- mineral oil (Sigma)
- 5 U/ml *Taq* polymerase (Promega, Madison, WI)
- 10x TBE: 89 mM Tris, 89 mM borate, 2 mM EDTA (pH 8.8)
- 10x loading buffer: 0.25 % bromphenol blue, 0.25 % xylene cynol FF, 0.25 % orange G, 30 % glycerol in water
- agarose
- ethidium bromide stock solution 10 mg/ml

4
Experimental Procedure

RAPD PCR can generally be optimized by looking at the interactions between primer, DNA, magnesium and deoxynucleotides. Investigation of these four variables requires a nine reaction array (Table 1).

Reaction matrix

1. Set up nine reaction tubes (1–9).

2. Add 2.5 ml of 10x reaction buffer to each of the nine tubes.

3. Add $MgCl_2$ to each reaction tube according to the concentrations A, B or C in Table 1.

4. Add dNTP mix to each reaction tube according to the concentrations A, B or C in Table 1.

5. Add primer to each reaction tube according to the concentrations A, B or C in Table 1.

6. Add genomic DNA to each reaction tube according to the concentrations A, B or C in Table 1.

7. Add deionized water to bring the reaction volume up to 23 µl.

8. Overlay each reaction with 30 µl of mineral oil.

9. Add 2 µl (0.5 U) of *Taq* polymerase to each reaction vessel.

10. Centrifuge tubes at full speed in a microcentrifuge for 10 s.

11. Transfer the reaction tubes to a thermocycler and amplify RAPDs   **PCR**
    through 2 template annealing cycles of 94 °C for 30 s, 32 °C for 30 s
    and 72 °C for 1 min 30 s, followed by 43 amplification cycles of
    94 °C for 5 s, 32 °C for 30 s and 72 °C for 1 min 30 s. Give reactions
    a final soak of 72 °C for 5 min.

12. Load 20 µl of each reaction with 1x loading buffer onto a 1.4 %
    agarose gel and separate the amplification products by electro-
    phoresis at 60 mA constant current in 1x TBE.

13. After electrophoresis stain the gel in 200 ml distilled water con-
    taining 100 µg ethidium bromide for 20 min; destain the gel in
    200 ml distilled water for 30 min.

14. Visualize the RAPD fingerprints under ultraviolet light (305 nm).

15. Score RAPD products in each reaction.   **Matrix analysis**

16. Calculate SN_L ratios for each component at each level.

17. Plot each SN_L ratio against its corresponding component level.

18. Set up optimized reactions using component concentrations equi-
    valent to the SN_L optima.

5
Results and Comments

The above protocol was used to optimize RAPD fingerprints from the
entomopathogenic fungus *Metarhizium anisopliae*. Figure 2 shows
the amplification profiles obtained for the nine reactions in the
orthogonal array. These were scored according to the number of
products in the profile $P = r + 1$ where r is the number of scorable
products (Table 2). The mean number of products (P) were then used
to calculate the SN_L ratio for each reaction component at each of the
three levels tested (Table 2). Component optima were calculated by
plotting the SN_L ratio against component concentration and using
2nd order polynomial regression to determine the component concentrations corresponding to peak SN_L ratios (Fig. 3).

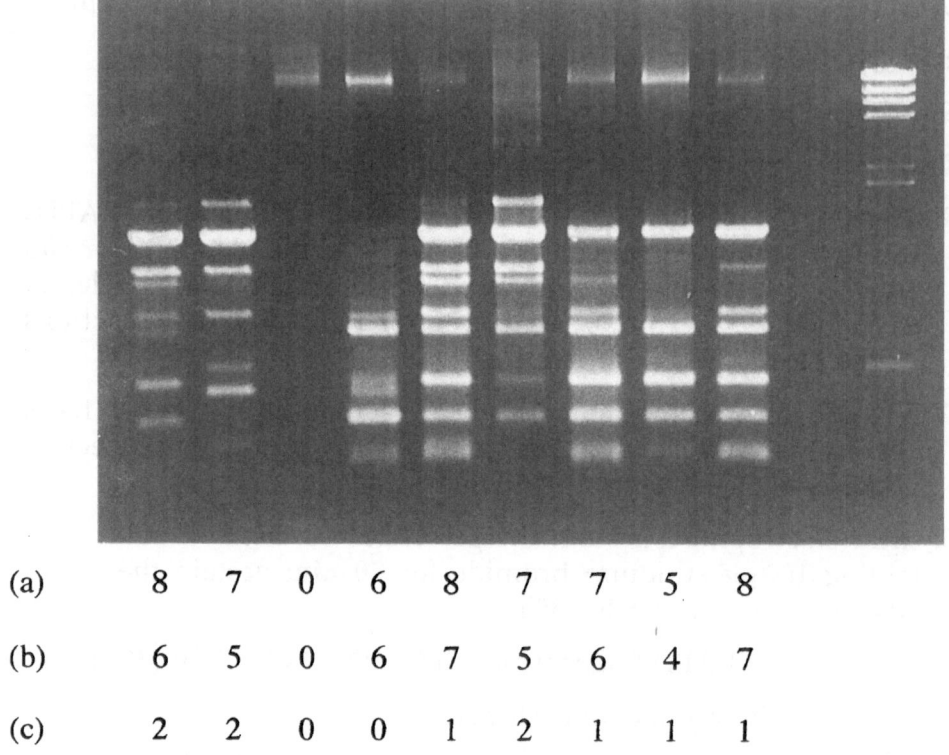

(a)	8	7	0	6	8	7	7	5	8
(b)	6	5	0	6	7	5	6	4	7
(c)	2	2	0	0	1	2	1	1	1

Fig. 2. RAPD fingerprints from the entomopathogenic fungus *Metarhizium aniso-pliae* obtained for an optimisation array testing four components each at three levels. Profiles were scored according to (*a*) the total number of amplification products, (*b*) products less than 1 kb in size, and (*c*) products greater than 1 kb

Figure 4 shows optimized RAPD fingerprints for *M. anisopliae*, *Rorippa nasturtium-aquaticum* and *Homo sapien*. Optimum conditions for fingerprints with the maximum number of scorable products were determined as 3 mM $MgCl_2$, 0.3 mM dNTP, 5 pmoles primer and 75 ng DNA (Fig. 4). This also corresponded to the optimal conditions required for generating fingerprints with amplification products larger than 1 kb.

Reactions yielding low molecular weight products (i.e., less than 1 kb) were optimal at 3.5 mM $MgCl_2$, 0.4 mM dNTP, 2.5 pmoles primer and 75 ng DNA. Magnesium ion availability has been shown to alter the profile of RAPD fingerprints (Wolff et al. 1993). Magnesium ion concentration required for optimal RAPD amplification is dependent on the sequence of the primer used (Wolff et al. 1993). Furthermore, magnesium and dNTPs affect the efficiency of priming and extension by altering the kinetics of association and dissociation of primer-template duplexes (Kwok et al. 1990; Huang et al. 1992). These com-

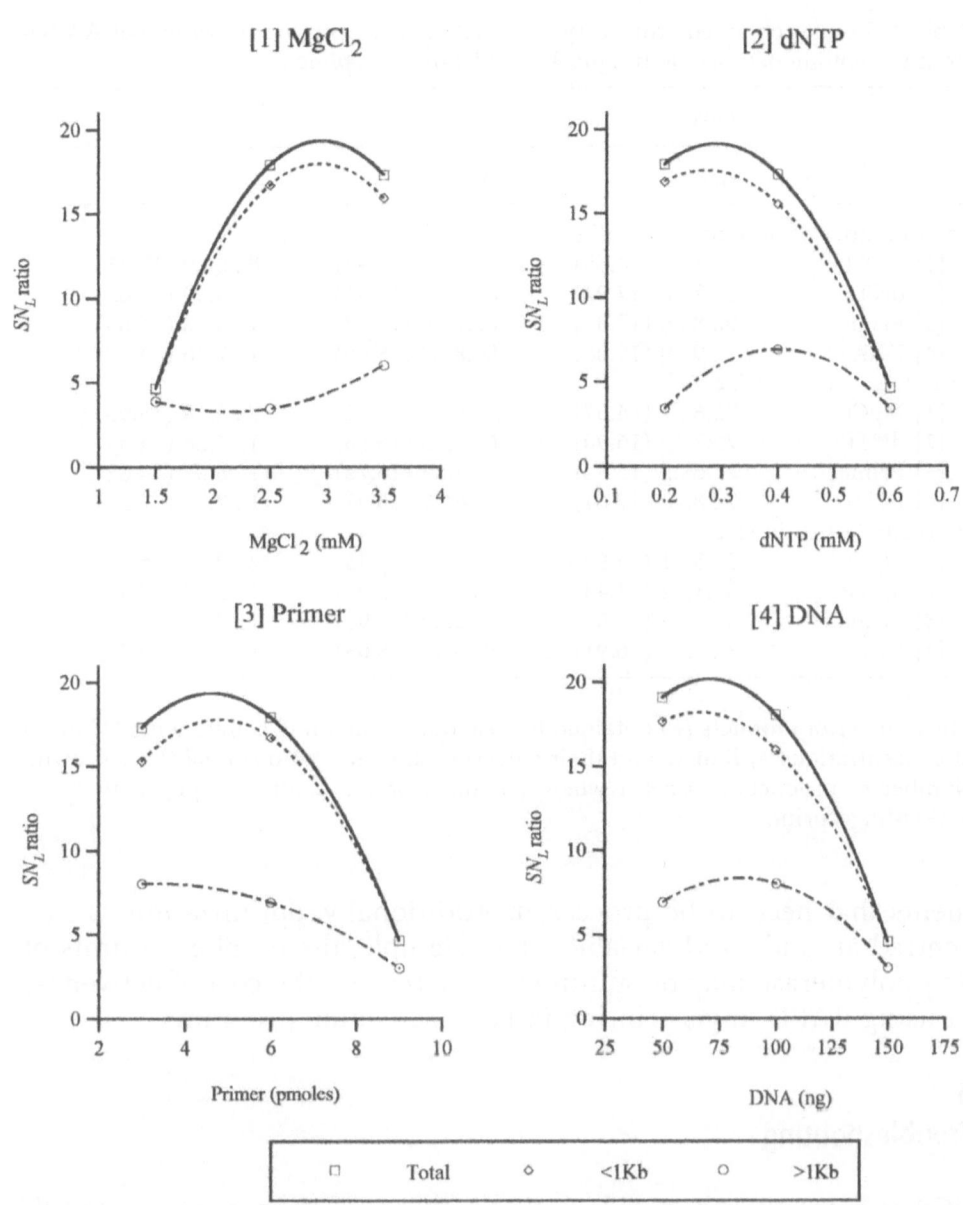

Fig. 3. Effect of reaction component concentration on the SN_L ratios for RAPD amplification

ponents are also important in determining the efficiency with which *Taq* polymerase recognizes and extends these duplexes (Wolff et al. 1993).

The complex interactions between components in PCR lends itself to optimization using orthogonal arrays. The advantages of using such strategies is the significant reduction in the number of experi-

Table 2. Results obtained from a typical orthogonal array optimization of RAPDs from the entomopathogenic fungus *Metarhizium anisopliae*

	Level		
	A	B	C
Total number of products in profile			
[1] MgCl$_2$	9...8...1 (4.65)	7...9...8 (17.92)	8...6...9 (17.31)
[2] dNTP	9...7...8 (17.92)	8...8...6 (17.31)	1...8...9 (4.65)
[3] Primer	9...8...6 (17.31)	8...7...9 (17.92)	1...9...8 (4.65)
[4] DNA	9...9...9 (19.08)	8...8...8 (18.06)	1...7...6 (4.57)
Products < IKb in size			
[1] MgCl$_2$	7...6...1 (4.57)	7...8...6 (16.72)	7...5...8 (15.96)
[2] dNTP	7...7...7 (16.90)	6...8...5 (15.56)	1...6...8 (4.59)
[3] Primer	7...6...5 (15.32)	6...7...8 (16.72)	1...8...7 (4.62)
[4] DNA	7...8...8 (17.64)	6...6...7 (15.97)	1...7...5 (4.52)
Products > IKb in size			
[1] MgCl$_2$	3...3...1 (3.90)	1...2...3 (3.45)	2...2...2 (6.02)
[2] dNTP	3...1...2 (3.43)	3...2...2 (6.91)	1...3...2 (3.43)
[3] Primer	3...3...2 (8.03)	3...2...2 (6.91)	1...2...2 (3.01)
[4] DNA	3...2...2 (6.91)	3...3...2 (8.03)	1...2...2 (3.01)

The number of products *(P)* obtained for reactions containing components *[1]* to *[4]* at concentrations *A, B* or *C*, and their corresponding SN_L ratio *(in bold)* are shown. Number or products $P = r + 1$ where r is the scorable number of products in the RAPD-fingerprint.

ments that need to be processed. Additionally, optimization can be centred around fixed variables. For example, the number of units of *Taq* polymerase may be minimized, increasing the cost effectiveness of using RAPD fingerprinting in large screening programs.

6
Troubleshooting

PCR is a notoriously problematic technique. Robust reaction conditions can be achieved using this protocol, however cycle to cycle variability during an amplification is the largest source of reaction failure. This is compounded in thermocyclers that use ambient air to cool the heating block. Running machines in constant temperature rooms (e.g., cold-rooms), enhances reaction reliability. Decreasing the denaturation times also significantly increases the amount of product amplified, presumably by increasing the half-life of *Taq* polymerase. We routinely find denaturation times of 5 s sufficient for all types of PCR.

Fig. 4. Optimized RAPD fingerprints obtained from *Metarhizium anisopliae, Rorippa nasturtium-aquaticum* and *Homo sapiens* DNA. Reactions were optimised using orthogonal arrays targeting profiles with the highest number of scorable products

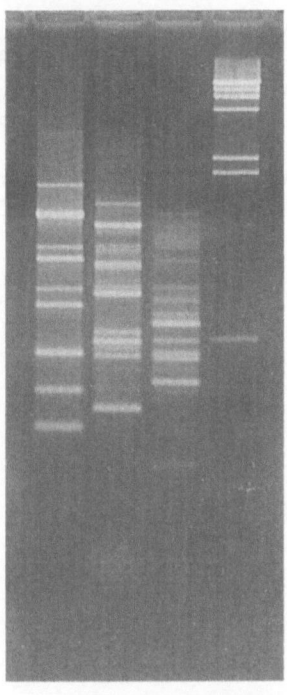

References

Charteris W (1992) Taguchi system of experimental design and data analysis – A quality engineering technology for the food industry. J Soc Dairy Technol 45:33–49

Cobb BD, Clarkson JM (1993) Detection of molecular variation in the insect pathogenic fungus *Metarhizium* using RAPD-PCR. FEMS Microbiol Lett 112:319–324

Cobb BD, Clarkson JM (1994) A simple procedure for optimising the polymerase chain reaction (PCR) using modified Taguchi methods. Nucleic Acids Res 22:3801–3805

Crowhurst RN, Hawthorn BT, Rikkerink EHA, Templeton MD (1991) Differentiation of *Fusarium solani* F-SP *cucurbitae* Race-1 and Race-2 by random amplification of polymorphic DNA. Curr Genet 20:391–396

Huang MM, Arnheim N, Goodman MF (1992) Extension of base mispairs by *Taq* polymerase – Implications for single nucleotide discrimination in PCR. Nucleic Acids Res 20:4567–4573

Kwok S, Kellogg DE, McKinney N, Spasic D, Goda L, Levenson C, Sninsky JJ (1990) Effects of primer template mismatches on the polymerase chain reaction – Human immunodeficiency virus type-1 model studies. Nucleic Acids Res 18:999–1005

Kresovich S, Williams JGK, McFerson JR, Routman EJ, Schaal BA (1992) Characterisation of genetic identities and relationships of *Brassica oleracea* L. via a random amplified polymorphic DNA assay. Theor Appl Genet 85:190–196

Taguchi G (1986) Introduction to quality engineering. Asian Productivity Organisation, UNIPUB, New York

Taguchi G, Wu Y (1980) Introduction to off-line quality control. Japan Quality Control Organisation. Nagoya, Japan

Welsh J, McClelland M (1990) Fingerprinting genomes using PCR with arbitrary primers. Nucleic Acids Res 18:7213–7218

Williams JGK, Kubelik AR, Livak KJ, Rafalski JA, Tingey SV (1990) DNA polymorphisms amplified by arbitrary primers are useful as genetic-markers. Nucleic Acids Res 18:6531–6535

Wolff K, Schoen ED, Petersvanrijn J (1993) Optimising the generation of random amplified polymorphic DNA's in *Chrysanthemum*. Theor Appl Genet 86:1033–1037

Chapter XII

Fingerprint Tailoring

G. CAETANO-ANOLLÉS

1
Introduction

The amplification of nucleic acid templates with one or more arbitrary oligonucleotide primer produces specific signatures that can be used to study virtually any nucleic acid whether anonymous in nature or previously characterized. This multiple arbitrary amplicon profiling (MAAP) strategy lends itself to the study of complex genomes in comparative and experimental biology applications that range from molecular systematics to genetic mapping (Caetano-Anollés 1994). It can also be applied to the fingerprinting of cDNA populations, PCR products, extrachromosomal nucleic acids, and cloned DNA. However, in many nucleic acid scanning applications there is a need to tailor the performance of the amplification reaction. For example, in molecular ecology and evolution there is sometimes a requirement to increase or decrease the ability to distinguish a group of organisms. Generally, this is done by using more than one fingerprinting technique capable of resolving genomes, for example, at the species or subspecies level. The versatility of nucleic acid scanning has overcome some of these limitations. The concept of "fingerprint tailoring" was introduced some time ago to depict those strategies which modify fingerprint pattern (Caetano-Anollés et al. 1991). Tailoring can target the number and range of amplification products, the level of polymorphic DNA detected, template complexity, and even the nature of amplified sites. It is based on improving: (a) analysis of amplification products, (b) primer design, and (c) amplification strategy (Caetano-Anollés 1996). The way amplification products are studied impacts on the outcome of fingerprint pattern. In DAF, some amplification products can only be detected when more sensitive separation techniques are used. For example, capillary electrophoresis promises to increase throughput but also detection of polymorphic DNA (Caetano-Anollés et al. 1995). While the capabilities of analyzing fingerprints are confined to the actual originating amplification process, the tailoring of primer design and of amplification strategy is much more versatile. This chapter describes some of these finger-printing alternatives.

2
Principles and Applications

While tailoring changes the complexity of the fingerprint pattern it can also increase the percentage of amplification products that are polymorphic within a particular group of organisms or templates under study. Two major mechanisms are responsible for such an effect: (a) an increase of the number of sites being probed in the template, for example during primer annealing, endonuclease restriction or postamplification manipulations, and (b) a change in the kinetics of the reaction due to novel primer-template interactions or changes in the stringency of amplification.

Improving Primer Design

A straightforward approach to tailor fingerprints is the improvement of primer design: primers can be tailored to produce DNA fingerprints of desired complexity by designing their sequence and ultimately the specificity of annealing to their targeted sites. The selection of primer sequence can be arbitrary and based on preliminary experience (i.e., choosing those primers that work best), or can be biased to introduce secondary structure within certain primer domains (see "DNA Amplification with Mini-hairpin Primers"), or to include degeneracy at particular locations within primer sequence (see "Oligonucleotide Base Substitution with Degenerate Bases"). Furthermore, the selection of primer sequence can be biased to recognize particular sequence motifs within the genome that represent dispersed sequences (see "Microsatellite primers" in "ASAP"), structural chromosomal domains or even consensus sequences complementary to gene families.

DNA Amplification with Mini-hairpin Primers

Amplification with very short primers is hampered by: (a) the existence of palindromic termini in amplification products capable of forming hairpin loops, as the short primer has great difficulty in displacing these hairpin-structures (Caetano-Anollés et al. 1992), and (b) an inherent decrease of primer annealing efficiency with decreasing primer length. However, these effects can be offset with the introduction of an extraordinarily stable and compact hairpin-turn structure at the 5' end of the primer that in some way interferes with the formation of hairpin loops in the resulting amplification products (Caetano-Anollés and Gresshoff 1994, 1996). These "mini-hairpin" structures consist of a loop of 3–4 nucleotides (nt) and a 2 nt stem, have high melting temperatures, unusually rapid mobilities during electrophoresis in polyacrylamide gels, and cause band compression during Maxam-Gilbert DNA sequencing (Hirao et al. 1988, 1989, 1992). Their extraordinary stability depends on the helical motif of the stem region, the loop closing base pair, and a balance between bendability and stacking stability of a B form structure in the loop (Hirao et al. 1994). Mini-hairpins have been observed in natural DNA, such as in the replication origin of phage G4 or in rRNA genes (Hirao et al. 1989, 1990; Antao and Tinoco 1991a,b).

Mini-hairpin primers are designed by attaching a 7–8 nt mini-hairpin to the 5' end of a "core" oligonucleotide of arbitrary sequence as short as 3 nt (Fig. 1). These oligonucleotides retain the characteristics of the mini-hairpin (high stability, high melting temperature and rapid electrophoretic mobility) and produce complex and reliable "sequence signatures," even from small template molecules such as plasmids, cloned DNA, and PCR amplified fragments (Caetano-Anollés and Gresshoff 1994, 1996). Their use increases detection of polymorphic DNA.

Amplification with mini-hairpin primers has been optimized (Caetano-Anollés and Gresshoff 1994) by using a buffer containing Triton X-100 and ammonium sulfate (Caetano-Anollés et al. 1994). Amplification requires 2–12 mM $MgSO_4$, with an optimal concentration range of 3–6 mM. Primers with long core regions are less tolerant of high magnesium levels. A minimum primer concentration of 1.5 μM is generally sufficient for reproducible amplification. However, higher primer concentrations (up to 30 μM) increase the efficiency of amplification but do not alter profile composition. In the study of PCR fragments, plasmids or viral DNA, reproducible patterns can only be obtained using high primer concentration (30 μM). Template DNA concentrations should be comparable to those used in standard DAF. Optimal template concentration is usually in the range of 0.1–5 ng/μl. For general considerations related to DAF optimization see Chap. IX.

Amplification parameters

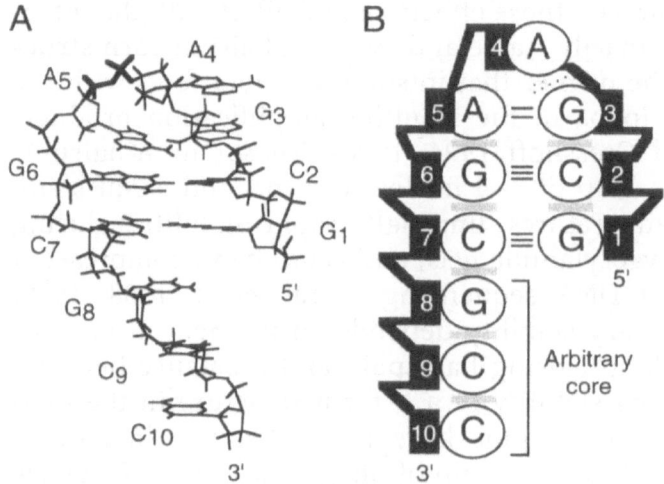

Fig. 1A,B. Three-dimensional structure of a mini-hairpin primer. Stereoview (**A**) and schematic diagram (**B**) of the mini-hairpin GCGAAGCGCC primer structure inferred from the nuclear magnetic resonance spectroscopy studies of Hirao et al. (1994). The final structure contains a hairpin-turn region composed of two B-form strands, G_1-to-A_4 and A_5-to-C_7, with two GC base pairs and one non-Watson-Crick G-A base pair. The G_3-A_5 nucleotides are shaded in the stereoview model. Three torsion angles between A_4 and A_5 differ from those of the typical B-form structure and allow a sharp turn region to occur. The sharp turn at the single phosphate intervening between A_4 and A_5 is also highlighted. Nucleotides take an anti-glycosidic angle conformation with A_5 wobbling around anti and high-anti conformations. All nucleotides adopt the C2'-*endo* conformation and form strong base-base stacking (*shaded small boxes*) ultimately responsible for the high thermal stability of these structures. *Circles* represent bases, *shaded boxes* C2'-*endo* sugars, and *lines* between bases hydrogen bond interactions

Oligonucleotide Base Substitution with Degenerate Bases

Another fingerprint tailoring strategy based on improving primer design involves the introduction of degeneracy in the primer sequence by base substitution with degenerate bases or analogs such as the "universal" base hypoxanthine (I) (Caetano-Anollés and Gresshoff 1994; Sakallah et al. 1995).

Introduction of nucleotides containing any of the four bases (N) or I substitutions within the sequence of a primer alter fingerprints by either simplifying or increasing the complexity of the patterns, the effect being quite marked if substitutions are close to the 3' terminus (Bassam and Caetano-Anollés, unpublished). For example, single I substitutions along the sequence of an octamer simplified fingerprints, probably by increasing the stringency of primer annealing. Hypoxanthine can base pair with low discrimination with all DNA bases, and in combination with other base analogues generates modified oligomers that can be used successfully as primers with reduced nonspecific priming in the PCR (Kong Thoo Lin and Brown 1992).

Improving Amplification Strategy

Tailoring can also be accomplished by improving the amplification strategy, such as in the cases of selective restriction fragment amplification (SRFA) (Vos et al. 1995), which will not be described here, ASAP analysis and tecMAAP.

ASAP (Arbitrary Signatures from Amplification Profiles)

As originally described, ASAPs are fingerprints generated from fingerprints by re-amplification of DAF profiles with mini-hairpin or standard arbitrary primers (Caetano-Anollés and Gresshoff 1996). ASAPs are also fingerprints obtained by reamplification of any amplification product, ranging from those generated with arbitrary primers to those produced in the PCR.

ASAP analysis is a dual-step amplification strategy that provides additional scanning of primary sequence within preselected amplicons. The procedure requires the use of one or more primers in each amplification step, allowing the combinatorial use of oligomers in fingerprinting. Provided the sequence of the primers differ substantially from each other, distinct fingerprints can be generated in each particular combination. For example a set of ten oligomers can be used in 100 different pairwise combinations, 90 of which should produce unique fingerprints. The power of the approach is further unveiled when multiplexing DAF reactions. If two primers are used in one of the amplification steps, or in both, 10^4 or 10^{16} different reactions are possible, respectively. However, priming during the second amplification occurs within the DAF fragments amplified in the first amplification. Therefore, the number of targeted locations in a genome depends on the number of original DAF reactions.

The arbitrary selection of primers in ASAP analysis can be biased to include recognition of particular sequence motifs or interspersed repetitive sequences. For example, primers complementary to simple sequence repeats (SSRs) present in microsatellite loci can be used to generate very simple ASAPs by reamplification of MAAP fingerprints (Caetano-Anollés and Gresshoff 1996). The advantage of targeting these loci is that they represent highly polymorphic regions, and in

Microsatellite primers

some cases can constitute codominant markers with many allelic forms.

In order to target consistently about 10–15 kb of DAF amplified sequence which contains only few SSR annealing sites, the SSR primers have to be anchored at their 5' termini with ambiguous (i.e., degenerate) nucleotides. Furthermore, the amplification reaction must occur under stringent conditions to avoid mismatch priming.

ASAP analysis using SSR primers generates very simple profiles containing in 50 % of the cases only one or two prevalent amplification products. These products represent SSR loci and in many cases are clearly codominant.

Optimization of primer and template levels

Being an extension of the MAAP strategy, ASAP analysis relies on finding appropriate "reproducibility windows" (see Chap. IX) for the many parameters in the amplification reaction, especially primer and template concentration. Primer concentration was found particularly important for reproducibility (Caetano-Anollés and Gresshoff 1996). Mini-hairpin decamers require 6–9 µM concentrations, while standard octamers require at least 9 µM primer levels. ASAP primers with sequences partially complementary to the termini of DAF products tolerate lower primer concentrations (about 3 µM) (Fig. 2). The reproducibility window for template concentration ranges from 0.001 to 1 ng/µl of DAF products, even if ASAP primers are mini-hairpins partially complementary to the DAF primers used in the first amplification reaction (Fig. 2). However, a decrease in primer concentration during ASAP amplification causes the window to narrow considerably. It is therefore important to consider the existence of primer-template interactions in new ASAP fingerprinting applications.

tecMAAP (Template Endonuclease Cleavage MAAP)

tecMAAP is a tailoring strategy which enhances polymorphic DNA detection by coupling MAAP to endonuclease cleavage of the template prior to amplification (Caetano-Anollés et al. 1993). Generally, one or more restriction endonucleases (preferably three 4 bp cutters) are used to restrict the DNA prior to amplification. This DNA then serves as a modified template for amplification with arbitrary primers. The strategy is straightforward but contingent on the assurance that template digestion has been complete.

During MAAP, many sites are targeted but only few are preferentially amplified in a dynamic reaction in which reaction kinetics and primer-template interaction determine the outcome of a particular fingerprint. In tecMAAP, digestion eliminates many of possible

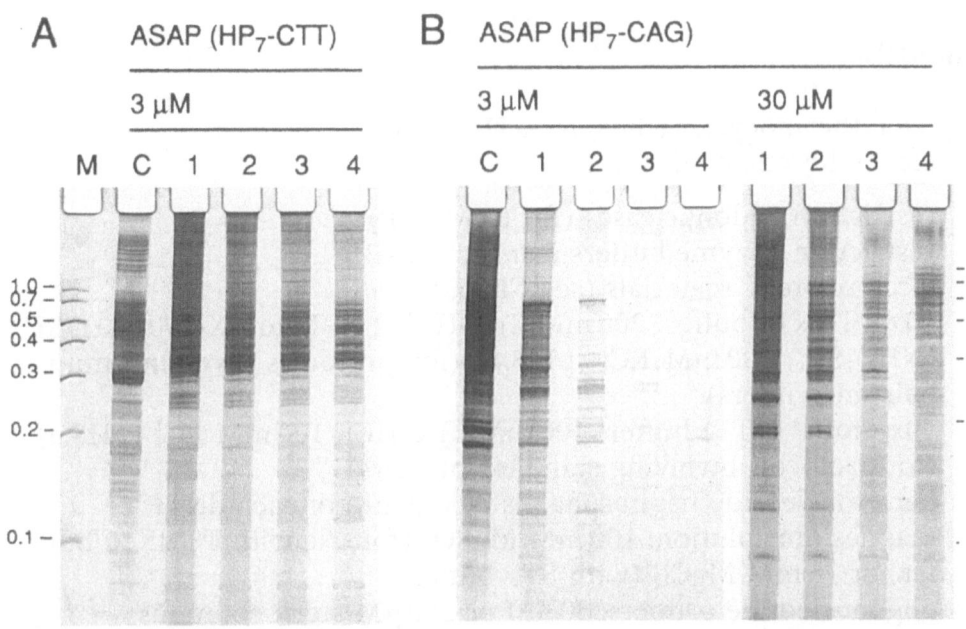

Fig. 2A,B. Effect of template concentration on ASAP analysis of bermudagrass. Reamplification of DAF profiles originally generated from bermudagrass cv. Tifway with primer HP$_7$-CTC (HP$_7$=GCGAAGC) diluted to 1 ng/µl (*lane 1*), 0.1 ng/µl (*lane 2*), 0.01 ng/µl (*lane 3*), and 0.001 ng/µl (*lane 4*). As control, DNA from cultivar Tifway was amplified during ASAP analysis (*lane C*). ASAP analysis was done using primers with one (**A**) or two (**B**) base substitutions in the arbitrary core sequence. Molecular weights standards (*M*) are given in kb

amplicons and ultimately reduces the effective length of template DNA. Restriction also increases the nucleotide sequence being probed. Sequence variation within restriction sites will either directly eliminate some of preferentially amplified products or will indirectly change the overall kinetics, creating new products or eliminating ones that were previously amplified. Computer simulation has been applied to this problem to show that, upon restriction, arbitrary octamer primers are forced to mismatch with the template in the last one or two 5' terminal nucleotides (Caetano-Anollés 1994). Ultimately, differential cleavage of template molecules enhances the detection of polymorphic DNA without changing appreciably the multiplex ratio.

tecMAAP has been shown to enhance considerably (up to 100-fold) the detection of polymorphic DNA, allowing the identification of closely related cultivars, plant accesions, and even near-isogenic lines.

3
Materials

Equipment
- oven thermocycler (Bios, New Haven, CT)
- electophoretic equipment

Reagents and solutions
- restriction endonucleases (for tecMAAP)
- restriction enzyme buffers (for tecMAAP)
- electophoretic materials (see Chap. XIII)
- 10x TTNK10 buffer: 200 mM Tris-HCl, 1 % Triton X-100, 40 mM $(NH_4)_2SO_4$, 100 mM KCl (pH 8.6) (for protocols involving mini-hairpin primers)
- 10x Stoffel (STF) buffer: 100 mM Tris-HCl, 100 mM KCl (pH 8.3) (for protocols involving standard primers)
- deoxynucleoside triphosphate stock: 2 mM of each dNTP
- magnesium solution: 100 mM $MgSO_4$ (for use with TTNK10 buffer) or 25 mM $MgCl_2$ (with STF buffer)
- oligonucleotide primers: 30 µM or 300 µM stock solutions
- AmpliTaq Stoffel fragment DNA polymerase (Perkin-Elmer/Cetus, Norwalk, CT)
- template: 1–50 ng/µl stock solutions
- mineral oil (when required)

4
Experimental Procedure

The following are protocols of MAAP techniques capable of increasing detection of polymorphic DNA; one of them, DAF analysis with mini-hairpin primers, can be applied to a wide variety of templates, including plasmids, cloned DNA, and PCR products. General precautions and considerations described for DAF analysis should also be applied to these techniques (see Chap. IX).

DAF Analysis with Mini-hairpin Primers

Primers containing hairpin-turn structures at their 5' termini (Fig. 1) can be used to obtain reliable fingerprints from almost any template nucleic acid. However, the design of the amplification reaction varies with template complexity.

1. Assemble an amplification cocktail in 10–20 µl total volume containing the following components: 0.1–5 ng/µl template DNA, 3–30 µM of primer(s), 0.3 U/µl of thermostable DNA polymerase,

200 µM of each deoxynucleoside triphosphate, 4 mM MgSO$_4$, and
TTNK10 buffer.

2. Cover the amplification cocktail with 1–2 drops of heavy mineral
   oil, when required.

3. Amplify in 35 cycles of 30 s at 96 °C, 30 s at 30 °C and 30 s at 72 °C
   in an oven thermocycler.

4. Retrieve the aqueous phase from reaction tubes.

5. Dilute samples five- to tenfold prior to electrophoresis.

Arbitrary Signatures from Amplification Profiles (ASAP)

ASAPs are fingerprints of fingerprints that result from the reamplification of DAF products with one or more arbitrary primer harboring a sequence that differs significantly from that of the primer used
to generate the original DAF reaction (Caetano-Anollés and Gresshoff 1996). Generally, primers used in the second round of amplification are mini-hairpin oligonucleotides, though unstructured DAF
primers or primers complementary to interspersed repetitive sequences can be used. The following protocol is designed for use with
mini-hairpin primers.

1. Dilute DAF reactions, usually containing 100–200 ng/µl of double
   stranded DNA, to about 1 ng/µl template (10x) stock solution.
   Typically, this corresponds to a 1:100 dilution of a standard
   amplification reaction.

2. Assemble the amplification cocktail in 10–20 µl total volume containing the following components: diluted amplification products
   (0.1 ng/µl final concentration), 9 µM of primer(s), 0.3 U/µl of thermostable DNA polymerase, 200 µM of each deoxynucleoside triphosphate, 4 mM MgSO$_4$, TTNK10 buffer.

3. Cover the amplification mixture with 1–2 drops of heavy mineral
   oil, when required.

4. Amplify in 35 cycles of 30 s at 96 °C, 30 s at 30 °C and 30 s at 72 °C
   in an oven thermocycler.

5. Retrieve the aqueous phase from reaction tubes.

6. Dilute samples five- to tenfold prior to analysis.

4.3
Template Endonuclease Cleavage MAAP (tecMAAP)

tecMAAP is based on a two-step reaction where the template nucleic acid is first subjected to enzymatic digestion with type II restriction endonucleases and then amplified with one or more arbitrary oligonucleotide primer.

1. Digest template DNA with one or more restriction endonucleases. Add 2 U of enzyme (usually 2 μl) and 2 μl of appropriate restriction enzyme buffer (Table 1) to 2 μg of DNA in a total 10 μl reaction volume. Incubate at the recommended temperature for 1 h to overnight. Both blunt-end or staggered-cutter enzymes can be used. However, enzymes that recognize 4 bp motifs are preferred.

2. Confirm complete digestion by electrophoresis in agarose or polyacrylamide gels.

3. Assemble an amplification cocktail in 10–20 μl total volume containing the following components: 0.1–5 ng/μl digested template, 3–30 μM of primer(s), 0.3 U/μl of thermostable DNA polymerase,

Table 1. Some restriction endonucleases with 4 bp specificities

Enzyme	Recognition sequence	Heat inactivation[a]	Buffer[b]	Temperature (°C)
AluI	AG'CT	+	A, B[c]	37
AsuI	G'GNCC	+	A, B[d]	37
HaeIII	GG'CC	–	B, A[c]	37
HhaI	GCG'C	+	B, A[c]	37
HinfI	G'ANTC	–	A, B[c]	37
HpaII	C'CGG	–	C, A[d], B[d]	37
MboI	'GATC	+	B, A[c]	37
MspI	C'CGG	+	A, B[c]	37
RsaI	GT'AC	+	B,.A[c]	37
Sau3A	'GATC	+	A, B[c]	37
TaqI	T'CGA	–	D, B[c], A[e]	65

[a] Inactivation (95 %) by incubation at 65 °C for 15 min.
[b] Buffers (10×): A (50 mM NaCl, 6 mM Tris-HCl, 6 mM MgCl$_2$, 1 mM dithiothreitol, pH 7.5), B (50 mM NaCl, 10 mM Tris-HCl, 10 mM MgCl$_2$, 1 mM dithiothreitol, pH 7.9), D (6 mM NaCl, 6 mM Tris-HCl, 6 mM MgCl$_2$, 1 mM dithiothreitol, pH 7.5), F (100 mM NaCl, 10 mM Tris-HCl, 10 mM MgCl$_2$, 1 mM dithiothreitol, pH 8.5).
[c] Buffer provides 75 %–100 % activity.
[d] Buffer provides 50 %–75 % activity.
[e] Buffer provides 15 %–50 % activity.

200 µM of each deoxynucleoside triphosphate, 1.5 mM $MgCl_2$, and STF buffer. Use $MgSO_4$ and TTNK10 buffer when amplifying with mini-hairpin primers.

4. Cover the amplification cocktail with 1–2 drops of heavy mineral oil, when required.

5. Amplify in 35 cycles of 30 s at 96 °C, 30 s at 30 °C and 30 s at 72 °C in an oven thermocycler.

6. Retrieve the aqueous phase from reaction tubes.

7. Dilute samples five- to tenfold prior to analysis.

Separation and Visualization of Amplification Products

Amplification products can be separated by polyacrylamide gel electrophoresis (PAGE) in vertical or open-faced slab gels, or using semi-automated miniaturized electrophoretic devices (see Chap. XIII). Gels are backed on polyester film, silver stained using the procedure of Bassam et al. (1991), and preserved by drying at room temperature.

5
Results and Comments

In an attempt to correlate abnormal electrophoretic mobility behavior of the mini-hairpin primers in denaturing polyacrylamide gels and amplification efficiency in DAF analysis, the decamer primer series HP_7-NNN was studied in detail. HP_7-NNN is a set of primers designed by attaching the constant mini-hairpin GCGAAGC (HP_7) at the 5' terminus to an arbitrary core of 3 nt containing one of 64 possible nucleotide sequences (NNN). The mini-hairpin HP_7 was chosen because it was the smallest and most stable (T_m=76.5) of described hairpins (Hirao et al. 1989, 1994). When these primers were used in DAF analysis of soybean (*Glycine max*) cv. Bragg and bermudagrass (*Cynodon dactylon*×*C. transvaalensis*) cv. Tifway, no correlation between electrophoretic mobility and efficiency of amplification was observed (Table 2). The majority of primers produced complex fingerprints with a balanced number of amplification products (range: 15–43 in the 50–700 bp interval) of high and low intensity. Only five out of the 64 primers tested failed to amplify soybean or bermudagrass DNA. However, it should be noted that the HP_7-CNN series contained several primers with very high amplification efficiencies that perhaps target specific sequences particularly abundant in tested genomes.

Mini-hairpin primers

Table 2. Electrophoretic and DAF analysis of GCGAAGC (HP$_7$)-derived mini-hairpin primers

Primer	L_{ap}[a]	N[b]	E[c]	Primer	L_{ap}	N	E
HP$_7$-CCC	6.53	31	4	HP$_7$-GCC	6.52	35	4
HP$_7$-CCT	6.85	33	3	HP$_7$-GCT	6.95	0	–
HP$_7$-CCG	7.00	0	–	HP$_7$-GCG	6.91	0	–
HP$_7$-CCA	6.62	34	3	HP$_7$-GCA	5.71	27	2
HP$_7$-CTC	6.76	43	4	HP$_7$-GTC	6.57	37	3
HP$_7$-CTT	7.19	30	3	HP$_7$-GTT	7.00	31	3
HP$_7$-CTG	7.24	28	3	HP$_7$-GTG	7.04	29	3
HP$_7$-CTA	6.90	30	3	HP$_7$-GTA	5.97	18	2
HP$_7$-CGC	6.92	29	2	HP$_7$-GGC	6.99	19	1
HP$_7$-CGT	6.20	36	2	HP$_7$-GGT	7.41	34	3
HP$_7$-CGG	7.30	0	–	HP$_7$-GGG	7.37	37	3
HP$_7$-CGA	6.92	34	4	HP$_7$-GGA	6.16	34	1.5
HP$_7$-CAC	6.64	36	4	HP$_7$-GAC	6.42	35	4
HP$_7$-CAT	7.00	39	4	HP$_7$-GAT	6.99	37	3
HP$_7$-CAG	7.00	40	4	HP$_7$-GAG	7.00	30	3
HP$_7$-CAA	6.64	39	4	HP$_7$-GAA	5.72	19	1
HP$_7$-TCC	6.81	34	3	HP$_7$-ACC	6.53	33	3
HP$_7$-TCT	7.09	27	2	HP$_7$-ACT	7.00	31	2
HP$_7$-TCG	7.18	23	2	HP$_7$-ACG	7.00	23	2
HP$_7$-TCA	6.68	38	3	HP$_7$-ACA	6.58	35	3
HP$_7$-TTC	6.95	0	–	HP$_7$-ATC	6.76	40	3
HP$_7$-TTT	7.23	25	2	HP$_7$-ATT	7.00	32	2
HP$_7$-TTG	7.23	36	3	HP$_7$-ATG	7.14	29	3
HP$_7$-TTA	7.00	15	1	HP$_7$-ATA	6.85	22	4
HP$_7$-TGC	7.41	29	3	HP$_7$-AGC	6.95	28	3
HP$_7$-TGT	7.51	32	2.5	HP$_7$-AGT	7.19	36	2
HP$_7$-TGG	7.68	29	3	HP$_7$-AGG	7.00	33	3
HP$_7$-TGA	7.84	33	3	HP$_7$-AGA	6.62	30	3
HP$_7$-TAC	6.80	31	3	HP$_7$-AAC	6.31	43	3
HP$_7$-TAT	7.25	30	3	HP$_7$-AAT	6.71	26	2
HP$_7$-TAG	7.51	20	1.5	HP$_7$-AAG	6.67	17	1
HP$_7$-TAA	7.00	18	2	HP$_7$-AAA	6.44	26	3

[a] Apparent oligonucleotide length (nt). The value was calculated by interpolation of the primer electrophoretic mobilities in 20% polyacrylamide-7M urea slab gels and the mobilities of unstructured oligonucleotide standards (CGAAGC, CCGAAGC, AGTCAGCCAC, and CGAAGCCCGAGCTG).

[b] Number of amplification products (within the 50–700 nt length range).

[c] Amplification efficiency, measured on an arbitrary scale, in which 1 depicts profiles with products of low intensity (mainly tertiary products), and 4 depicts profiles with products of high intensity (mainly primary products), and sometimes on an overall background of amplification.

Mini-hairpin primers have the property of increasing detection of polymorphic DNA when fingerprinting genomic DNA from a variety of organisms, including plant species such as centipedegrass (Caetano-Anollés and Gresshoff 1994), bermudagrass (Caetano-Anollés et al. 1995), and soybean (Caetano-Anollés and Gresshoff 1996). For example, the ability of mini-hairpin primers to differentiate bermudagrass cultivar Tifway from its γ-irradiation mutant, Tifway II was increased almost fivefold when compared to the use of standard primers (Caetano-Anollés et al. 1995).

Degenerate octamers containing a tail of two N or I substituted nucleotides at their 3' or 5' termini were tested by amplifying a wide range of templates (from bacterial to mammal origin). The 5' N-tailed primers produced profiles almost indistinguishable from those generated by the related octamers. In contrast, 5' I-tailed primers produced highly variant fingerprints. Our results indicate that 5' extension of primer length over the 8 nt 3'-terminal domain defined in previous studies (Caetano-Anollés et al. 1992) has little influence on the distribution of amplification products. However, if this extension is made with nucleotides containing hypoxanthine a major effect on amplification is observed. This may result from a differential discrimination of I in the formation of base pairs with A and C over T (Corfield et al. 1987, Kong Thoo Lin and Brown 1992), or from steric hindrance during annealing due to conformational changes. The addition of degenerate tails at the 3' termini altered completely the DAF patterns. However, N-primers produced profiles of increased complexity, while I-primers resulted in simplified fingerprints. Therefore, hypoxanthine substitutions at the 3' termini can be used effectively to simplify fingerprint pattern, instead of invoking other alternatives such as reducing primer length or increasing annealing temperatures (Caetano-Anollés et al. 1992).

Oligonucleotide base substitution

The general observation, that N-substituted primers generate DNA profiles of increased complexity, was extended to mini-hairpin primers by substituting the 5' nucleotide of the primer core with a degenerate N base (Caetano-Anollés and Gresshoff 1994). As expected, such substitutions increased fingerprint complexity and altered amplification profiles. Substitution experiments of mini-hairpin primers with varying arbitrary cores suggested that mini-hairpin primers with short cores provide a more defined tool for DNA fingerprinting.

References

Antao VP, Tinoco Jr I (1991a) Thermodynamic parameters for loop formation in RNA and DNA hairpin tetraloops. Nucleic Acids Res 20:819–824

Antao VP, Lai SY, Tinoco Jr I (1991b) A thermodynamic study of unusually stable RNA and DNA hairpins. Nucleic Acids Res 19:5901–5905

Bassam BJ, Caetano-Anollés G, Gresshoff PM (1991) Fast and sensitive silver staining of DNA in polyacrylamide gels. Anal Biochem 196:81–84

Caetano-Anollés G (1994) MAAP: a versatile and universal tool for genome analysis. Plant Mol Biol 25:1011–1026

Caetano-Anollés G (1996) Fingerprinting nucleic acids with arbitrary oligonucleotide primers. Agro-Food-Industry High-Tech 7:26–35

Caetano-Anollés G, Gresshoff PM (1994) DNA amplification fingerprinting using arbitrary mini-hairpin oligonucleotide primers. Bio/Technology 12:619–623

Caetano-Anollés G, Gresshoff PM (1996) Generation of sequence signatures from DNA amplification fingerprints with mini-hairpin and microsatellite primers. Bio-Techniques 20:1044–1056

Caetano-Anollés G, Bassam BJ, Gresshoff PM (1991) DNA amplification fingerprinting: a strategy for genome analysis. Plant Mol Biol Rep 9:294–307

Caetano-Anollés G, Bassam BJ, Gresshoff PM (1992) Primer-template interactions during DNA amplification fingerprinting with single arbitrary oligonucleotides. Mol Gen Genet 235:157–165

Caetano-Anollés G, Bassam BJ, Gresshoff PM (1993) Enhanced detection of polymorphic DNA by multiple arbitrary amplicon profiling of endonuclease digested DNA: identification of markers linked to the supernodulation locus in soybean. Mol Gen Genet 241:57–64

Caetano-Anollés G, Bassam BJ, Gresshoff PM (1994) Buffer components tailor DNA amplification with arbitrary primers. PCR Methods Applic 4:59–61.

Caetano-Anollés G, Callahan LM, Williams PE, Weaver KR, Gresshoff PM (1995) DNA amplification fingerprinting analysis of bermudagrass (*Cynodon*): genetic relationships between species and interspecific crosses. Theor Appl Genet 91:228–235

Corfield PWR, Hunter WN, Brown T, Robinson P, Kennard O (1987) Inosine-adenine base pairs in B-DNA duplex. Nucleic Acid Res 15:7935–7945

Hirao I, Naraoka T, Kanamori S, Nakamura M, Miura K (1988) Synthetic oligodeoxyribonucleotides showing abnormal mobilities on polyacrylamide gel electrophoresis. Biochem Int 16:157–162

Hirao I, Nishimura Y, Naraoka Y, Watanabe K, Arata Y, Miura K (1989) Extraordinary stable structure of short single-stranded DNA fragments containing a specific base sequence: d(GCGAAAGC). Nucleic Acids Res 17:2223–2231

Hirao I, Ishida M, Watanabe K, Miura K (1990) Unique hairpin structures occurring at the replication origin of phage G4 DNA. Biochem Biophys Acta 1087:199–204

Hirao I, NishimuraY, Tagawa Y, Watanabe K, Miura K (1992) Extraordinarily stable mini-hairpins: electrophoretical and thermal properties of the various sequence variants of d(GCGAAAGC) and their effect on DNA sequencing. Nucleic Acids Res 20:3891–3896

Hirao I, Kawai G, Yoshizawa S, NishimuraY, Ishido Y, Watanabe K, Miura K (1994) Most compact hairpin-turn structure exerted by a short DNA fragment, d(GCGA-AGC) in solution: an extraordinarily stable structure resistant to nucleases and heat. Nucleic Acids Res 22:576–582

Kong Thoo Lin P, Brown DM (1992) Synthesis of oligodeoxiribonucleotides containing degenerate bases and their use as primers in the polymerase chain reaction. Nucleic Acids Res 20:5149–5152.

Sakallah SA, Lanning RW, Cooper DL (1995) DNA fingerprinting of crude bacterial lysates using degenerate RAPD primers. PCR Meth Applic 4:265–268

Vos P, Hogers R, Bleeker M, Reijans M, van de Lee T, Hornes M, Frijters A, Pot J, Peleman J, Kuiper M, Zabeau M (1995) AFLP: a new technique for DNA fingerprinting. Nucl Acids Res 23:4407–4414

Resolving DNA Amplification Products Using Polyacrylamide Gel Electrophoresis and Silver Staining

G. Caetano-Anollés

1
Introduction

The scanning of nucleic acids by amplification produces an array of DNA products that can be resolved using a variety of methods. As originally described, DNA amplification fingerprinting (DAF) (Caetano-Anollés et al. 1991) and arbitrarily primed PCR (AP-PCR) (Welsh and McClelland 1990) use polyacrylamide gel electrophoresis (PAGE), while random amplified polymorphic DNA (RAPD) analysis (Williams et al. 1990) separates amplification products by electrophoresis in agarose gels. These techniques differ also in the way nucleic acid fragments are detected: DAF uses silver staining, AP-PCR uses autoradiography, and RAPD uses staining with ethidium bromide. These procedures provide simple, fast and low-cost analysis of amplification products. However, many researchers would prefer avoiding the use of radioactivity, acrylamide, and ethidium bromide. Despite its low resolving power, the familiarity and simplicity of agarose gel electrophoresis has made RAPD popular. However, it is a poor strategy for the analysis of complex fingerprints. Alternatively, PAGE has traditionally offered a superior separation of nucleic acids, proteins and polysaccharides. Acrylamide polymerization onto backing supports such as polyester films or glass has simplified gel handling and improved the overall performance of the technique. In particular, the coupling of PAGE and optimized silver staining protocols that detect nucleic acids at the picogram level provides the speed, simplicity and high resolution needed in nucleic acid scanning applications (Bassam et al. 1991; Bassam and Caetano-Anollés 1993; Caetano-Anollés and Bassam 1993; Caetano-Anollés and Gresshoff 1994; Bassam and Bentley 1995).

Amplification products can also be separated by several alternative techniques. Fluorophore-labeled DAF products can be sized by laser scanning with automated DNA sequencers (Caetano-Anollés et al.

1992) (protocols for detection and analysis of fluorophore-labeled amplicons are described in Chap. X) or by capillary electrophoresis (CE) (Caetano-Anollés et al. 1995). These real-time analysis techniques increase resolution but lack throughput and are limited by availability of proper instrumentation. Recently, other approaches that rely on the separation of single-stranded DNA in polyacrylamide gels according to size and base composition have been used: denaturing gradient gel electrophoresis (DGGE) (see Chap. XIV), temperature sweep gel electrophoresis (TSGE) (see Chap. XV), and analysis of single strand conformation polymorphisms (SSCPs) (McClelland et al. 1994) appear to increase resolution and detection of polymorphic DNA. An emerging strategy able to probe DNA sequences without resorting to any separation of DNA is based on short oligonucleotide arrays immobilized on solid supports and used to survey by hybridization the sequence of amplification products (Chetverin and Kramer 1994; Beattie et al. 1995). This strategy, profiling-by-hybridization (PHB, in analogy to sequence-by-hybridization, SBH; Strezoska et al. 1991), can detect single nucleotide polymorphisms through multiple pairwise comparisons (G. Caetano-Anollés, patent pending).

This chapter focuses on several polyacrylamide-based separation techniques that can be used to separate DAF or differentially displayed RNA profiles. It also describes the staining of nucleic acid fragments with silver.

2
Principles and Applications

Resolving DAF Fingerprints

DAF amplification products can be electrophoretically separated in polyester-backed polyacrylamide gels. DNA separation using vertical, open-faced discontinuous (isotachophoresis) or miniaturized electrophoretic units produces reliable and clearly resolved fingerprints after silver staining (Fig. 1). Polyester backing films constitute an improved support for polyacrylamide gels. These films bind covalently to the acrylamide monomers during polymerization and provide better handling, staining and preservation of the gels. If these films are unavailable, the gels can be attached to glass or plastic via a silane bridge using bind-silane (γ-methacryloxy-propyl-trimethoxysilane) solutions.

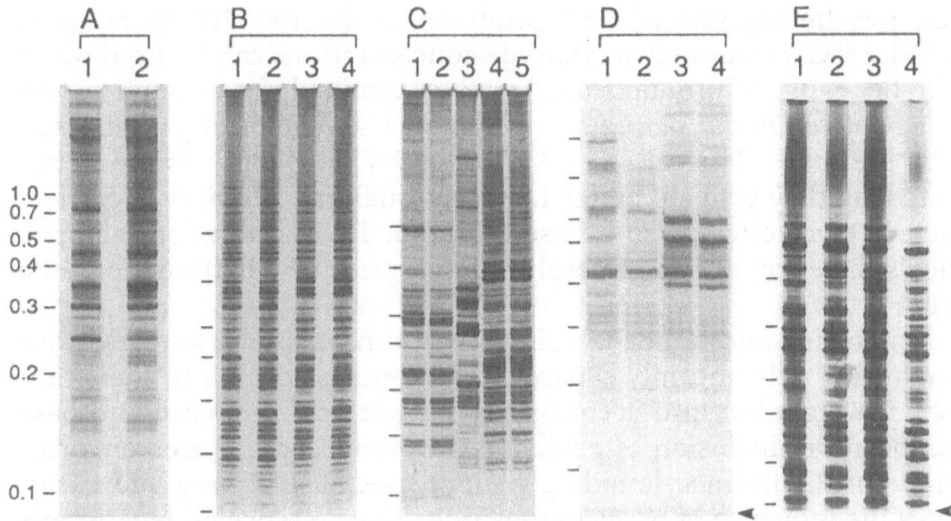

Fig. 1A–E. DNA amplification fingerprinting (DAF) profiles resolved by electrophoresis in vertical, open-faced horizontal and miniature polyacrylamide gels. DAF profiles were generated from bermudagrass (*Cynodon dactylon×C. transvaalensis*) cultivar Tifway and its γ-irradiation-induced mutant Tifway II (**A**), several isolates of *Fusarium oxysporum* f. sp. *cubense* (**B**) and *Phytium* (**C**), African bermudagrass (*C. transvaalensis*) selections Ctr2570 and Ctr2747 (*lanes 1, 2*) and interspecific hybrid bermudagrass cultivars Tifgreen and Tifdwarf (*lanes 3, 4*) (**D**), and related Indonesian fruit bat (*Pteropus hypomelanus*) individuals (**E**), using standard octamer (**A, B, C, E**) or decamer mini-hairpin (**D**) primers. The profiles were separated using vertical denaturing 0.45 mm thick 4 %T:5 %C polyacrylamide gels (8×10 cm) and standard 8-prong combs (**A**), denaturing 10 %T:2 %C polyacrylamide gels and standard 15-prong (**B**) or customized 20-prong (**C**) combs, open-faced discontinuous 6 %T:5 %C polyacrylamide gels (**D**), and miniature 10 %–15 % gradient PhastGels using reconstituted buffer strips (**E**). *Arrowheads* indicate the presence of a buffer front. Please note the stacking gel in the top of **E** and staining artifacts generated during handling (surface blemishes on the center of the gel) on the magnified view of the PhastGel. Molecular weight markers are indicated in kilobase pairs. (Data from G. Caetano-Anollés and B.J. Bassam, unpublished)

Polyacrylamide concentration

A wide range of gel configurations are possible. Pore size in polyacrylamide gels can be increased by either diluting total monomer components (i.e., decreasing polyacrylamide concentration, %T) or by increasing the percent of cross-linker (%C). At a fixed %T, porosity also increases with both higher and lower %C levels following a nearly parabolic function with a minimum at about 5 %C. We have routinely resolved DAF patterns using 4 %T:5 %C polyacrylamide gels (Caetano-Anollés and Bassam 1993). However, lower cross-linker levels permit the use of higher %T gels without the risk of cracking or peeling away from the backed gels and appear to increase the quality of the silver stain (Bassam and Bentley 1995). An optimal formulation for the analysis of DAF products in the 100–700 bp range is 10 %T:2 %C. Lower polyacrylamide concentrations can be used when a higher range of products needs to be examined, for example, when using native *Thermus aquaticus* polymerases. When using nonbacked gels, the more classical 4 %–5 %T:3 %–5 %C formulations are better suited, as they provide better handling qualities. Mobility modifiers such as glycerol can also be used to tailor DNA separation. Figure 2 shows the effect of 5 % glycerol on fingerprint distribution along the gels.

Denaturing and nondenaturing PAGE can be used. However, nondenaturing gels not only separate fragments according to their molecular weight but also according to nucleic acid sequence and base composition. Inclusion of passive denaturants such as urea or formamide in the polyacrylamide gel suppresses base pairing and eliminates the influence of base composition. The presence of denaturants enhances band resolution.

PhastSystem   Nucleic acid scanning techniques have the potential of analyzing thousands of individuals in plant and animal breeding applications. While DNA extraction remains the limiting step, automated work stations that use pipetting robots to assemble amplification reactions (Harrison et al. 1993, Garner et al. 1993) and real time DNA separation of amplification products using fluorophores or CE promise to speed pre- and postamplification steps. Currently, there is a need for

Fig. 2. Effect of loading buffer, loading procedure and glycerol on polyacrylamide gel electrophoresis (PAGE). Arbitrary signatures from amplification profiles (ASAP) generated from two independent DNA extractions of an isolate of the dogwood-pathogen *Discula destructiva* (NY329) were separated in 10 %T:2 %C polyacrylamide gels in the absence (**A**) or presence (**B**) of 5 % glycerol. Samples were loaded in urea (LB_a, formulation A) or urea-Ficoll (LB_b, formulation B) loading buffer using small-bore flat pipette tips (*FT*) or by sample "dropping" using standard pipette tips (*T*). DAF profiles originally amplified with primer GTAACGCC were reamplified with the mini-hairpin primer GCGAAGC-CAG. Molecular weight markers are indicated in kilobase pairs. **C** The most common artifacts observed during PAGE of DAF products: *Lane 1*: dust particle (*arrowhead*) embedded in the gel during improper gel casting; *Lane 2*: sample overloading and surface staining blemishes (arrowhead) due to poor gel handling; *Lane 3*: gel streaking caused by contaminants introduced during loading (*left arrowhead*) and typical "fuzzy" bands from sample overloading (*right arrowhead*); *Lane 4*: artifacts due to air bubbles introduced during gel casting (*arrowhead*) and band distortion and localized overdevelopment during silver staining caused by acrylamide deposition behind the polyester film (due to the incorrect application of the backing sheet to the glass plate); *Lane 5*: excessive tightening of clamp assembly on one of its sides causing the bottom edge of the film to lift away from the glass and ultimately band distortion. Also note the use of old acrylamide solutions and poor loading in *lanes 3, 4*

a relatively inexpensive DNA separation alternative that would still fulfill high throughput analysis. Semi-automatic separation and staining of DAF fragments in miniature polyacrylamide gels using the PhastSystem can resolve DNA fingerprints with high throughput and simplify detection of diagnostic DNA polymorphisms. The procedure combines our ultrasensitive silver staining technique with vitually complete automation of sample loading, electrophoresis and staining (Baum et al. 1994; G. Caetano-Anollés and B.J. Bassam, unpublished),

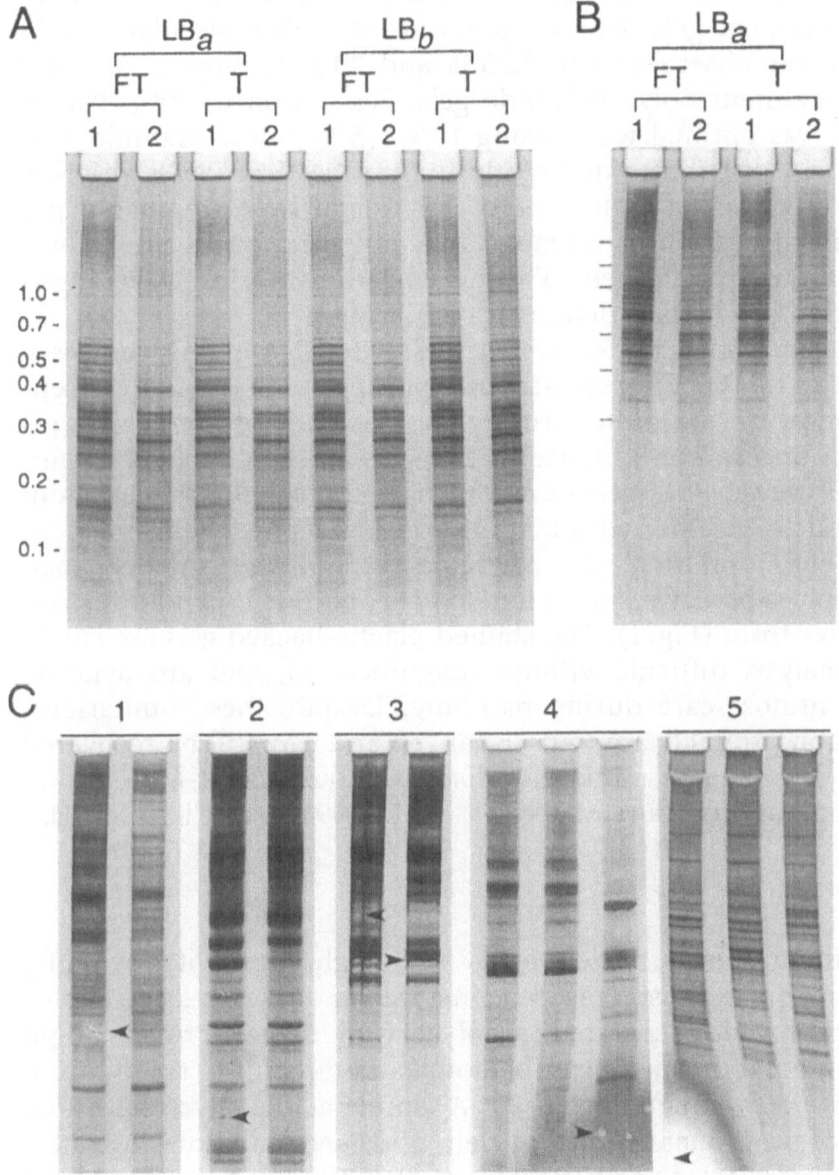

allowing processing of over 180 samples/day per unit with minimal manipulation.

Although requiring the added expense of proprietary reagents and consumables (US $8/gel), the PhastSystem eliminates gel preparation (with the use of precast gels polymerized onto clear backing film), permits fast electrophoretic performance (30 samples/2 gels/30 min using one unit), high throughput (30–60 samples/h vs 15 samples/h using vertical PAGE), and reproducible and sensitive semi-automated DNA separation and staining due to preprogrammed conditions).

Mobility shifts vary with gel concentration and the use of homogeneous or gradient gels. Good separation of high molecular weight fragments was observed with 12.5% and 20% homogeneous, and 8%–25% gradient polyacrylamide gels. Separation of 200–1000 bp fragments was optimal when using 10%–15% polyacrylamide gels. Overall, DNA distributes differently than in vertical PAGE, offering superior resolution of higher molecular weight DNA fragments and allowing identification of valuable DNA polymorphisms unresolved using more traditional gels. The visualization of closely arranged amplification products is demonstrated in Fig. 1.

The discontinuous buffer system resolves DNA poorly unless a long pre-run of 100 Vh generates an appropriate in-gel ion concentration following the buffer front. The existence of a stacking gel (Fig. 1), an uncommon practice in DNA separation, appears advantageous. However, Phastgels have a restricted resolving range (about 20 mm) which, together with loading constraints, results in the general loss of resolution. In particular, fragments smaller than 100–150 bp are poorly resolved due to gel compression and difussion of the buffer front (Fig. 1). The stained plastic-backed gels are small, making analysis difficult without magnification, and are delicate, requiring utmost care during handling. Despite these limitations, amplification products are well visualized and can still be recovered succesfully from these miniature, discontinuous plastic-backed matrices using the procedure of Weaver et al. (1994) (see Chap. XVII).

Silver Staining

Silver staining detects nucleic acids with high sensitivity, avoiding fluorophore or radioisotopic labeling. Silver staining of complex nucleic acid profiles separated in polyacrylamide gels provides high band resolution but can be cumbersome and difficult to reproduce if an adequate technique is not used. A simple acidic silver stain that detects picogram quantities of nucleic acids separated on polyester-backed polyacrylamide gels (Bassam et al. 1991) has been developed recently to provide unsurpassed sensitivity and reproducibility. The technique has few steps and reagents, is fast, and produces the least number of staining artifacts (Bassam and Caetano-Anollés 1993; Caetano-Anollés and Gresshoff 1994), and has even been used in DNA sequencing applications (Storts et al. 1993). The silver stained polyester-backed gels can be preserved for many years by drying and act as safe repositories of electrophoresed DNA amplification products. These gels permit the retrospective recovery and examination of nucleic acids, constitute experimental records, avoid the costs of photography, and allow easy densitometrical scanning of DNA profiles for data analysis. A desired band, whether part of a simple or complex array of DNA fragments in a profile, can be directly excised and used as template for further amplification. For a detailed description of band recovery from silver stained gels see Chap. XVII.

The silver staining protocol is based on a photochemically derived silver stain originally designed for the staining of proteins (Goldman and Merril 1982). The technique is based on the use of silver nitrate as the impregnating agent and formaldehyde in an alkaline environment as the reducer. Impregnation is done with relatively low concentrations of silver in the presence of formaldehyde. Image development occurs by reduction of silver to metallic silver by formaldehyde at low temperature (8°–12 °C) and alkaline pH. The metallic silver deposits in the immediate vicinity of the staining substratum while the complexant sodium thiosulfate keeps silver reduction in the polyacrylamide matrix to a minimum. Silver complexation alters the kinetics of silver reduction and helps minimize background staining. Image development is finally stopped by decreasing pH, preferably with the use of weak acids such as acetic or citric acid. A detailed discussion of the many parameters influencing the silver staining reaction can be found in Caetano-Anollés and Gresshoff (1994).

Data Analysis

A fingerprint pattern is only informative if it can be compared to other patterns. To do so the pattern has to be evaluated for matching components. For example, when DAF samples are separated by electrophoresis, individual bands within each lane have to be **sized** relative to molecular weight standards and then compared (**matched**) in the search for comigrating bands.

Band scoring can be arduous if done by the eye and hand. Bands are sized and matched directly on gels, autoradiographic or photographic films, or photocopies on transparency overlays. The presence, absence and intensity of a band in a particular location is noted relative to the weight standards, usually with the help of rulers or alignment devices and a lightbox. Scoring by the eye and hand is demanding and may suffer from error and bias from the investigator. However, image analysis can now be automated with appropriate hardware and software tools. The image can be recorded via a high-resolution video camera, scanning device, or phosphorimage analyzer, edited to reduce background and allow band alignment, and then analyzed for band matching. The analysis usually takes into account the mobility and intensity of the bands permitting the investigator to set thresholds of tolerance for each of these parameters. Automated image analysis has increased the level of accuracy with which fingerprints are analyzed and has simplified enormously the task (Gill et al. 1991).

3
Materials

Equipment
– electrophoresis apparatus: vertical PAGE unit for 0.45 mm-thick slab gels (Mini-Protean II cell, Bio-Rad, Hercules, CA), horizontal isothermally controlled electrophoresis unit (EC1001, E-C Apparatus Corp., St. Petersburg, FL) or semi-automated miniaturized electrophoresis equipment (PhastSystem, Pharmacia LKB, Piscataway, NJ)
– image recording device, for example, Apple Color One scanner (Apple Computer, Cupertino, CA) and Ofoto program (version 2.02; Light Source Computer Images Inc., Salinas, CA)
– image analysis device, for example, Apple Macintosh computer (Apple Computer, Cupertino, CA) and Think Pascal program Image for the Macintosh computer (Version 1.45; Wayne Rasband, NIH; Internet, wayne@helix.nih.gov)

Supplies
– polyester gel-backing film (GelBond PAG, FMC Bioproducts, Rockland, ME; or other silanized plastic sheets)
– membrane syringe filters
– precast backed gels and buffer strips for use with the PhastSystem (Pharmacia LKB)
– staining trays with flat bottoms and straight sides (if desired, use clear plastic lids from 1000 µl pipette tip racks)

Reagents and solutions
Reagents used in DNA separation must be electrophoresis grade and those used in silver staining of high purity analytical grade.

– Tris-borate-EDTA (TBE) buffer stock (10x): 1 M Tris-HCl, 0.83 M boric acid, and 10 mM $Na_2EDTA \cdot H_2O$ (pH 8.3)

- leading ion buffer: 50 mM formate and 130 mM Tris-HCl (pH 8.5)
- trailing ion buffer: 200 mM of an appropriate trailing acid species (such as proline, glycine or tricine; Doktycz 1993) in 100 mM NaOH
- acrylamide/cross-linker/urea stock solution (1x): 9.8 % acrylamide, 0.2 % piperazine diacrylamide, and 10 % urea in 1x TBE.

Note: Acrylamide solutions are light sensitive and should be stored in the dark at 4 °C.

Caution: Acrylamide is a potent neurotoxin. Handle this solution with care and dispose of it appropriately.

- ammonium persulfate (10 % stock)
- N,N,N',N'-tetramethylethylenediamine (TEMED)
- loading buffer: 30 % urea, 0.08 % xylene cyanol FF (formulation A; Caetano-Anollés et al. 1991) or 40 % urea, 3 % Ficoll, 0.02 % xylene cyanol, 0.02 % bromphenol blue, in 1x TBE (formulation B; Bassam and Bentley 1995)
- fixative/stop solution: 7.5 % (v/v) glacial acetic acid
- silver impregnating solution: 1 g/l silver nitrate, 1.5 ml/l 37 % formaldehyde. Recycle the used silver by precipitation with NaCl.

Caution: Silver is toxic. Handle this solution with care and dispose of it appropriately.

- developer: 30 g/l sodium carbonate, 3 ml/l formaldehyde, 2 mg/l sodium thiosulfate

Note: Use relatively fresh formaldehyde and never store it in the cold. Prepare impregnation and developer solutions in advance and store at room temperature. Add formaldehyde and freshly prepared (at least weekly) sodium thiosulfate prior to staining.

4
Experimental Procedure

Amplification products can be separated in vertical or open-faced polyacrylamide slab gels using a variety of protocols. I will describe DNA separation by vertical denaturing PAGE using 0.45 mm thick polyacrylamide gels (8×10 cm) backed on polyester film (Caetano-Anollés and Bassam 1993), open-faced discontinuous denaturing or nondenaturing polyacrylamide gels (Allen et al. 1989; Doktycz 1993), and by using a semi-automated miniaturized electrophoretic and staining device.

Vertical PAGE

1. Assemble each electrophoretic rig under running distilled water to avoid dust particles. Place the large glass plate on clamp assembly, then the polyester backing sheet (hydrophilic side up) in tight apposition to the large glass plate, spacers, and finally the small glass plate. Make sure all rig components are flush against the bottom before tightening assembly screws.

2. Dry the gel rig in a dust-free area in the dark (backing sheets are light-sensitive) overnight. If desired, use an oven at 40°–50 °C to speed the process. Alternatively, subject the gel rigs to an 85 % ethanol rinse before drying.

3. Assemble gel rigs in casting stand.

4. Prepare to cast 4 %–15 %T and 2 %C polyacrylamide-urea gels. Mix 10 ml of the acrylamide/cross-linker/urea stock solution with 150 μl ammonium persulfate and 15 μl TEMED.

   Note: Use relatively fresh acrylamide stock solutions to attain optimal fingerprint quality. De-gassing of the solution is not necessary.

5. Deliver the gel mix by injection through a 0.45 μm pore size membrane (this eliminates dust particles and the possibility of staining artifacts) and insert Teflon comb. The gel mix begins to set in about 2 min and should fully polymerize in 30 min.

6. Attach gel rigs to electrode core, place into buffer tank, and fill buffer reservoirs with 1x TBE.

7. Remove combs, rinse wells with buffer using a fine-gauge syringe needle, and pre-electrophorese gels at 150 V for at least 5 min.

8. Rinse and then load wells with 3 μl of a dilution of each amplification reaction (usually containing 30–40 ng DNA) mixed with 3 μl of loading buffer.

9. Electrophorese at 150 V for about 60–90 min.

   Note: Voltages as high as 300 V can be used; depending on running buffer, gel composition and gel thickness, higher voltages can produce band distortion.

10. Disassemble gel rigs and remove backed gels

Open-faced discontinuous PAGE

1. Pour the gel mix, prepared as described in step 4 of Vertical PAGE and filtered through 0.2–0.45 μm filters, by capillary action between a glass plate and backing polyester film separated by 0.2–0.4 mm spacers.

2. Wash the gel with abundant water and air dry.

3. Before electrophoresis, rehydrate the gel with leading ion buffer.

4. Place the gel on the horizontal plate of an isothermal controlled electrophoresis apparatus maintained at 10 °C.

5. Carefully place electrodes on top of cathodal and anodal buffer strips containing trailing ion buffer and leading ion buffer, respectively. Buffer strips are made of filter paper and are stacked vertically on the gel.

6. Apply DNA samples directly or diluted 1:3 in loading buffer to the gel surface, 1 cm from the cathode, with a sample applicator or by applying the sample on the surface of squares made of polypropylene filter mesh.

7. Electrophorese for about 1 h at 300 V or until the bromophenol blue reaches the anode.

Note: The use of ultrathin gels diminishes heat generation and permits the use of higher electrical loads. Similarly, the use of long gels in the system increases the tolerance for higher differences in potential (as electrical loads vary with potential drop/cm).

1. Load precast, polyester-backed, nondenaturing, 5 %–15 % polyacrylamide gels on the horizontal plate of a PhastSystem apparatus using a forceps, and place buffer strips into position. **Miniaturized semi-automated PAGE**

2. Place sample applicator containing 0.5 µl of amplification reactions on applicator holding arm.

3. Separate amplification products at 15 °C using the following steps: pre-electrophoresis for 100 Vh (400 V, 10 mA, 2.5 W), sample application for 2 Vh (400 V, 1 mA, 2.5 W); sample electrophoresis for 100 Vh (200 V, 10 mA, 2.5 W), and slow hold for 100 Vh (50 V, 1 mA, 0.5 W).

4. Remove gels at the holding phase (about 200 Vh total).

The polyester-backed gels can be fixed and stained with silver using the procedure of Bassam et al. (1991). The stain detects about 1 pg DNA/mm^2 band cross-section, and is commercially available (Promega Corp., Madison, WI). During steps 2–6 of the following protocol, subject gels to agitation on a shaker at 50 rpm. **Silver staining**

1. Place gels in staining trays (gel facing up): lift one corner of the backing film and peel away the backed gel from the rear glass plate. Handle gels with care; they should not touch any surface until image development.

2. Fix the gels in fixative (acetic acid solution) for 10 min.

3. Wash the gels three times with distilled water, 2 min each time.

4. Impregnate with silver solution for 20 min. During staining, the gel remains attached to the tray by surface tension facilitating removal of spent staining solutions by tipping out the liquid.

5. Rinse with distilled water for 5–20 s.

6. Develop the image with developer solution kept at 8 °C. Optimum image contrast and minimum background is generally attained in 4 min.

7. Stop image development in stop (acetic acid) solution for at least 1 min. Wash extensively with water.

8. Dry stained gels at room temperature and store in photographic albums.

Automated silver staining

1. Place gels on the chamber of the PhastSystem automated silver staining unit.

2. Program and run the following routine: step 1, 10 min fixation; steps 2–4, 2 min washes; step 5, 20 min silver impregnation; step 6, 0.1 min wash; step 7, 3.5 min image development; step 8, stop reaction with acetic acid (1 min); and step 9, final wash for 4 min.

   Note: The steps use four staining reagents (fixative-stop, impregnation, and developer solutions and water) delivered through six in-ports and discarded through two out-ports (silver discarded separately), and are done at 15 °C, except for steps 7 and 8 which use ice-cold solutions. The use of ice-cold developer compensates for temperature shifts during image development due to time delays in the automated delivery and evacuation of developer solution, resulting in a highly reliable stain.

3. Silver stained backed gels are preserved by drying at room temperature.

Image processing and data analysis

Silver stained profiles can be scanned and the images stored in electronic files. For example, fingerprints can be scanned with an Apple Color One scanner and the Ofoto program, and the images subsequently analyzed using the program Image for Macintosh computer using a gel analysis macro (see Fig. 2 in Chap. IX). Scanned images can be analyzed with a variety of computer programs. For example, Dendron (Solltech Inc., Oakdale, IA) permits manipulation of lanes and gels for proper alignment with molecular weights, constructs composite images from different gels, finds and assigns polymorphisms, computes similarity coefficients, and generates dendrograms that permit, for example, the analysis of microbial isolates.

- Gel casting: To avoid staining artifacts, it is important to wear gloves during gel rig assembly and silver staining. The use of some detergents has been shown to leave residues that increase background staining (Krutchinina and Gresshoff 1995). Clean glass plates with distilled water only; if necessary use ethanol or an hydrochloric acid wash. **Tips**

- Gel loading: To avoid production of irregular wells and aberrant fingerprints, rinse the wells with running buffer immediately after removing the Teflon combs and before loading. This will eliminate residual acrylamide and the chances of polymerization in the wells.

The efficacy of loading depends on selecting a suitable loading buffer and on simplifying experimental manipulations. Loading buffers should contain the sample within the well or the gel surface, undiluted, until DNA fragments have entered the polyacrylamide matrix. Moreover, these buffers should not interfere with normal electrophoresis of the samples. When using denaturing urea-polyacrylamide gels, avoid sucrose or glycine containing buffers. Urea, Ficoll, formamide, and even glycerol can be used if added in suitable concentrations. Samples can be applied with the help of small-bore flat pipette tips by deposition on the bottom of the well. Alternatively, normal pipette tips (for 20 µl or 200 µl pipettes) can be used. In this case, the wells should be cast only about 5 mm deep and the pipette tip should be placed vertical immediately on top of the well before sample "dropping." The two alternative loading procedures produce identical results. An example of these procedures used in combination with two alternative loading buffers can be found in Fig. 2.

In general, the samples should be loaded swiftly but with a steady hand to permit uniform placement on the bottom of the wells.

- Data analysis: During electrophoresis some bands are not as well resolved as others. It is good practice to score bands within a region of the gel where fingerprint patterns are well discernable. For example, using a standard PAGE protocol for DAF analysis, only those products that are <500–700 bp in length are generally scored. Bands should be scored consistently throughout the different samples to be compared. In some cases, comigrating bands exhibit differences in intensity which could represent the dosage of the amplification products (perhaps indicative of the homozygous or heterozygous state) or existence of more than one comigrating band (probably representing more than one locus). Exclude from the analysis those products that exhibit wide but unexplainable variations in intensity or lend themselves to ambiguous interpretation.

5
Troubleshooting

- Polyester film alignment and application: The application of polyester backing film can introduce artifacts during both DNA separation and silver staining. Incorrect film alignment on the bottom of the gel rigs is responsible for the majority of acrylamide leaks during gel pouring. Edge sealing can be accomplished by simply leveling the glass plates, film and spacers along the bottom edge during gel rig assembly. When doing this do not tighten screws forcefully and make sure that the rig components are correctly assembled. If problems persist, Bassam and Bentley (1995) have suggested the use of malleable plasticine adhesive (Blu-Tack, Bostick America, Middleton, MA) as a sealant. Simply press the adhesive along the bottom of the gel rig assembly and snap into position in the casting stand at the time of gel pouring. Remember to remove the original rubber gasket of the stand to make space for the plasticine adhesive.

  Incorrect application of the backing sheet to the glass plate can also cause the bottom edge of the film to lift away from the glass. This results in acrylamide depositing behind the film causing uneven gel thickness and irregularities in the electric field during electrophoresis (Fig. 2). Avoid this problem by assembling gel rig components under running water and by pressing the backing film firmly onto the large glass plate. This can be done by rubbing a finger on the film surface and then aligning the rig components. Rig assembly under running water also facilitates alignment of film, glass plates and spacers into position, and eliminates dust and contamination. In particular, avoidance of dust particles (during assembly and gel pouring) eliminates a common type of silver staining artifact, the appearance of spots and streaks in the gels (Fig. 2). The small (front) glass plate makes contact with the gel and should therefore be kept particularly clean and free of scratches.

- Gel loading: Avoid introducing contaminants in the wells; they can produce streaking after electrophoresis and staining (Fig. 2). Nucleic acid overloading can also cause streaking and fuzzy bands (Fig. 2). Dilute samples appropriately (usually 1:3–1:10) to avoid these artifacts.

- Gel electrophoresis: During electrophoresis, mobility of DNA fragments can shift due to irregularities of the electric field or protein or polysaccharide impurities in the samples. These problems can be usually corrected by running molecular weight standards every few lanes or by including them together with the samples, by using monomorphic products as internal references, and by running samples at least in duplicate.

  Scanned images of gels can also be manipulated to correct artifactual electrophoretic irregularities, such as the common "smiling" effect on PAGE in which centrally located bands exhibit higher mobilities, and warping due to variation in electric field.

References

Allen RC, Graves G, Budowle B (1989) Polymerase chain reaction amplification products separated on rehydratable polyacrylamide gels and stained with silver. BioTechniques 7:736–744

Bassam BJ, Bantley S (1995) Electrophoresis using polyester-backed polyacrylamide gels. BioTechniques 19:568–573

Bassam BJ, Caetano-Anollés G (1993) Silver staining of DNA in polyacrylamide gels. Appl Biochem Biotechnol 42:181–188

Bassam BJ, Caetano-Anollés G, Gresshoff PM (1991) Fast and sensitive silver staining of DNA in polyacrylamide gels. Anal Biochem 196:81–84

Baum TJ, Gresshoff PM, Lewis SA, Dean RA (1994) Characterization and phylogenetic analysis of four root-knot nematode species using DNA amplification fingerprinting and automated polyacrylamide gel electrophoresis. Mol Plant-Microbe Interact 7:39–47

Beattie KL, Beattie WG, Meng L, Turner SL, Coral-Vazquez R, Smith DD, McIntyre PM, Dao DD (1995) Advances in genosensor research. Clin Chem 41:700–706

Caetano-Anollés G, Bassam BJ (1993) DNA amplification fingerprinting using arbitrary oligonucleotide primers. Appl Biochem Biotechnol 42:189–200

Caetano-Anollés G, Gresshoff PM (1994) Staining nucleic acids with silver. An alternative to radiosotopic and fluorescent labeling. Promega Notes 45:13–18

Caetano-Anollés G, Bassam BJ, Gresshoff PM (1991) DNA amplification fingerprinting using very short arbitrary oligonucleotide primers. Bio/Technology 9:553–557

Caetano-Anollés, G., Bassam, B.J., and Gresshoff, P.M. (1992) DNA amplification fingerprinting with very short primers. In Applications of RAPD technology to plant breeding. Joint Plant Breeding Symposia Series. Crop Science Society of America, Minneapolis, pp. 18–25

Caetano-Anollés G, Callahan LM, Williams PE, Weaver KR, Gresshoff PM (1995) DNA amplification fingerprinting analysis of bermudagrass (*Cynodon*): genetic relationships between species and interspecific crosses. Theor Appl Genet 91:228–235

Chetverin AB, Kramer FR (1994) Oligonucleotide arrays: new concepts and possibilities. Bio/Technology 12:1093–1099

Doktycz MJ (1993) Discontinuous electrophoresis of DNA: adjusting DNA mobility by trailing ion net mobility. Anal Biochem 213:400–406

Garner HR, Armstrong B, Lininger DM (1993) High-throughput PCR. BioTechniques 14:112–115

Gill P, Evett IW, Woodroffe S, Lygo JE, Millican E, Webster M (1991) Databases, quality control and interpretation of DNA profiling in the home-office-forensic science service. Electrophoresis 12:204–209

Goldman D, Merrill CR (1982) Silver staining of DNA in polyacrylamide gels: linearity and effect of fragment size. Electrophoresis 3:24–26

Harrison D, Baldwin C, Prockop DJ (1993) Use of an automated workstation to facilitate PCR amplification, loading agarose gels and sequencing of DNA templates. BioTechniques 14:88–97

Jayarao BM, Bassam BJ, Caetano-Anollés G, Gresshoff PM, Oliver SP (1992) Subtyping of *Streptococcus uberis* by DNA amplification fingerprinting. J. Clin. Microbiol. 30:1347–1350

Kruchinina NG, Gresshoff PM (1994) Detergent affects silver sequencing. BioTechniques 17:280–282

McClelland M, Arensdorf H, Cheng R, Welsh J (1994) Arbitrarily primed PCR fingerprints resolved on SSCP gels. Nucleic Acids Res 22:1770–1771

Storts DR, Wu LC, Mendoza L, Oler JK (1993) Silver staining: a new approach to nonradioactive DNA sequencing. Promega Notes 42:10–14

Strezoska Z, Paunesku T, Radosavljevic D, Labat I, Drmanac R, Crkvenjakov R (1991) DNA sequencing by hybridization: 100 bases read by a non-gel-based method. Proc Natl Acad Sci USA 88:10089–10093

Weaver KR, Caetano-Anollés G, Gresshoff PM, Callahan LM (1994) Isolation and cloning of DNA amplification products from silver stained polyacrylamide gels. BioTechniques 16:226–227

Welsh J, McClelland M (1990) Fingerprinting genomes using PCR with arbitrary primers. Nucleic Acids Res 18:7213–7218

Williams JGK, Kubelik AR, Livak KJ, Rafalski JA, Tingey SV (1990) DNA polymorphisms amplified by arbitrary primers are useful as genetic markers. Nucleic Acids Res 18:6531–6535

Denaturing Gradient Gel Electrophoresis for Enhanced Detection of DNA Polymorphisms

I. DWEIKAT and S. MACKENZIE

1
Introduction

DNA-based marker systems have, over the past 5 years, become much more refined as adaptations are incorporated to address needs specific to different genetic systems. PCR-based methodologies have found important applications in genomic mapping efforts due to the small amounts of tissue required and the rapidity of the procedure; these advantages allow the investigator to conduct studies with much larger population sizes than were previously feasible. Elaborations on the standard PCR procedure have facilitated the simultaneous evaluation of exceedingly large numbers of loci per experiment (Vos et al. 1995) as well as the rapid screening for linkage disequilibrium in segregating or heterogeneous populations (Michelmore et al. 1991). Such approaches have proven invaluable for high density mapping of a target genomic region.

One important constraint that has presented constant challenge to the investigator interested in DNA-based mapping and genotypic screening is the lack of sufficient DNA polymorphism detectable in some species and population structures (e.g., those derived from highly related parental strains) using standard methodologies. One effective means of circumventing these problems is the application of denaturing gradient gel electrophoresis (DGGE). The adoption of DGGE for the resolution of arbitrarily primed PCR products (AP-PCR, RAPD, DAF) has proven particularly effective for the enhanced detection of DNA polymorphism.

DNA fragments subjected to DGGE migrate based not only on size but on melting properties of the fragment (Myers et al. 1985). Consequently, use of this system allows the simultaneous resolution of DNA fragment length polymorphism as well as DNA sequence variations existing between two like-sized fragments. Accordingly, DGGE has proven especially useful in the monitoring of allelic variants, permitting detection of even single base pair substitutions (Myers et al. 1985; Abrams et al. 1990), pedigree assessments in highly selected populations comprised of closely related individuals (He et al. 1992; Dweikat et al. 1993), and gene mapping in species that are particularly low in DNA fragment length polymorphisms by conventional methodologies (Dweikat et al. 1994; Procunier et al. 1995). Here we describe the DGGE procedure, its basis, and its application to genome analysis.

2
Principles and Applications

DNA fragment migration under DGGE conditions is dependent on both fragment size and melting properties. This is by virtue of a gradient of denaturant (generally formamide plus urea) within the polyacrylamide gel that runs from low (0 %–40 %), at the origin to high (40 %–80 %) at the gel base. As a DNA fragment migrates through the gel, it reaches a denaturant concentration at which melting is initiated within the fragment. At this point, the primary melting domain of each fragment gives rise to a "bubble" or "fork" configuration, producing a partially single-stranded molecule that slows fragment migration (Myers et al. 1985). Slight differences in sequence, resulting from as few as one or two base pair changes, are reflected in a corresponding difference in melting properties and a potentially dramatic shift in fragment migration (see Fig. 1). Resolution of the polymorphism arising due to sequence differences can be further enhanced by the development of GC-rich primers that act as "clamps" to further stabilize the bubble structure (Sheffield et al. 1989; Abrams et al. 1990)

The DGGE procedure is well-suited for the detection of DNA polymorphism in situations in which restriction fragment length polymorphism is low but DNA sequence variations are expected. Thus, DGGE has been successfully applied to the detection of intragenic (allelic) variation in humans, accounting for over 75 % of the reported uses of DGGE. The monitoring of intragenic sequence variation using DGGE has facilitated population studies in complex microbial (Muyzer et al. 1993; Wawer and Muyzer 1995), domestic animal (Johnston and Fernando 1995), and marine animal (Norman et al. 1994) populations. DGGE has been successfully applied to gene mapping studies in wheat (*Triticum aestivum*) (Dweikat et al. 1994; Procunier et al. 1995), a species of unusually large genome size and low DNA polymorphism, by significantly enhancing the resolution of heritable polymorphism (He at al. 1992; Procunier et al. 1994). DGGE

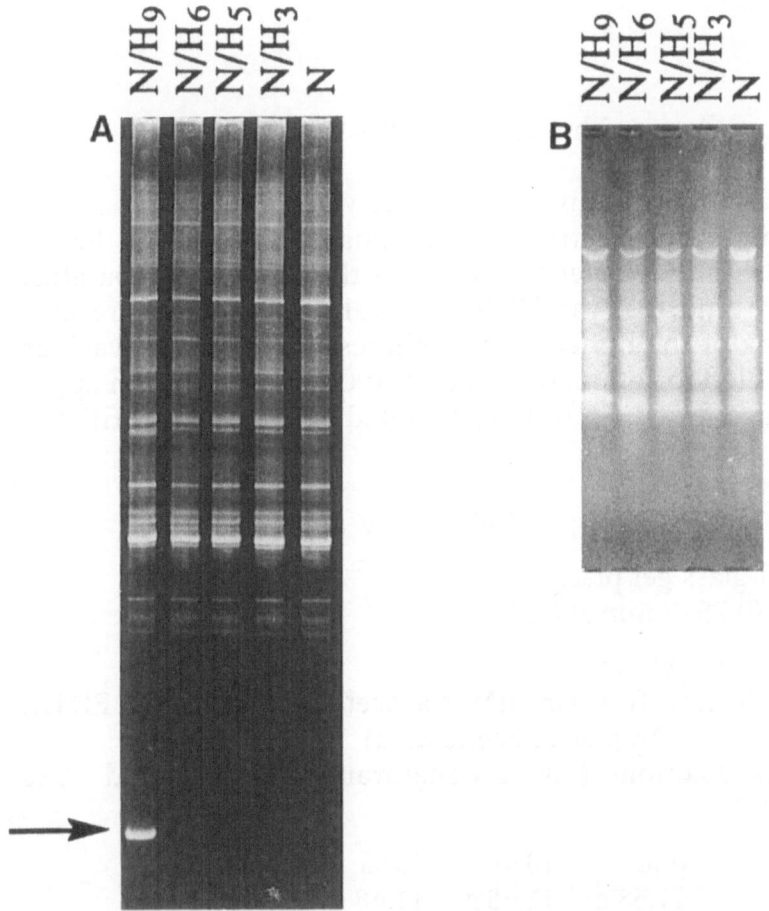

Fig. 1A,B. Comparison of denaturing gradient gel electrophoresis (DGGE) and agarose gel electrophoresis in resolving a DNA fragment polymorphism associated with
Hessian fly resistance gene *H9* in wheat. A PCR amplification of wheat genomic DNA
using an oligonucleotide (10-mer) primer and fractionated by 10%–50% denaturing
gradient gel electrophoresis. B PCR amplified product identical to that used in A but
fractionated in 1.2% agarose, 0.5x TBE. Wheat line Newton (*N*), susceptible to all
known Hessian fly biotypes, is shown with four near-isogenic lines, each containing
a different gene for resistance. *Arrow* indicates a DNA polymorphism associated with
the *H9* resistance gene. (From Dweikat et al. 1994)

has also been applied to the discrimination of highly related germ-
plasm in oat, wheat and barley (Dweikat et al. 1993). One additional
and logical use of DGGE that has, surprisingly, not yet been fully
exploited is as a rapid, initial screening procedure for allelic variants
derived by chemical mutagenesis (Hovig et al. 1992; Keohavong et al.
1995).

3
Materials

Equipment
- waterbath circulator with a range of 10°–100 °C
- stir plate
- gradient maker (10 ml minimum capacity)
- polyacrylamide gel electrophoresis apparatus (e.g., the SE600 vertical PAGE unit, Hoefer Scientific; or the Protein II apparatus, BioRad). We find that the SE600 vertical unit allows more uniform heating of the buffer. The usefulness of DGGE for various analyses has led to commercialization of the procedure. An apparatus designed for DGGE is now marketed by BioRad as D GENE.
- peristaltic pump
- Hamilton syringe
- power supply capable of providing 200 V and 150 mA

Supplies
- heat-treated glass gel plates
- gel spacers (0.75–1 mm thick)

Reagents and solutions
- 1 M KOH in 50 % ethanol
- 20x TAE (800 mM Tris, 400 mM Na acetate, 20 mM Na_2 EDTA, adjust pH to 7.6 with glacial acetic acid)
- DGGE stock solutions (100 % denaturant=7 M urea and 40 % formamide)

	0 %	10 %	50 %
acrylamide	11.68 g	11.68 g	11.68 g
bis-acrylamide	0.32 g	0.32 g	0.32 g
20x TAE	5 ml	5 ml	5 ml
formamide	0	4 ml	20 ml
urea	0	4.2 g	21.0 g
distilled water to	100 ml	100 ml	100 ml

- 10 % ammonium persulfate (prepare fresh daily)
- TEMED
- loading buffer (20 % sucrose or Ficoll, 0.1 % bromphenol blue in 1x TAE buffer)
- ethidium bromide:10 mg/ml in water

4
Experimental Procedure

Gel electrophoresis
This protocol is based on the procedure originally described by Myers et al. (1985, 1987).

1. Clean glass plates by soaking in 1 M KOH dissolved in 50 % ethanol for at least 1 h and rinsing in hot water followed by deionized water. Allow to air dry.

2. Assemble the plates with spacers and brackets for casting.

3. Place small (3 mm) stirbar in high density solution chamber of the gradient maker (chamber proximal to gel apparatus) and attach gradient maker tubing to a gel loading or 0.23 gauge needle.

4. Prepare two small beakers. To beaker 1 add 10 ml of 10 % denaturant stock solution, 100 μl of 10 % ammonium persulfate, mix well, and 3 μl TEMED, mix well. To beaker 2 add 10 ml of 50 % denaturant stock solution, 100 μl of 10 % ammonium persulfate, mix well, and 3 μl TEMED, mix well. No de-gassing is necessary.

5. Pour the beaker 1 (10 % denaturant) solution into the gradient maker low density chamber (distal to the gel apparatus) and open inter-chamber valve briefly, allowing a very small amount of the solution to pass between chambers, to eliminate bubbles between the chambers.

6. Pour the beaker 2 (50 % denaturant) solution into the gradient maker high density chamber (proximal to the gel apparatus). Turn on the stir plate to allow mixing in the 50 % denaturant chamber and simultaneously open the outlet valve.

7. Immediately open the inter-chamber valve to allow mixing between chambers and turn on the peristaltic pump for a pump rate of approximately 3–4 ml/min.

8. During gel pouring, hold needle still to avoid disruption of the gradient and fill plates to approximately 1 cm from the rim.

9. Insert comb at an angle to avoid introduction of air bubbles.

10. Allow gel polymerization for a minimum of 2 h at room temperature.

11. Prepare 4.5–5 l of 1x TAE running buffer and fill buffer chamber of gel apparatus.

12. Set circulating water bath to 65 °C, attach tubing to heat exchange unit of the gel apparatus, and begin water circulation at least 30 min prior to gel loading to prewarm running buffer.

13. Gently remove gel comb and fill wells with prewarmed 1x TAE running buffer.

14. Add loading buffer to the DNA samples in a ratio of 1:5 (loading buffer:sample) and underlay DNA samples into the gel wells using a Hamilton syringe.

15. Fill upper buffer chamber with warmed 1x TAE buffer, attach electrodes to power supply. Gel should be run at 150 V, constant voltage, for 5 h.

16. After run, gel can be stained in ethidium bromide (1 μl stock solution/10 ml 1x TAE buffer). Stain with gentle shaking for 30 min, followed by destaining in deionized water for at least 2 h.

Important tips

- Glass plates, spacers and clamps must be very clean.

- Standard glass plates are sometimes susceptible to cracking during a DGGE gel run. Consequently, it is advisable to use heat-treated glass plates.

- A minimum of 2 h are required for complete gel polymerization. Incomplete gel polymerization will significantly affect gel resolution. After polymerization, a gel can be stored overnight at 4 °C prior to running.

- When the gel clamps are not properly tightened, buffer seepage will result in distortion of the outer gel lanes.

- The running buffer and gel pH must be equal. When this is not the case, glass cracking or poor gel resolution can occur.

- The running buffer can be reused two or three times without affecting gel quality.

- Greater resolution of DNA bands is achieved with smaller DNA sample volumes.

- Once the gel run is complete, or if it is unexpectedly interrupted due to power failure, the undisturbed gel can be retained in the gel apparatus for as long as 48 h without significant loss of resolution.

References

Abrams ES, Murdaugh SE, Lerman LS (1990) Comprehensive detection of single base changes in human genomic DNA using denaturing gradient gel electrophoresis and a GC clamp. Genomics 7:463–475

Dweikat IM, Mackenzie S, Levy M, Ohm H (1993) Pedigree assessment using RAPD-DGGE in cereal species. Theor Appl Genet 85:497–505

Dweikat IM, Ohm H, Mackenzie S, Patterson F, Cambron S, Ratcliffe R (1994) Association of a DNA marker with Hessian fly resistance gene *H9* in wheat. Theor Appl Genet 89:964–968

He S, Ohm H, Mackenzie S (1992) Detection of DNA sequence polymorphisms among wheat varieties. Theor Appl Genet 84:573–578

Hovig E, Smith-Sorensen B, Uitterlinden AG, Borresen AL (1992) Detection of DNA variation in cancer. Pharmacogenetics 2:317–328

Johnston DA, Fernando MA (1995) *Eimeria* spp of the domestic fowl: Analysis of genetic variability between species and strains using DNA polymorphisms amplified by arbitrary primers and denaturing gradient gel electrophoresis. Parasit Res 81:91–97

Keohavong P, Melacrinos A, Shukla R (1995) In vitro mutational spectrum of cyclopenta(cd)pyrene in the human *HPRT* gene. Carcinogenesis 16:855–860

Michelmore R, Paran I, Kesseli R (1991) Identification of markers linked to disease-resistance gene by bulked segregant analysis: A rapid method to detect markers in specific genomic regions by using segregating populations. Proc Natl Acad Sci USA 88:9828–9832

Muyzer G, DeWaal ED, Witterlinden AG (1993) Profiling of complex microbial populations by denaturing gradient gel electrophoresis analysis of polymerase chain reaction-amplified genes coding for 16S rRNA. Appl Environ Microbiol 59:695–700

Myers RM, Lumelsky N, Lerman LS, Maniatis T (1985) Detection of single base substitutions in total genomic DNA. Nature 313:495–497

Myers RM, Maniatis T, Lerman LS (1987) Detection and localization of single base changes by denaturing gradient gel electrophoresis. Methods Enzymol 155:501–527

Norman JA, Moritz C, Limpus CJ (1994) Mitochondrial DNA control region polymorphisms: Genetic markers for ecological studies of marine turtles. Molec Ecology 3:363–373

Procunier JD, Wolf M, Howes NK (1994) Increase of inheritable polymorphisms of arbitrary primed PCR products on DGGE gels. Biotech Techniq 8:707–710

Procunier JD, Townley-Smith TF, Fox S, Proshar S, Gray M, Kim WK, Czarnecki E, Dyck PL (1995) PCR-based RAPD/DGGE markers linked to leaf rust resistance genes *Lr29* and *Lr25* in wheat (*Triticum aestivum* L) J Genet Breed 49:87–91

Sheffield VC, Cox DR, Lerman LS, Myers RM (1989) Attachment of a 40-base-pair G+C-rich sequence (GC-clamp) to genomic DNA fragments by the polymerase chain reaction results in improved detection of single base-pair changes. Proc Natl Acad Sci USA 86:232–236

Vos P, Hogers R, Bleeker M, Reijans M, van de Lee T, Hornes M, Frijters A, Pot J, Peleman J, Kuiper M, Zabeau M (1995) AFLP: a new technique for DNA fingerprinting. Nucl Acids Res 23:4407–4414

Wawer C, Muyzer G (1995) Genetic diversity of desulfovibrio spp. in environmental samples analyzed by denaturing gradient gel electrophoresis of (NiFe) hydrogenase gene fragments. Appl Environ Microbiol 61:2203–2210

Modified Temperature Sweep Gel Electrophoresis for the Separation of Arbitrarily Amplified DNA Fragments

G. A. PENNER

1
Introduction

Early suggestions by Williams et al. (1990), that RAPD analysis tended to yield the same number of fragments regardless of the size of the genome analyzed, appear to have been based on the level of resolution applied. Initial attempts to detect chromosome-specific amplified fragments through the application of arbitrary primers to the nullisomic/tetrasomic aneuploid set available for all wheat chromosomes (except chromosome 5B) were largely unsuccessful (Devos and Gale 1992). It appears that by far the majority of bands identified on agarose gels were the result of the comigration of fragments amplified from different sites in the genome. If a band is composed of a single fragment corresponding to a single site in the genome, then we would expect to be able to detect its absence in one of a series of nullisomic lines.

Agarose gels are capable of differentiating fragments on the basis of size only. It is possible through the use of acrylamide gels to increase the resolution based on size by maintaining tighter control on deviation from average pore size. The DNA duplex is held together by hydrogen bonds that differ in strength depending on the base pairs involved. Adenine (A)-thymine (T) attraction is based on only two hydrogen bonds, while guanine (G)-cytosine (C) attraction is based on three hydrogen bonds. By exerting an increasing denaturation force in the same direction as the fragment migrating through a gel, sequence differences can be manifested by differences in mobility. A – T bonds will be broken at lower denaturation conditions than C – G bonds and cause a local site of denaturation in the double-stranded helix causing the fragment's mobility to be retarded. In practice, it is postulated that this effect would be most apparent in regions that are AT-rich, and hence susceptible to denaturation, rather than in regions that are GC-rich.

Dweikat et al. (1993) applied this theory by introducing denaturing gradient gel electrophoresis, a technique that involved the construction of a physical gradient of denaturant within an acrylamide gel (see Chap. XIV). Yoshino et al. (1991) developed a denaturating acrylamide technique that relies on an increase in temperature over time to create a denaturation gradient. This technique, termed temperature sweep gel electrophoresis (TSGE), was developed for the separation of single-stranded DNA or RNA fragments. Penner and Bezte (1994) modified the protocol such that the migrating DNA fragments were maintained in at least a partially double-stranded state, thus facilitating rapid staining with ethidium bromide. By imposing the denaturation gradient over time, all fragments, regardless of their size, are exposed to the same denaturing conditions. This results in equally sharp bands throughout the gel.

2
Materials

Equipment
- gel apparatus (BioRad Protean II)
- power supply (BioRad model 550/200)
- peristaltic pump (Buchler)
- circulating water bath (Haake D8)
- UV transilluminator
- photographic apparatus

Reagents and solutions
- 50x TAE: 2.0 M Tris-acetate, 0.05 M EDTA (pH 8.0)
- 40 % acrylamide: 37.5 g acrylamide: 1 g bis-acrylamide
- 100 % denaturant solution: 7 M urea, 40 % deionized formamide
- APS: 20 % (w/v) ammonium persulfate dissolved in water (made fresh every day)
- TEMED
- 6x loading buffer. For a 10 ml solution: 4 g sucrose, 25 mg bromophenol blue, 2.4 ml 0.5 M EDTA (pH 8.0)
- ethidium bromide stock solution: 10 mg/ml

3
Experimental Procedure

Preparation of TSGE gels
1. Start with clean, dry glass plates, spacers, combs and clamps, otherwise gel polymerization and/or sealing may not be adequate. Assemble glass plates using 0.75 mm spacers.

2. Prepare gel as follows for 22 ml of an 8 % acrylamide/50 % denaturant gel. Combine the following in a small, side-arm Erlenmeyer flask:
   - 4.4 ml 40 % acrylamide
   - 11 ml 100 % denaturant
   - 6 ml ddH$_2$O
   - 440 µl 50x TAE

3. Mix with gentle swirling.

4. Warm solution to room temperature by running hot water over
   the flask or by immersion in a water bath (optional to decrease
   polymerization time).

5. De-gas the solution for about 10 min using a vaccum trap
   (optional to decrease polymerization time).

6. Add 55 µl APS and 22 µl TEMED and gently mix.

7. Pour solution into preassembled glass plates using a syringe. The
   solution should be injected slowly to avoid the formation of tiny
   air bubbles that inhibit polymerization.

8. Insert comb (0.75 mm) being careful not to trap any air bubbles.

9. Allow at least 50 min for gel to polymerize. (If steps 4 and 5 are
   performed allow 10–20 min.)

10. Prepare 4 l of 1x TAE buffer for the lower buffer chamber and pre-
    warm to 25 °C through the use of the circulating water bath.

11. Prepare an additional 500 ml of 1x TAE buffer for the upper buffer
    chamber and for rinsing wells.

12. Following gel polymerization remove the comb carefully and
    purge the wells with 1x TAE. Use enough force to displace all the
    unpolymerized acrylamide. Remove all liquid from the wells.

13. Rinse the wells twice with ddH$_2$O, removing all liquid after each
    rinse, then fill the wells with 1x TAE for loading.

14. Add sufficient 6x loading buffer to PCR reaction samples such
    that the final concentration is 1x, prior to loading. Load 10 µl of
    sample into each well with a 50 µl glass syringe. Rinse the syringe
    thoroughly with ddH$_2$O after loading each sample.

15. Prior to attaching the plates to the cooling core, wet the seals of
    the core with buffer. Attach the plates.

16. Pour remaining buffer into upper buffer chamber.

17. Attach tubing through peristaltic pump and begin circulating buffer from the lower to the upper chamber at a rate of 80 ml/h.

Running TSGE separations

1. Attach a power supply to the electrophoresis unit and apply constant voltage at 300 V. The current should be in the range of 30–50 mA.

2. Gradually increase the temperature in the cooling core by increasing the temperature of the circulating water bath in the following manner.
   Prewarm running buffer to 25 °C.
   Start electrophoresis:1 h at 25 °C; 1 h at 35 °C; 1 h at 45 °C; 1.5 h at 55 °C.
   End electrophoresis.

Temperature conditions are for Haake circulating water bath, the temperature in the buffer tank should increase from 25 °C to 43 °C.

Visualizing separations

1. After running electrophoresis for 4.5 h remove the gel apparatus from the buffer tank and cool for 10 min before separating glass plates.

2. Remove one spacer and carefully pry plates apart with a plastic wedge. Do not use the spacers for prying. The gel should adhere to one plate or the other.

3. Invert the glass plate with the gel adhered to it and gently tease the gel off the plate into a dilute ethidium bromide solution (0.5 μg/ml) and stain for 10 min with gentle agitation.

4. Decant the ethidium bromide solution into a dark glass bottle and store in the dark at room temperature for subsequent use.

5. Wash the gel briefly with water (less than 5 min) and view on a UV transilluminator.

6. Photograph as described for agarose gels.

4
Results and Comments

A TSGE separation of arbitrarily amplified wheat DNA is compared to an agarose separation in Fig. 1. It is clear that considerably more bands are detected with TSGE than with agarose. We have found that silver staining does not result in the identification of significantly more information than staining with ethidium bromide. DNA frag-

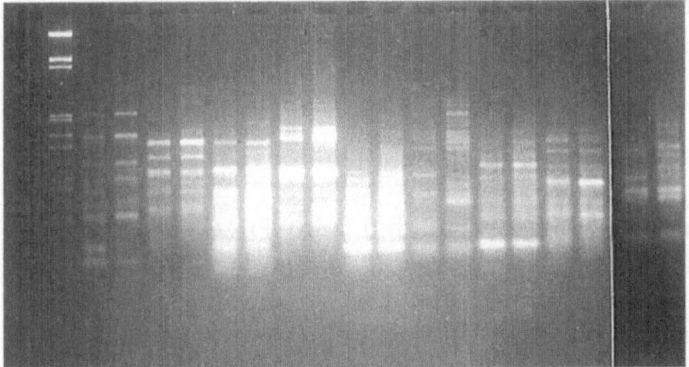

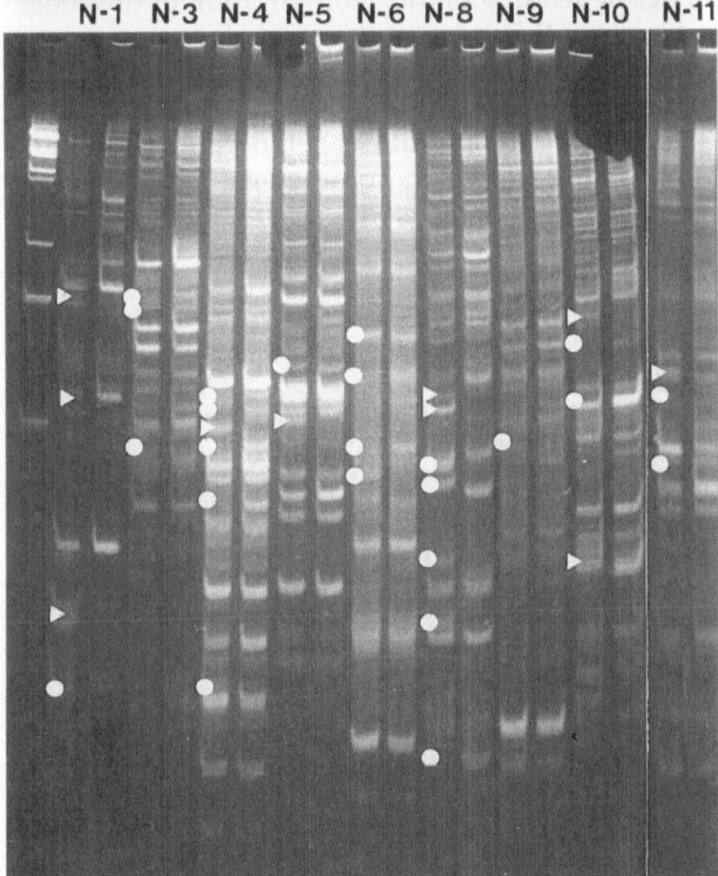

Fig. 1A,B. Comparison of resolving power between agarose gels and TSGE for RAPD fragments. **A** Agarose: *Left-most lane*, molecular weight marker (λ digested with *Eco*RI and *Hind*III); Primers used to amplify genomic DNA are listed across the *top*; *C*, Chinese Spring genomic DNA; *K*, Kenya Farmer genomic DNA. **B** TSGE: As above. *Circles* denote fragments that are chromosome-specific; *Triangles/Circles* denote fragments that were not designated as chromosome-specific but are polymorphic between these two wheat lines

ments can be readily purified from TSGE separations simply by soaking excised fragments in 0.1x TE overnight followed by ethanol precipitation.

5
Troubleshooting

TSGE separations can easily be overloaded with DNA, resulting in streaking within a lane and poorly resolved bands. Denaturation solutions must be reasonably fresh, and the formamide adequately deionized to ensure sharp bands. Different gel apparatuses have different heating properties depending on the configuration of the cooling core. The Bio-Rad Protean II has one of the most efficient temperature transfer systems available. The temperature curve with other apparatuses should be adjusted such that the run starts at 25 °C and ends at 43 °C 4.5 h later. If the running buffer temperature becomes warmer than 45 °C the fragments will become single-stranded and much more difficult to detect.

References

Devos KM, Gale MD (1992) The use of random amplified polymorphic DNA markers in wheat. Theor Appl Genet 84:567–570

Dweikat L, MacKenzie S, Levy M, Ohm H (1993) Pedigree assessment using RAPD-DGGE in cereal crop species. Theor Appl Genet 85:497–505

Penner GA, Bezte LJ (1994) Increased detection of polymorphism among randomly amplified wheat DNA fragments using a modified temperature sweep gel electrophoresis (TSGE) technique. Nucleic Acids Res 22:1780–1781

Williams JGK, Kubelik AR, Livak KJ, Rafalski JA, Tingey SV (1990) DNA polymorphisms amplified by arbitrary primers are useful as genetic markers. Nucleic Acids Res 18:6531–6535

Yoshino K, Nishigaki K, Husimi Y (1991) Temperature sweep gel electrophoresis: a simple method to detect point mutations. Nucleic Acids Res 19:3153

High Throughput Scoring of RAPD Fragments Through the Use of Dot-Blot Hybridization

G. A. PENNER

1
Introduction

The application of marker assisted selection to plant breeding is constrained by the cost of the technology employed and throughput capacity. At present, labor is a much more significant component of the cost per reaction than consumables. The reliance of PCR based marker methods on gel electrophoresis limits the number of samples that can be reasonably processed per unit of labor. Recent improvements in the detection of polymorphic fragments through denaturing acrylamide techniques such as denaturing gradient gel electrophoresis (DGGE) (see Chap. XIV) and temperature sweep gel electrophoresis (TSGE) (see Chap. XV) are not readily amenable to large scale, marker screening programs.

Product detection techniques that do not involve electrophoresis have been developed (Gu et al. 1995), but their lack of specificity makes them unsuitable for RAPD analysis. The conversion of RAPD fragments into allele-specific amplicons leads to increased reliability of amplification, increased allele specificity (increased information potential) and the facility to multiplex markers. In hexaploid species such as wheat, it has been difficult to convert RAPD markers due to cross-specificity with homologous loci in the other subgenomes (Penner, unpublished results). In these situations a more reliable and rapid means of scoring RAPD markers would be useful.

We recently developed a dot-blot hybridization technique (rapid RAPDs) that involves labeling the diagnostic RAPD fragment and hybridizing it to total amplified products immobilized in dot-blots on a nylon membrane (Penner et al. 1996). The use of chemiluminescence reduces X-ray film exposure times to under 1 h. Dot-blot hybridization is more time consuming than agarose electrophoresis with small sample sizes (2 days of analysis time with rapid RAPDs compared to 4 h with agarose electrophoresis). Marker-assisted selection requires analysis of thousands of samples with the same marker. Hybridization analysis of several membranes (each containing a minimum of 96 samples) can be performed in the same time frame as a single membrane. It has been our experience that the rapid RAPD detection method also increases the sensitivity of product detection and hence the reliability of sample scoring.

2
Materials

Equipment
– thermocycler
– dot-blot apparatus (Bio-Rad)

Supplies
– nylon membrane (Hybond N, Amersham)
– X-ray films (X-Omat 5, Kodak)

Kits
– chemiluminescent detection kit (Boehringer Mannheim)

Reagents and solutions
– phenol:chloroform (1:1)
– 95 % ethanol
– PCR labeling reaction: 50 µM of each of dATP, dCTP, and dGTP, 40 µM dTTP, 0.25 µM DIG-11-dUTP, 1.25 U *Taq* DNA polymerase, 1x reaction buffer, 1.5 mM $MgCl_2$
– 20x SSC: 3 M NaCl, 0.3 M Na citrate (pH 7.0)
– 10 % SDS
– prehybridization solution: 5x SSC, 1 % (w/v) blocking reagent (Boehringer Mannheim), 0.1 % (w/v) N-lauroylsarcosine, 0.02 % (w/v) SDS
– hybridization solution: the same as prehybridization solution plus the labeled diagnostic fragment

3
Experimental Procedure

Labeling of diagnostic fragments

1. Excise diagnostic RAPD fragments from agarose gels using long wavelength UV with a scalpel.

2. With your gloved thumb, apply firm pressure to chunk of agarose containing diagnostic fragment and wrapped in Parafilm

3. Pipette liquid extruded into a fresh Eppendorf tube. Extract with phenol:chloroform (1:1).

4. Precipitate DNA with 95 % ethanol. Resuspend in sterile water.

5. PCR label diagnostic RAPD fragments using the PCR labeling reaction at an annealing temperature of 38 °C.

6. Test the specific activity of the labeled fragments on test blots with 1/25 to 3/5 of total PCR product.

Dot blotting of arbitrarily amplified product

1. Insert a Hybond N membrane into a Bio-Rad dot-blot apparatus.

2. Pipette 5 µl aliquots from the 25 µl PCR reactions into each dot-blot hole.

3. Apply vacuum pressure to immobilize DNA fragments onto membrane.

4. Bake membrane at 80 °C for 1.5 h.

Hybridization of diagnostic fragment to total PCR product

1. Perform prehybridization at 65°–68 °C for 1 h. Use at least 20 ml of prehybridization solution/100 square cm of filter.

2. Perform hybridization at 65°–68 °C for 1.5 h. Use 2.5 ml of hybridization solution/100 square cm of filter; the amount of labeled diagnostic fragment to be used is determined empirically as described in Sect. 3.

3. Wash membranes twice at room temperature in 2x SSC/0.1 % SDS for 5 min.

4. Wash membranes twice in 0.1x SSC/0.1 % SDS for 20 min at 68 °C.

5. Detect chemiluminescence according to Boehringer Mannheim protocols.

6. Expose membranes to X-ray film for periods varying from 5 to 90 min.

4
Results and Comments

Clear differentiation of the presence or absence of diagnostic fragments is demonstrated in Fig. 1. This technique has been shown to work with chromosome-specific fragments in wheat, and polymorphic fragments in wheat and barley.

5
Troubleshooting

Amplification failure may lead to miscoring of results. Positive controls can be performed through the application of labeled monomorphic, nonchromosome-specific fragments.

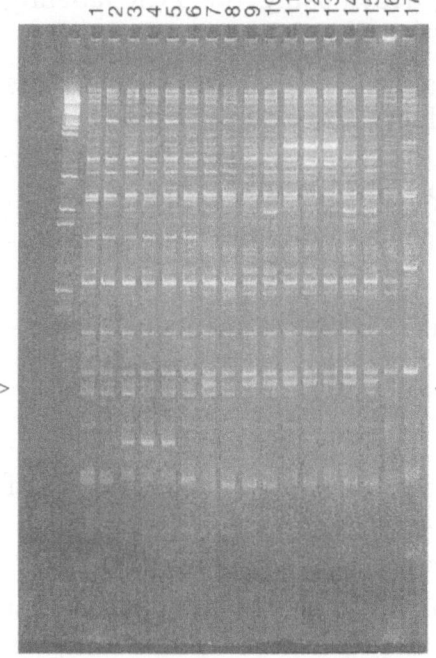

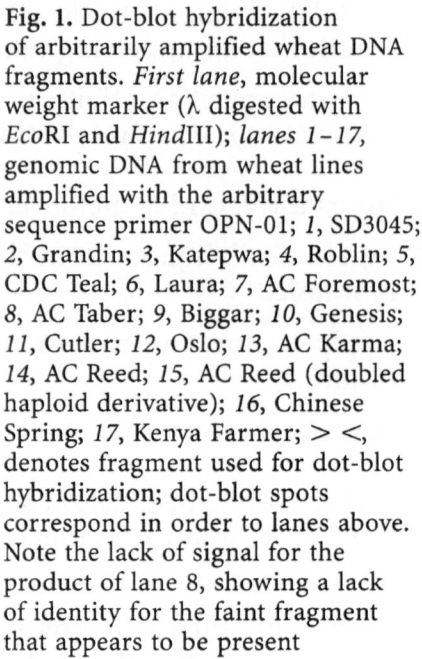

Fig. 1. Dot-blot hybridization of arbitrarily amplified wheat DNA fragments. *First lane,* molecular weight marker (λ digested with *Eco*RI and *Hind*III); *lanes 1–17,* genomic DNA from wheat lines amplified with the arbitrary sequence primer OPN-01; *1,* SD3045; *2,* Grandin; *3,* Katepwa; *4,* Roblin; *5,* CDC Teal; *6,* Laura; *7,* AC Foremost; *8,* AC Taber; *9,* Biggar; *10,* Genesis; *11,* Cutler; *12,* Oslo; *13,* AC Karma; *14,* AC Reed; *15,* AC Reed (doubled haploid derivative); *16,* Chinese Spring; *17,* Kenya Farmer; > <, denotes fragment used for dot-blot hybridization; dot-blot spots correspond in order to lanes above. Note the lack of signal for the product of lane 8, showing a lack of identity for the faint fragment that appears to be present

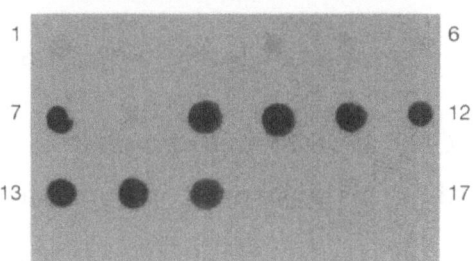

References

Gu WK, Weeden NF, Yu J, Wallace DH (1995) Large-scale, cost-effective screening of PCR products in marker-assisted selection applications. Theor Appl Genet 91:465–470

Penner GA, Lee SJ, Bezte LJ, Ugali E (1996) Rapid RAPD screening of plant DNA using dot blot hybridization. Molecular Breeding 2:7–10

Chapter XVII

Recovering Amplified DNA from Silver Stained Gels

G. Caetano-Anollés

1
Introduction

Individual nucleic acid fragments can be recovered from agarose or polyacrylamide gels by elution from the electrophoretic matrix (Smith 1980). Table 1 describes some of the methods that have proved effective in eluting nucleic acids into liquid or solid supports. These methods usually require a further purification step by phenol and chloroform extraction, sometimes followed by concentration before subcloning. Fragment isolation is particularly complicated when a complex assortment of nucleic acids has been resolved. This is the

Table 1. Elution of nucleic acids into liquid or solid supports

Method	Reference
Blotting	Southern 1975
Centrifugal filtration through filter membranes	Zhu et al. 1985
	Krowczynska et al. 1995
Centrifugal filtration through siliconized sterile glass	Heery et al. 1990
wool or Sephadex beads	Wang and Rossman 1994
Electroelution into dialysis bags	McDonell et al. 1977
Electroelution into DEAE cellulose	Dretzen et al. 1981
	Sylvers and Beresten 1993
Use of low-melting point agarose	Wieslander 1979
	Zintz and Beebe 1991
	Ausubel et al. 1992
Gel compression by freeze-squeeze	Thuring et al. 1975
	Tautz and Renz 1983
Mechanical gel extrusion	Li and Ownby 1993
Flush DNA extraction	Grey and Brendel 1992
Centrifugal DNA extraction	Schwartz and Whitton 1992
Passive overnight diffusion into buffer	Ausubel et al. 1992

case in a number of applications in which polyacrylamide gel electro-phoresis (PAGE) was coupled to silver staining, including DNA sequencing (Doktycz 1993; Storts et al. 1993), single strand conforma-tion polymorphism (SSCP) analysis (Ainsworth et al. 1991; Dockhorn-Dworniczak et al. 1991; Mohabeer et al. 1991; Maekawa et al. 1993; Sugano et al. 1993; Ainsworth et al. 1993), DNA profiling (Allen et al. 1989; Budowle et al. 1991), DNA amplification fingerprinting (DAF) with arbitrary oligonucleotide primers (Caetano-Anollés et al. 1991), and differential display (DD) of messenger RNA (see Chap. XXVIII). The majority of these applications are DNA amplification-based and produce a collection of amplification products generally representing one or more discrete portions of a genome. The close proximity of bands in complex DNA profiles, such as those obtained in DAF or DD analysis, makes the isolation of DNA fragments physically demanding. In this chapter I will describe a simple procedure to recover DNA amplification products from silver stained polyacrylamide gels that we have recently developed (Weaver et al. 1994). The method uses one or more rounds of amplification of DNA diffusing passively from gel seg-ments during thermal cycling. Isolated DNA fragments can be used directly without further purification or subcloning. They can be used as probes for Southern hybridization either to confirm their mono-morphic or polymorphic nature when hybridized to DAF profiles, or to determine if the originating DAF products represent single or multi-locus sites in the genome when hybridized to genomic DNA. Alterna-tively, isolated fragments can be cloned without further purification using either "blunt-ended" or "overhanging-ended" cloning strategies (for the latter, see Chap. XVIII), especially when purity of the selected fragment is mandated or when isolated products are to be sequenced. Isolated DAF products can be used as hybridization probes in many applications (such as marker-assisted breeding, genetic mapping, gen-eral fingerprinting, molecular ecology and evolution). They can also be sequenced and converted into landmarks for genome mapping applications, the so-called sequence characterized amplified regions (SCAR) (Paran and Michelmore 1993).

2
Materials

Equipment
- oven thermocycler (Bios, New Haven, CT)
- benchtop centrifuge

Supplies
- heavy mineral oil
- scalpel

- template nucleic acids (DNA amplification products, including complex mixtures from DAF, DD or sequencing analysis) or silver stained gels
- reaction buffer (10x): 100 mM Tris-HCl, 100 mM KCl (pH 8.3)
- deoxynucleoside triphosphate stock solution: 2 mM of each dNTP
- magnesium: 25 or 100 mM MgCl$_2$ stock solution
- thermostable DNA polymerase enzyme. When using nonstringent amplification conditions, use truncated derivates such as AmpliTaq Stoffel DNA polymerase (Perkin-Elmer/Cetus, Norwalk, CT).
- oligonucleotide primer(s): 3 μM or 30 μM stocks

Reagents and solutions

3
Experimental Procedure

Isolation of DNA amplification fragments involves their initial separation by PAGE and silver staining (see Chap. XIII). Selected DNA fragments embedded in polyacrylamide are isolated directly from the silver stained gels by one or more cycles of isolation and amplification.

Recovering DNA from gels

1. Carefully excise a small piece of gel containing the desired DNA fragment with a flamed scalpel or dissection probe. When using a preserved gel, clean the surface of the gel with 95 % ethanol, use the scalpel to sharply delimit the segment of interest, and then rehydrate the excised band with a drop of sterile water.

2. Place the gel segment in 10–100 μl of standard amplification mixture containing 0.3–3 μM of primer(s), 0.3 U/μl of AmpliTaq Stoffel fragment, 1.5 mM MgCl$_2$, 200 μM of each dNTP, and buffer.

3. Centrifuge and, if necessary, cover the amplification mix with 1–2 drops of heavy mineral oil.

4. Heat for a minimum of 5 min at 95 °C if the fragment to isolate is larger than 500 bp.

5. Generally, amplify in 35 cycles of 30 s at 96 °C, 30 s at 50 °C and 30 s at 72 °C in an oven thermocycler. Several amplification regimens can be used depending on primers and stringency required.

6. Resolve amplification products using PAGE and silver staining (see Chap. XIII).

7. Alternatively, bands can be eluted from the gel segments at 95 °C for 20–30 min and the recovered eluate used as template in step 3.

4
Results and Comments

We have shown that silver stained polyester-backed polyacrylamide gels act as safe repositories of nucleic acids (Weaver et al. 1994). The gels can be preserved dried for many years without suffering distortion or image loss, allowing the retrospective analysis of the embedded nucleic acid molecules.

We have isolated many bands representing interesting monomorphic or polymorphic DAF products from bacterial, fungal and plant fingerprints (see Caetano-Anollés et al. 1992, 1993; Weaver et al. 1994, 1995; Caetano-Anollés and Trigiano 1996). Silver stained products were recovered from profiles separated using vertical and horizontal PAGE, discontinuous polyacrylamide isotachophoresis, and the PhastSystem, under both denaturing and nondenaturing conditions. The procedure was efficient and selective enough to remove the unwanted contaminants, even if the isolated bands were relatively minor amplification products and were very close in molecular weight to abundant products in the profile.

Number of recovery cycles

The number of amplification recovery cycles depends mostly on the complexity of the DNA pattern from where a particular fragment is isolated. Recovery occurs in a single cycle when bands are isolated from PCR products, even if resulting from multiplex PCR reactions (Fig. 1). In contrast, at least two cycles are necessary to recover bands from complex profiles such as those generated in DAF or DD analysis. During recovery, and in a few instances, minor contaminating products persist at very low levels even after several isolation cycles. These contaminants usually represent only a small fraction (about 5 %) of recovered DNA and are highlighted by the sensitivity of nucleic acid silver staining.

Factors affecting recovery

Several factors influence the recovery of nucleic acids, especially polyacrylamide concentration, DNA polymerase and annealing temperature used during amplification, and the length of isolated fragments. Higher polyacrylamide concentrations appear to decrease the number of recovery cycles, most probably by decreasing product carryover during electrophoresis (Weaver et al. 1994). For example, the isolation of a soybean DAF marker required five cycles when using 4.5 % polyacrylamide gels, but only three cycles when using 6 % gels. In SSCP analysis, band isolation from 20 % polyacrylamide gels demanded thorough washing of the silver stained gel prior to band excision and the use of a high number of temperature cycles during recovery (Calvert et al. 1995).

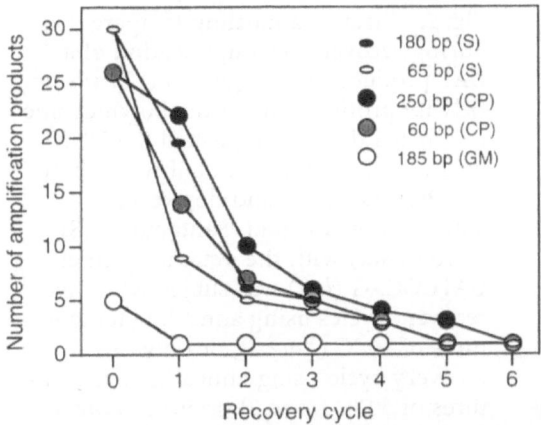

Fig. 1. Isolation of amplification products from dogwood (*Cornus florida*) and soybean (*Glycine max*) DNA fingerprints. Polymorphic DAF fragments generated from dogwood cultivars 'Cherokee Princess' (CP) and 'Santamour' (S) with primer GATCGCAG (see Fig. 2) and from soybean cv. Bragg (GM) DAF CTAACGCC profile with primer NN(AG)$_6$ (where N is A, T G or C), were subjected to a series of recovery cycles that comprised subsequent rounds of band excision, amplification, DNA separation by polyacrylamide gel electrophoresis, and silver staining of the gels. Amplification products were scored in the 50–1000 bp range

Band recovery also appears strongly conditioned by the DNA polymerase used during amplification (Weaver et al. 1994). When using some variants of *Thermus aquaticus* DNA polymerase, such as the truncated Klentaq LA (AB Peptides, St. Louis, MO), succesive rounds of band recovery and amplification were unable to eliminate contaminating products. One of possible explanations is the very efficient amplification of contaminating template DNA, even if present at very low levels in the excised gel segments.

High annealing temperature and therefore high stringency amplification helps eliminate many contaminants during band recovery. For example, the number of contaminants still remaining after a first cycle of isolation was halved if annealing temperature was raised from 30 °C to 50 °C (Fig. 2). If higher stringency is required, a "hot-start" (D'Aquila et al. 1991) can enhance relative enrichment and fragment recovery (for a simple hot-start procedure, see Bassam and Caetano-Anollés 1993).

Finally, the length of the amplification fragments to be isolated is an important consideration. Generally, small fragments are readily isolated, the shorter products requiring a lower number of recovery cycles (Fig. 1). In contrast, DNA fragments of more than 600 bp can only be isolated if gel segments are incubated at 95 °C for at least 20 min prior to amplification (Sanguinetti et al. 1994). Obviously,

G. Caetano-Anollés

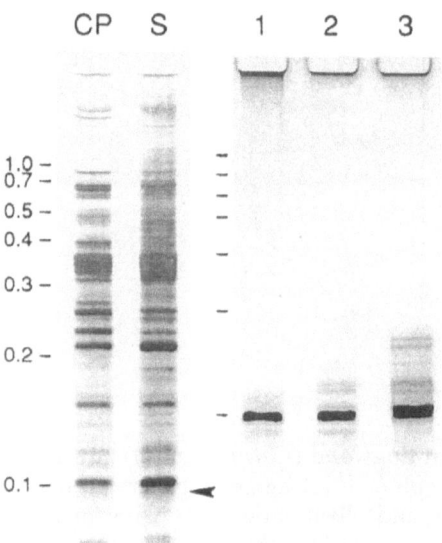

Fig. 2. Effect of annealing temperature during recovery by amplification of a DAF product from dogwood (*Cornus florida* L.) profiles. One of the polymorphic fragments that distinguished the DNA patterns generated from cultivar 'Cherokee Princess' (CP) and dogwood anthracnose-resistant 'Santamour' (S) (*arrowhead*) with the octamer primer GATCGCAG (*left*) was subjected to five recovery cycles using annealing temperatures of 30 °C (*lane 1*), or only one recovery cycle using annealing temperatures of 50 °C (*lane 2*) or 30 °C (*lane 3*) (*right*). Molecular weight markers (*lane M*) are given in kb

long fragments should be given enough time to diffuse out of the gel segments. Furthermore, it is recommended that only a small gel segment be excised (to minimize inhibitory effects of silver or silver-staining components; Calvert et al. 1995) and that each round of isolation be given at least 25–30 temperature cycles to maximize amplification.

References

Ainsworth PJ, Surh LC, Coultier-Mackie MB (1991) Diagnostic single strand conformational polymorphism (SSCP): a simplified non-radioisotopic method as applied to a Tay-Sachs B1 variant. Nucleic Acids Res 19:405–406

Ainsworth PJ, Rodenhiser D, Costa M (1993) Identification and characterization of sporadic and inherited mutations in exon 31 of the neurofibromatosis (NF1) gene. Hum Genet 91:151–156

Allen RC, Graves G, Budowle B (1989) Polymerase chain reaction amplification products separated on rehydratable polyacrylamide gels and stained with silver. Bio-Techniques 7:736–744

Ausubel FM, Brent R, Kingston RE, Moore DD, Seidman JG, Smith JA, Struhl K (1992) Current Protocols in Molecular Biology, 2nd Edition. J. Wiley & Sons, New York.

Bassam BJ, Caetano-Anollés G (1993) Automated "hot start" PCR using mineral oil and paraffin wax. BioTechniques 14:30–33

Budowle B, Chakraborty R, Giusti AM, Eisenberg AJ, Allen RC (1991) Analysis of VNTR locus DIS80 by the PCR followed by high-resolution PAGE. Am J Hum Genet 48:137–144

Caetano-Anollés G, Trigiano RN (1996) Recovery of DNA amplification products from silver stained polyacrylamide gels: applications in nucleic acid fingerprinting and genetic mapping. In White BA (ed), Methods in molecular medicine: preparative PCR protocols. Humana Press, Totowa, New Jersey. In press

Caetano-Anollés G, Bassam BJ, Gresshoff PM (1991) DNA amplification fingerprinting using short arbitrary oligonucleotide primers. Bio/Technology 9:553–557

Caetano-Anollés G, Bassam BJ, Gresshoff PM (1992) Primer-template interactions during DNA amplification fingerprinting with single arbitrary oligonucleotide. Mol Gen Genet 235:157–165

Caetano-Anollés G, Bassam BJ, Gresshoff PM (1993) Enhanced detection of polymorphic DNA by multiple arbitrary amplicon profiling of endonuclease digested DNA: identification of markers tightly linked to the supernodulation locus in soybean. Mol Gen Genet 241:57–64

Calvert RJ, Weghorst CM, Buzard GS (1995) PCR amplification of silver-stained bands from cold SSCP gels. BioTechniques 18:782–786

D'Aquila RT, Bechtel LJ, Videler JA, Eron JJ, Gorczyca P, Kaplan JC (1991) Maximizing sensitivity and specificity of PCR by pre-amplification heating. Nucleic Acids Res 19:3749

Dockhorn-Dworniczak B, Dworniczak B, Brommelkamp L, Bulles J, Horst J, Bocker WW (1991) Non-isotopic detection of single strand conformation polymorphism (PCR-SSCP): a rapid and sensitive technique in diagnosis of phenylketonuria. Nucleic Acids Res 19:2500

Doktycz MJ (1993) Discontinuous electrophoresis of DNA: adjusting DNA mobility by trailing ion net mobility. Anal Biochem 213:400–406

Dretzen G, Bellard M, Sassone-Corsi P, Chambon P. (1981) A reliable method for the recovery of DNA fragments from agarose and acrylamide gels. Anal Biochem 112:295–298

Grey M, Brendel M (1992) Rapid and simple isolation of DNA from agarose gels. Curr Genet 22:83–84

Heery DM, Gannon F, Powell R (1990) A simple method for subcloning DNA fragments from gel slices. Trends Genet. 6:173

Krowczynska AM, Donoghue K, Hughes L (1995) Recovery of DNA, RNA and protein from gels with microconcentrators. BioTechniques 18:698–703

Li Q, Ownby CL (1993) A rapid method for extraction of DNA from agarose gels using a syringe. BioTechniques 15:976–978

Maekawa, M., Sudo, K., Kitajima, M., Matsuura, Y., Li, S. and Kanno, T. (1993) Detection and characterization of new genetic mutations in individuals heterozygous for lactate dehydrogenase-B(H) deficiency using DNA conformation polymorphism analysis and silver staining. Hum Genet 91:163–168

McDonell MW, Simon MN, Studier FW (1977) Analysis of restriction fragments of T7 DNA and determination of molecular weights by electrophoresis in neutral and alkaline gels. J Mol Biol 110:119–146

Mohabeer AJ, Hiti AL, Martin WJ (1991) Non-radioactive single strand conformation polymorphism (SSCP) using the Pharmacia PhastSystem. Nucleic Acid Res 19:3154

Paran I, Michelmore RW (1993) Development of reliable PCR-based markers linked to downy mildew resistance genes in lettuce. Theor Appl Genet 85:985–993

Sanguinetti CJ, Diaz-Neto E, Simpson AJG (1994) Rapid silver staining and recovery of PCR products separated on polyacrylamide gels. BioTechniques 17:15–19

Schwarz H, Whitton JL (1992) A rapid, inexpensive method for eluting DNA from agarose or acrylamide gel slices without using toxic or chaotropic materials. BioTechniques 13:205–206

Smith HO (1980) Recovery of DNA from gels. Methods Enzymol 65:371–380

Southern EM (1975) Detection of specific sequences among DNA fragments detected by gel electrophoresis. J Mol Biol 98:503–517

Storts DR, Wu LC, Mendoza L, Oler JK (1993) Silver staining: a new approach to non-radioactive DNA sequencing. Promega Notes 42:10–14

Sugano, K., Kyogoku, A., Fukayama, N., Ohkura, H., Shimosato, Y., Sekiya, T. and Hayashi, K. (1993) Rapid and simple detection of c-Ki-*ras*-2 gene codon 12 mutations by nonradioisotopic single-strand conformation polymorphism analysis. Lab Invest 68:361–366

Sylvers LA, Beresten S (1993) A rapid automated method for simultaneous elution and purification of RNA from polyacrylamide gels. BioTechniques 14:378–380

Tautz D, Renz M (1983) An optimized freeze-squeeze method for the recovery of DNA fragments from agarose gels. Anal Biochem 132:14–19

Thuring RW, Sanders JB, Borst PA (1975) Freeze-squeeze method for recovering long DNA from agarose gels. Anal Biochem 66:213–220

Wang Z, Rossman TG (1994) Isolation of DNA fragments from agarose gel by centrifugation. Nucleic Acids Res 22:2862–2863

Weaver KR, Caetano-Anollés G, Gresshoff PM, Callahan LM (1994) Isolation and cloning of DNA amplification products from silver stained polyacrylamide gels. BioTechniques 16:226–227

Weaver KR, Callahan LM, Caetano-Anollés G, Gresshoff PM (1995) DNA amplification fingerprinting and hybridization analysis of centipedegrass. Crop Sci 35:881–885

Wieslander L (1979) A simple method to recover intact high molecular weight RNA and DNA after electrophoretic separation in low gelling temperature agarose gels. Anal. Biochem. 98:305–309

Zhu J, Kempenaers W, Van der Straeten D, Contreras R, Fiers W (1985) A method for fast and pure DNA elution from agarose gels by centrifugal filtration. Bio/Technology 3:1014–1016

Zintz CA, Beebe DC (1991) Rapid re-amplification of PCR products purified in low melting point agarose gels. BioTechniques 11:158–162

Chapter XVIII

Cloning of RAPD Markers

E. Mori and R. Fani

1
Introduction

Cloning of RAPD markers is a valuable technique for the study and utilization of RAPD amplification products. It can contribute to the characterization of a DNA region that is species- or group-specific, allowing the construction of probes and oligonucleotides to be used for the detection of microorganisms (see Chap. XXV). Moreover cloning of RAPD markers is an important step in procedures to determine the nucleotide sequence of the marker (see Chap. XIX). Sequencing could be used to identify and study the regions of a target genome amplified in RAPD experiments. This may help in determining whether there are some regions of the bacterial genome that are preferentially amplified during RAPD. In addition, the identification of RAPD fragments may highlight the nucleotide sequence of primer annealing sites and may help in understanding the molecular mechanism generating RAPD patterns.

In general, high intensity RAPD bands are the most reproducible ones in amplification patterns, as they probably lie between two strong annealing sites. Thus, in the selection of RAPD markers to be cloned and according to the purpose of the experiment, it is convenient to choose the sharpest and most intense amplification fragments. Furthermore, these types of RAPD bands probably do not contain contaminating fragments or secondary products.

The insertion of an internal restriction site in the 3' end of the primer used in the RAPD amplification may be useful in the subsequent cloning of the RAPD band. In fact, cleavage of primer sequences from the ends of fragments might be necessary when dealing with short markers amplified by long primers (over 20 nucleotides) (Fani et al. 1993).

2
Experimental Strategy

Restriction Digestion Strategy

The experimental procedure begins with the analysis of amplification products by agarose gel electrophoresis in order to identify an appropriate RAPD band. The selected fragment is excised from the agarose gel and amplified again using the same primer. This step is aimed at increasing the abundance of DNA molecules to be cloned and reducing the fraction of secondary amplification products. The RAPD product is then digested with a restriction endonuclease that cuts inside the primer sequence. It is important in this step to try and avoid restriction enzymes that cut multiple sites within the fragment to be cloned. Finally, the digested DNA is ligated into a plasmid vector previously treated with the same restriction enzyme or with an enzyme leaving compatible ends. Successive steps involve the transformation of *E. coli* competent cells and the selection and identification of recombinant plasmids by restriction analysis and/or sequencing.

One-Step Strategy

A one-step cloning strategy for the direct insertion of a RAPD product into a plasmid vector is provided by the TA cloning kit (Invitrogen,USA) or by other similar kits available from molecular biology companies. The use of this cloning strategy eliminates the need for enzymatic modifications of the amplification product and does not require restriction site-containing primers. This procedure is based on a nontemplate-dependent activity of *Taq* polymerase which adds a single deoxyadenosine (A) to the 3' ends of the amplified molecules. The linearized vector pCR II (Fig. 1) supplied in the TA cloning kit has a single 3' protruding deoxythymidine (T) residue; this allows RAPD inserts to ligate efficiently with the vector (Fig. 1). Products generated by *Taq* polymerase clone with high efficiency into the vector pCR II, as the 3' A-overhangs are not removed. By contrast, other thermostable polymerases containing extensive 3' to 5' exonuclease activity do not leave 3' A-overhangs. The vector is modified at the unique *Eco*RI site so that the inserted RAPD fragment is flanked by two *Eco*RI sites. Features of plasmid pCR II include kanamycin and ampicillin resistance genes and M13 forward and reverse priming sites for the sequencing of inserts.

CLONING OF RAPD BANDS

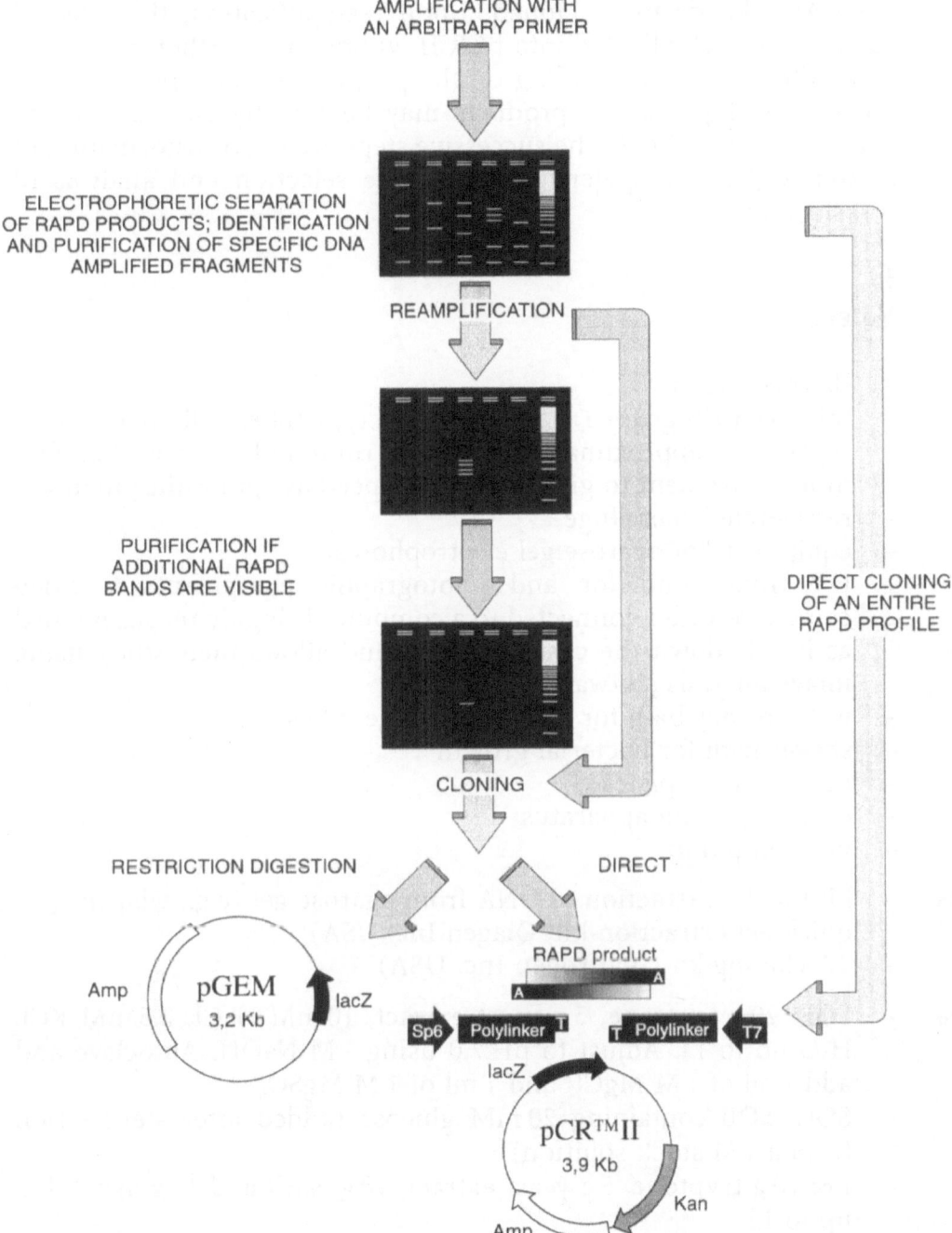

Fig. 1. Outline of the strategies for cloning RAPD bands by restriction digestion (*left*) or direct one-step protocol (*right*)

The experimental procedure involves the generation of RAPD products and their analysis by gel electrophoresis in order to select the RAPD bands to be cloned. Upon reamplification, the selected bands are directly ligated into pCR II without any further manipulation. Alternatively, depending on the purpose of the experiment, all the RAPD amplification products may be directly cloned into the plasmid vector pCR II. The successive steps are the transformation of *E. coli* INVαF' competent cells and the selection and analysis of transformants.

3
Materials

Equipment
- thermal cycler
- microcentrifuge for 1.5 ml Eppendorf type tubes. All microcentrifuges have approximately the same rotor radius; it is therefore more convenient to give centrifuge speed as rpm rather than *g*
- refrigerated centrifuge
- equipment for agarose gel electrophoresis
- UV transilluminator and photographic equipment. A video camera recorder connected to a computer is highly recommended as it cuts down the cost of photos and allows their study using image analysis software
- water or dry bath for microcentrifuge tubes
- shaker bath for bacterial growth
- incubator for plates
- electroporation apparatus
- vacuum pump

Kits
- kit for the extraction of DNA from agarose gel (e.g., Qiagen Qiaquick gel extraction kit, Qiagen Inc. USA)
- TA cloning kit (Invitrogen Inc. USA)

Media
- SOB: 20 g tryptone, 5 g yeast extract, 10 mM NaCl, 2.5 mM KCl, H_2O up to 1 l. Adjust to pH 7.0 using 1 M NaOH. Autoclave and add 1 ml of 1 M $MgCl_2$ and 1 ml of 1 M $MgSO_4$
- SOC: SOB containing 20 mM glucose (added after sterilization from a 1 M stock solution)
- LB: 10 g tryptone, 5 g yeast extract, 10 g NaCl and 15 g agar, H_2O up to 1 l

Bacterial strains and plasmids
- *Escherichia coli* DH5α, genotype: *sup*E44 *lac*U169 (φ80 *lacZ*ΔM15) *hsd*R17 *rec*A1 *end*A1 *gyr*A96 *thi*–1 *rel*A1. Other transformable *E. coli* strains allowing white-blue selection are also suitable

- plasmid pGEM3 (Promega, USA), or an equivalent cloning vector with ampicillin resistance and blue-white selection
- *Escherichia coli* One Shot INVαF' (provided with Invitrogen TA cloning system), genotype: F' *end*A1 *rec*A1 *hsd*R17 (r_K^-, m_K^-) *sup*E44 *thi*–1 *gyr*A96 *rel*A1 φ80 *lacZ*ΔM15 Δ(*lacZYA-arg*F) U169
- plasmid pCR II (provided with TA cloning system, Invitrogen)

- TE: 10 mM Tris-HCl (pH 8.0), 1 mM EDTA
- agarose: multi-purpose agarose from Boehringer-Mannheim (Germany) or any equivalent agarose at 2 % in TAE buffer for PCR products; at 0.6 % for analysis of plasmid DNA
- 50x TAE buffer: 242 g Tris base, 57.1 ml glacial acetic acid, 100 ml 0.5 M EDTA (pH 8.0), distilled H_2O up to 1 l
- ethidium bromide: prepared in sterile water at 10 mg/ml, stored at 4 °C and used at a final concentration of 0.5 µg/ml (10 µl for 200 ml agarose gel)

Reagents and solutions

Caution: Ethidium bromide is a strong mutagen; never mouth pipette, avoid skin contact, and wear disposable gloves.
- restriction enzymes with their 10x buffers
- 5 M NaCl
- 95 % ethanol
- 10x ligation buffer and T4 DNA ligase
- glycerol: 10 % (v/v) in distilled water, autoclave to sterilize
- ampicillin: stock solution at 50 mg/ml in distilled water and filter sterilized
- X-gal: 20 mg/ml in dimethylformamide
- IPTG: 20 mg/ml in sterile water
- 0.5 M β-mercaptoethanol: make a fresh solution just before use by adding 500 µl of the 14.4 M solution supplied by the manufacturer to 14.1 ml of water
- kanamycin: stock solution at 50 mg/ml in distilled water and filter sterilized

4
Experimental Procedure

4.1 Cloning of RAPD Markers Using Primers Containing a Restriction Site

Obtain RAPD products using one of the available amplification protocols, amplifying with a single primer containing a restriction site at its 3' end. Analyze the RAPD products on agarose gel to select the desired band(s).

Production of RAPD markers

Reamplification of RAPD bands

1. Cut the chosen band from the gel with a clean scalpel.

2. Place the agarose plug containing the DNA in a 1.5 ml microcentrifuge tube with 100 µl of TE.

3. Heat the sample at 94 °C for 15 min (mixing by inversion every 3–4 min) to dissolve the agarose.

4. Dilute the sample (1:5; 1:10; 1:50 in sterile water) and reamplify 1 µl aliquots of each dilution under standard conditions with the same primer used to generate the RAPD band. The dilution series reduces the probability of finding contaminating bands that will be amplified along with the RAPD marker.

Analysis of amplification products

5. An aliquot of the amplification products from each dilution is loaded on a 2 % agarose gel in TAE buffer containing 0.5 µg/ml of ethidium bromide and electrophoresed at 10 V/cm for 2 h.

Purification of RAPD products from agarose gel

6. RAPD fragments are purified from agarose gel using the Qiagen Qiaquick gel extraction kit. The experimental procedure is described in the instruction manual provided with the kit. Any other procedure or kit giving highly purified DNA is also suitable.

Note: The simple extraction procedure described before for reamplification of the RAPD band (steps 1–3) is suitable for amplification but not for restriction digestion.

Digestion with restriction endonuclease

7. Digest 200 ng of the purified DNA for 3 h with 5 U of the restriction enzyme and 2 µl of 10x buffer, in 20 µl volume. Also digest 50–100 ng of the plasmid vector (pGEM3 or equivalent) with the same enzyme (or with a different restriction enzyme generating compatible ends) in 20 µl.

8. Mix together the two digestion reactions and precipitate the DNA by adding 56 µl of TE, 4 µl of 5 M NaCl and 200 µl of 95 % ethanol. Mix well. Keep 30 min at 0 °C, centrifuge 30 min at 4 °C at 12000 rpm. Dry the pellet under vacuum and dissolve in 8 µl of TE.

Cloning into a plasmid vector

9. Add 1 µl of 10x ligation buffer and 1 µl (1 U) of T4 DNA ligase.

10. Incubate overnight at the optimal temperature, depending on the presence of sticky (16°–20 °C) or blunt ends (4 °C) in the ligation mixture.

Transformation

11. Transform *E. coli* competent cells of strain DH5α via electroporation, following the procedure described below (any other protocol that gives high efficiency of transformation can also be used).

12. Inoculate a fresh colony in 100 ml SOB medium in a 1 l flask and grow at 37 °C with vigorous shaking up to O.D.$_{660}$=0.5.

13. Keep the culture on ice for 30 min.

14. Centrifuge at 4 °C 15 min at 4000 g.

15. Remove the supernatant and suspend the cells in 50 ml of cold sterile water.

16. Centrifuge again as in step 14.

17. Suspend the bacteria in 2 ml of sterile cold glycerol.

18. Centrifuge again as in step 14.

19. Suspend the bacteria in 0.2 ml of sterile cold glycerol.

20. In a microcentrifuge tube put 40 μl of the glycerol suspended bacteria and 2 μl of ligated DNA. Mix and keep on ice for 10 min.

21. Set the output of the electroporation instrument (adopt the values suggested by the manufacturer of the instrument).

22. Transfer the DNA-bacteria mixture into a sterile cold electroporation cell and place the cell in the appropriate slot of the instrument.

23. Pulse the current and quickly remove the cell from the instrument.

24. Transfer the mixture into a fresh microcentrifuge tube and add 0.2 ml of SOC. Mix well and incubate at 37 °C for 1 h.

25. Select for transformants plating 5, 20, 100 and 200 μl on LB plates with 50 μg/ml ampicillin, 20 μg/ml X-Gal and 20 μg/ml IPTG.

26. Incubate the plates at 37 °C for at least 18 h. Shift the plates to 4 °C for 2–3 h for color development.

27. Purify recombinant plasmids from a few white transformants and determine the insert size by restriction analysis.

    Note: The size of the insert could be smaller than the original fragment. This is due to the possible presence of additional restriction sites inside the original fragment.

4.2
Cloning of RAPD Markers by the Invitrogen TA Cloning System

Production of RAPD markers

Obtain RAPD products using a *Taq* polymerase leaving 3' A-overhangs and following one of the protocols reported in this manual.

After the amplification all the RAPD products can be directly cloned (using 0.5–1 µl of the reaction), or an aliquot of the reaction can be analyzed by agarose gel electrophoresis in order to select the product(s) to be cloned. Proceed by following steps 1–6 of the section "Cloning of RAPD markers using primers containing a restriction site."

Note: Purification of a RAPD band from agarose gel prior to cloning into pCR II plasmid vector may decrease the ligation efficiency. This is probably due to the loss of 3 'A-overhangs during the purification step and could result in the absence (or low number) of transformants. We therefore recommend, whenever possible, not to purify the RAPD fragment(s) from agarose gel. If smearing of RAPD products occurs, we suggest purifying the desired band, reamplifying it (see "Cloning of RAPD markers using primers containing a restriction site") and finally cloning the reamplified band into pCR II vector.

Cloning into pCR II

1. Ligation reaction (10 µl):

Sterile water	5 µl
10x ligation buffer	1 µl
pCR II vector (25 ng/µl)	2 µl
RAPD product	1 µl (about 10 ng)
T4 DNA ligase	1 µl

2. Incubate the ligation reactions at 14 °C overnight.

3. Centrifuge 1 min and place on ice.

Transformation

4. Thaw an appropriate number of vials of One Shot INVαF' cells on ice.

5. Pipette 2 µl of 0.5 M β-ME into each vial of the One Shot INVαF' cells and mix by stirring gently with pipette tip.

6. Pipette 1–2 µl of each ligation reaction into cells and stir gently with pipette tip to mix.

7. Incubate the vials on ice for 30 min.

8. Heat shock for exactly 30 s in a 42 °C water bath. Do not mix.

9. Place the vials on ice for 2 min.

10. Add 450 µl of SOC medium to each vial.

11. Shake the vials at 37 °C in a shaker at 225 rpm for exactly 1 h. Place the vials with the transformed cells on ice.

12. Plate 50 µl and 200 µl from each transformation vial on LB plates with 50 µg/ml kanamycin or ampicillin and 20 µg/ml X-Gal. (IPTG is not necessary with INVαF' cells).

Analysis of transformants

13. Incubate at 37 °C for at least 18 h. Shift plates to 4 °C for 2–3 h for color development.

14. Analyze plasmid DNA from some white transformants for the presence of inserts by restriction digestion.

5
Troubleshooting

Smearing of RAPD products on the gel may require purification of the desired band from the gel. This step may decrease ligation efficiency resulting in absence of transformants. To avoid smearing and secondary products, it is recommended that amplification be optimized using no more than 30 cycles or a hot start before the cycling program.

Another reason for failure to obtain transformants could be the incorrect ratio between plasmid and insert in the ligation mixture. In general, 0.5–1 µl of a typical amplification product of about 400–700 bp in length gives the proper vector-insert ratio of 1:1. We recommend not to use more than 2–3 µl of the amplification product in the ligation reaction as the T4 DNA ligase may be inhibited by salts present in the RAPD sample.

References

Fani R, Damiani G, Di Serio C, Gallori E, Grifoni A, Bazzicalupo M (1993) Use of random amplified polymorphic DNA (RAPD) for generating specific DNA probes for microorganisms. Mol Ecol 2:243–250

Chapter XIX

Sequencing of RAPD Markers

E. Mori and R. Fani

1
Introduction

Although the random amplified polymorphic DNA (RAPD) methodology, described by Williams et al. (1990) and Welsh and McClelland (1990), has been extensively used for many purposes, very little is known about the nucleotide sequence of RAPD markers and the primer binding sites within the target genome. It may be assumed that RAPD amplification is initiated at many sites which are imperfectly complementary to the primer sequence and which form short inverted repeats separated in the target genome by 0.2–3 kb. Moreover, some RAPD bands could result from self-priming events due to hairpin loop formation at the 3' terminus of the initially amplified products. Such events should produce large inverted repeats. Determination of the nucleotide sequence of RAPD products could help in understanding the molecular mechanisms generating RAPD patterns and highlight novel applications of this methodology. The molecular nature of the polymorphisms detected in RAPD amplification has yet to be described. This would require DNA sequencing of several genomic primer binding sites from a number of polymorphic bands. Determining the nucleotide sequence of RAPD products might permit identification and study of those DNA regions of target genomes that were preferentially amplified. Determination of the nucleotide sequence of RAPD markers that are species- or group-specific could also permit design of specific oligonucleotides for the identification, detection and monitoring of bacteria by selective amplification via PCR (see Chap. XXV).

Sequencing of RAPD markers can be performed by two different strategies (Fig. 1): using standard techniques (Sanger et al. 1977) on DNA fragments cloned into appropriate vectors (see Chap. XVIII) or directly on the amplification product via PCR. A drawback of the first procedure is that it could lead to enrichment of molecules with nucle-

Sequencing strategies

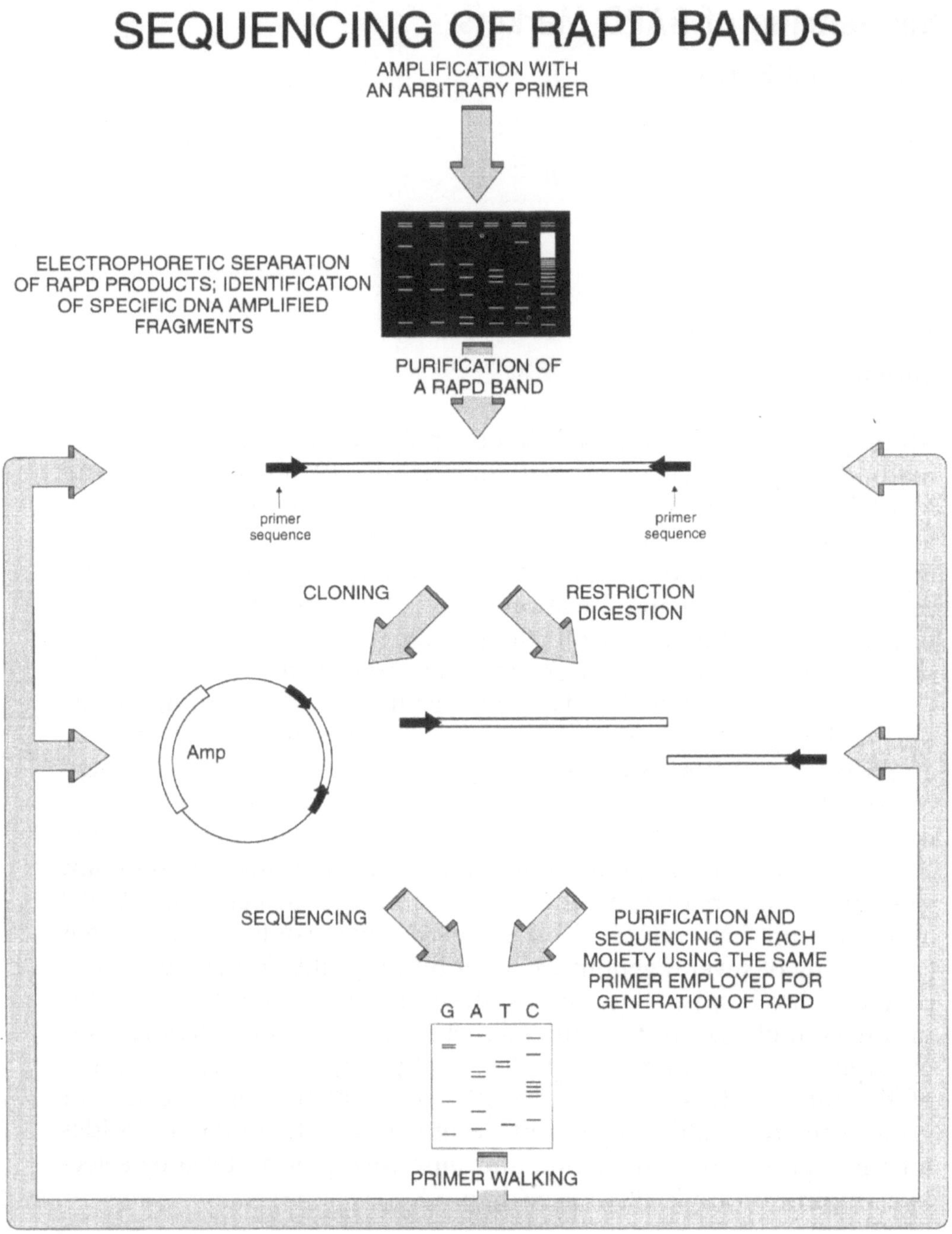

SEQUENCING OF RAPD BANDS

Fig. 1. Outline of the strategies for sequencing RAPD markers. Standard sequencing on DNA fragments cloned into appropriate vectors (*left*) or cycle sequencing of the amplification product (*right*)

otide misincorporation, produced by *Taq* polymerase. Furthermore, cloning may require synthesis of primers containing restriction sites and postamplification restriction digestion of products. The amplification product may need to be purified from a gel prior to ligation into a plasmid. Recently, however, a variety of methods have been developed for cloning PCR products directly without the need for modifying enzymes, purification or restriction digestion (see Chap. XVIII).

The second strategy, cyclic sequencing, works directly on the amplification products without any further manipulation (Sears et al. 1992). Besides there is also the advantage of avoiding the errors present in a single cloned sequence, since very few misincorporated nucleotides are present at each position in the pool of amplified molecules. Moreover, direct sequencing requires very small amounts of amplification products. Nevertheless, the direct sequencing of RAPD products via PCR presents a problem: RAPD amplification is generally primed by a single oligonucleotide and each amplified fragment contains the same nucleotide sequence at both ends. Therefore it is not possible to sequence the DNA fragment using the same oligonucleotides employed in the generation of RAPD. Sequencing of a RAPD product can be performed directly when two or three different primers have been used to generate RAPD; in this situation it is likely that the majority of RAPD fragments results from amplification primed by two different oligonucleotides. The problem may also be solved by digesting the RAPD band with a restriction endonuclease which recognizes a single site within the fragment, generating two fragments, each containing one single annealing site for the primer (Fig. 1). After part of the sequence has been obtained, it is necessary to design a new specific primer for direct sequencing of the remaining portion of the fragment.

2
Materials

- agarose gel electrophoresis equipment
- UV transilluminator and photographic equipment. A video camera recorder connected to a computer is highly recommended as it cuts down the cost of photos and allows their study using image analysis software
- water or dry bath for microcentrifuge tubes
- sequencing gel apparatus with power supply
- thermal cycler

Equipment

- microcentrifuge for 1.5 ml Eppendorf-type tubes. All microcentrifuges have approximately the same rotor radius; it is therefore more convenient to give centrifuge speed as rpm rather than g

Kits
- kit for PCR sequence with *Taq* polymerase (CircumVent DNA sequencing kit, New England Biolabs, Beverly, MA)

Reagents and solutions
- restriction enzymes with their 10x buffers
- 50x TAE buffer: 242 g Tris base, 57.1 ml glacial acetic acid, 100 ml 0.5 M EDTA (pH 8.0), distilled H_2O up to 1 l
- agarose: multi-purpose agarose is used at 2 % in TAE buffer when analyzing RAPD products
- low-melting temperature agarose: prepare according to the instructions supplied by the manufacturer at a concentration of 1 % in TAE buffer. During the preparation, do not heat the gel over 70 °C
- ethidium bromide: prepared in sterile water at 10 mg/ml, stored at 4 °C and used at a final concentration of 0.5 µg/ml (10 µl for 200 ml agarose gel)

Caution: Ethidium bromide is a strong mutagen; never mouth pipette, avoid skin contact, and wear disposable gloves.
- liquid nitrogen (or CO_2 + ethanol)
- ^{35}S-labeled dATP or [^{32}P]dATP

3
Experimental Procedure

1. Produce RAPD patterns with any of the available protocols.

2. Reamplify and purify the selected RAPD band as described in Chap. XVIII.

Restriction digestion
3. Digest the selected RAPD band using different restriction endonucleases in order to find an enzyme that recognizes a single site within the fragment. Perform restriction digestion using 2 µl aliquots (containing 100–300 ng of DNA) of the reamplified product with 3 U of the selected restriction endonuclease, 2 µl of 10x buffer in a 20 µl final volume.

4. Incubate the reactions 3 h at optimal temperature.

Analysis of digestion products
5. Analyze the products (from 5 to 10 µl of the restriction reaction) by gel electrophoresis on 1 % low-melting agarose in 1x TAE buffer containing ethidium bromide. This separates the two fragments generated after treating the RAPD band with the proper restriction endonuclease.

Note: DNA can be purified by the Qiagen Qiaquick gel extraction kit as described in Chap. XVIII; however as cycle sequencing does not require highly pure template DNA, it is simpler and faster to use the following protocol.

6. Excise the DNA fragment from the gel with a clean, sharp scalpel.

7. Place the gel slice into a 1.5 ml microcentrifuge tube.

Gel purification of RAPD products

8. Freeze the contents of the tube by immersing the tube for 1 min in liquid nitrogen (or in CO_2 + ethanol bath) and then heat at 50 °C for 1 min. Repeat this step four times.

9. Centrifuge 10 min at 12000 rpm.

10. Draw 8 µl of the supernatant (which usually contains at least 30 ng of DNA), and use it for sequencing reactions.

11. Sequence RAPD fragments via PCR with the same primer used for generating the RAPD pattern. Prepare reaction mixtures according to the instruction manual of CircumVent DNA sequencing kit with the $Vent_R$ (exo⁻) DNA polymerase. The manual provides a set of methods for sequencing nanogram amounts of DNA templates by thermal cycle sequencing using ^{35}S-labeled dATP or [^{32}P]dATP radiolabeled incorporation. The annealing temperature is 25 °C for 10-mer primers and 45 °C for 17-mer primers. The thermal program is:
    - 90 °C for 1 min
    - 95 °C for 1 min, 30 s
    - 95 °C for 30 s
    - 25 °C(or 45 °C) for 1 min
    - 75 °C for 2 min
    - Repeat steps 3–5 a further 44 times.
    - 75 °C for 10 min
    - 60 °C for 10 min
    - Store at 4 °C

Determination of nucleotide sequence

Note: There are several other kits available for cycle sequencing with *Taq* polymerase. Any of them can be used alternatively.

12. Resolve amplification products on a sequencing gel.

13. Based on the obtained sequence, design new primer(s) for sequencing the remaining part of the fragment.

Note: It is advisable to use a computer program for the design of sequencing primers.

14. Determine nucleotide sequence via PCR (as described above) on the nondigested RAPD fragment using the new primers.

4
Troubleshooting

When DNA sequencing by cyclic PCR results are polymorphic (presence of more than one band in the same position), this may be due to the comigration and co-purification of different RAPD bands (or of the products of their restriction digestion) with similar size with the subsequent simultaneous sequencing of two different templates. In this case it is necessary to change the restriction enzyme chosen.

Problems may also be encountered when sequencing short RAPD markers that contain two restriction sites for the same enzyme. In this case only two of the three fragments generated by restriction digestion can be sequenced by cycle sequencing because of the lack of primer annealing sites on the central fragment. However in this case it is necessary to use as template the nondigested marker, as should be done for all the large markers.

References

Sanger F, Nicklen S, Coulson AR (1977) DNA sequencing with chain terminating inhibitors. Proc Nat Acad Sci USA 74:5463–5466

Sears LE, Moran LS, Kissinger C (1992) Circumvent™ thermal cycle sequencing and alternative DNA sequencing protocols using the highly thermostable Vent$_R$™ (exo⁻) DNA polymerase. Bio/Techniques 13:626–633

Welsh J, McClelland M (1990) Fingerprinting genomes using PCR with arbitrary primers. Nucleic Acids Res 18:7213–7218

Williams JGK, Kubelik AR, Livak KJ, Rafalski JA, Tingey SV (1990) DNA polymorphisms amplified by arbitrary primers are useful as genetic markers. Nucleic Acids Res 18:6531–6535

Chapter XX

Analysis of Tumor-Specific Genetic Alterations by Arbitrarily Primed PCR

R. Arribas, S. Tòrtola, J. Welsh, M. McClelland, and M. A. Peinado

1
Introduction

It has been demonstrated that arbitrarily primed PCR (AP-PCR) is useful for the detection and isolation of tumor specific allelic losses and gains (Peinado et al. 1992; Kohno et al. 1994), thus providing a molecular alternative to cancer cytogenetics. This is based predominantly on a plethora of propitious properties of the procedure: (1) the amplified bands usually originate from single copy sequences, rather than from repetitive elements; (2) there is no apparent bias for the chromosomal origins of the amplified bands, and therefore, fingerprints representative of the full chromosomal complement can be obtained by the use of a few arbitrary primers; and (3) the amplification is semi-quantitative in that the intensity of an amplified band is almost proportional to the concentration of its corresponding template sequence.

The possibility of detecting moderate gains of genetic material, such as those corresponding to triploidy and tetraploidy, represents a significant technical development because such genomic changes cannot be detected by conventional allelotyping by restriction fragment length polymorphism (RFLP) or by typing of polymorphic microsatellites.

Here we describe the procedure for analyzing genetic alterations that occur during the tumorigenic process by AP-PCR. This procedure involves the following steps: (1) comparative AP-PCR analysis of matching normal-tumor tissue; (2) isolation and identification of altered amplified bands; (3) chromosomal localization of altered amplified bands. Protocols for isolation and characterization of an AP-PCR amplified sequence are equally useful in other applications of the technology including applications to RNA (Welsh et al. 1992; Liang and Pardee 1992).

2
Materials

Equipment
- thermal cycler (e.g., MJ Research model PTC100, Perkin Elmer model 9600)
- sequencing gel electrophoresis apparatus (40 cm long/30 cm wide/0.4 mm thick)
- gel dryer
- UV cross-linker (e.g., Amersham Life Sciences) (optional)

Supplies
- PCR tubes
- autoradiogram markers
- X-ray film and exposure cassettes
- nitrocellulose or nylon blotting membrane (e.g., BIO-RAD Zeta-Probe or AMERSHAM Hybond)
- 3MM Whatman paper

Media
- LB: 10 g bacto-tryptone, 5 g bacto-yeast extract, 10 g NaCl, H_2O up to 1 l, pH 7.0. Sterilize by autoclaving.

Reagents and solutions
- TE: 10 mM Tris-HCl (pH 7.5), 1 mM EDTA
- 10x PCR buffer: 100 mM Tris (pH 8.0), 500 mM KCl, 15 mM $MgCl_2$, 0.01 % gelatin
- 100 mM $MgCl_2$
- 10x dNTPs: 1.25 mM each dNTP
- arbitrary primer 62R1 (5'-TTTCTCGTTAACTTATTTCATCTTG-3'): 100 pmol/μl
- $[\alpha^{33}P]$dATP >2000 Ci/mmol (2 μCi/reaction tube) or $[\alpha^{32}P]$dCTP >3000 Ci/mmol (1–2 μCi/reaction tube)
- *Taq* polymerase 5 U/μl (Perkin Elmer Cetus or Boehringer Mannheim)
- genomic DNA from frozen tissue diluted to 20 and 10 ng/μl
- 1x TBE buffer: 0.089 M Tris-borate, 0.025 M disodium EDTA, pH 8.3
- 6 % polyacrylamide/8 M urea gel in 1x TBE buffer
- denaturing loading buffer: 95 % formamide, 0.1 % bromophenol blue, 0.1 % xylene cyanol, 10 mM EDTA
- ampicillin stock solution: 50 mg/ml in H_2O. Sterilize by filtration
- glycerol
- 3 M NaCl, 0.5 M Tris-HCl (pH 7.4)
- 2x SSPE: 0.036 M NaCl, 0.02 M sodium phosphate (pH 7.7), 2 mM EDTA

3
Experimental Procedure

Arbitrarily Primed PCR of Genomic DNA

Genomic DNAs of normal-tumor tissue pairs from a number of individuals are analyzed according to the standard AP-PCR protocol described in Chap. VII.

Isolation of an AP-PCR Amplified Sequence

To confirm that observed differences correspond to changes in the genome of the tumor cells, it is necessary to isolate and identify altered amplified bands. Although direct reamplification may be sufficient, it is highly recommended to clone the product in order to obtain a unique sequence.

1. Align the autoradiogram markers on the gel with their exposed images. Use a needle to mark in the dried gel the exact position of the band, then excise with a scalpel or razor blade. Reexposure of the gel will confirm the accuracy of excision of the band.

2. Place the excised portion of the gel (approx. $0.5-1 \times 2-3$ mm) in $50-100 \,\mu$l of water and incubate at 60 °C for $10-20$ min to elute the DNA.

3. Reamplify 1 μl of the eluted DNA with the same primer and using the same conditions as described in Chap. VII, except do not add extra $MgCl_2$ (final concentration of $MgCl_2$ is 1.5 mM instead of 2.5 mM). Perform 30 high stringency cycles as described in Chap. VII.

4. Analyze the PCR product by gel electrophoresis running the sample next to an arbitrarily-primed PCR product to verify its size and purity. If other bands are co-amplified, the desired band can again be cut from the gel and reamplified by PCR. If the major product of the PCR is the appropriate band, this product can then be cloned. We have also experimented successfully with using SSCP to separate the desired band from other products of the same size (Mathieu-Daude and McClelland, manuscript submitted).

Gel excision and reamplification of the band.

Some arbitrarily chosen primers will contain recognition sites for restriction endonucleases that digest DNA to produce fragments with staggered ends compatible with the cloning sites of commercially

Cloning of reamplified bands

available plasmid or phagemid vectors. In this case, one can digest both the vector and the reamplified band DNA with the appropriate restriction enzyme(s), then ligate and transform using standard protocols (see Chap. XVIII). The restriction site should lie four or more bases from the end of the product. Frequently, however, arbitrarily chosen oligonucleotides do not contain recognition sites for restriction enzymes. Commercially available vectors allow the cloning of any PCR amplified product with no need of enzymatic digestion. We use the Invitrogen TA cloning kit for this purpose, but other manufacturers supply equivalent products. Ligation of the PCR product into the vector, transfection and colony selection is performed according to manufacturer instructions. Blue/white selection of transformed cells (X-Gal/IPTG) is recommended. To analyze the cloned product perform the following steps:

Analysis of the cloned product

1. Pick white colonies into culture tubes containing 2–5 ml LB + ampicillin (50 µg/ml). Grow overnight at 37 °C with aeration.

2. Add glycerol to 15 % to an aliquot of each culture and freeze at –70 °C for long-term storage.

3. Mix 10 µl of the bacterial culture from step 1 above with 90 µl of 1x PCR buffer.

4. Boil for 30 s, then pellet the cells in a microcentrifuge (13 000 g).

5. Perform high stringency PCR as described in "Gel excision and reamplification of the band" (step 3) using the original primer.

Characterization of the Cloned Fragment

Confirmation that the cloned band corresponds to the band visualized in the AP-PCR gel is performed by Southern blot hybridization analysis. A dried AP-PCR gel can be transferred to a blotting membrane. Then it is possible to demonstrate if the cloned band, used as probe, hybridizes to the expected band.

Transfer of AP-PCR gel DNA to a blotting membrane

1. Cut out the part of the dried gel that is going to be transferred to the nitrocellulose or nylon membrane. Cut at least ten sheets of Whatman 3MM paper and the blotting membrane to the same size as the gel segment.

2. Dip four sheets of paper and the dried gel in 3 M NaCl, 0.5 M Tris-HCl (pH 7.4) and lay them on a glass plate with the gel on top.

3. Soak the blotting membrane with water and lay this on top of the gel. Place one wet sheet of 3MM paper and five or more sheets of dry 3MM paper on top of the membrane.

4. Cover with another glass plate and a weight (0.5 kg). Leave for 2 h to overnight.

5. Wash the blotting membrane with water and 2x SSPE. Irradiate with UV to cross-link the DNA. To do this, either use a UV cross-linker or place the membrane face down on clear plastic wrap on a transilluminator for 15 min.

6. Dry the membrane (a few minutes at 40°–70 °C) and expose to X-ray film. With this procedure, 10 %–50 % of the radioactive material is transferred to the blotting membrane. The transfer efficiency is not the same for all the bands and depends on their size; bands bigger than about 1500 base pairs will transfer with lower efficiency.

Preparation of the probe

The probe for the hybridization step is prepared directly from the PCR product (obtained in step 5 of "Analysis of the cloned product") or from a plasmid obtained by a miniprep. If the PCR product is to be used, it must first be extracted with any of the conventional methods used for the purification of PCR products. Most DNA precipitation methods or commercially available kits can be used for this purpose (including Bio101 Geneclean, Millipore 10000 MW cartridges, DS Primer remover, Qiagen QIAquick Spin PCR purification kit). Purified DNA fragment is labeled by standard methods; we use the random primer method with commercially available kits (e.g., Promega, Prime-a-Gene). Unincorporated nucleotides are removed by elution of the probe using resin columns such as QIAquick nucleotide removal kit (Qiagen).

Hybridization

Hybridization is performed using standard methods (Sambrook et al. 1989). Specific hybridization by the probe to the band of the correct size is evidence of successful cloning. The presence of background due to the radioactive nature of the AP-PCR bands (although with a low signal) facilitates the identification of the band.

Sequencing of the cloned AP-PCR product

Once the identity of the cloned fragment with the AP-PCR band has been confirmed, it can be sequenced directly from a plasmid miniprep or rescued phagemid using standard DNA sequencing methods. Different types of sequencing kits are available from many manufacturers, including Perkin Elmer, Promega, and USB. In our hands, cycle sequencing methods work very well, especially when sequencing PCR products. In general, regular PCR amplification from the transformed bacteria (see "Analysis of the cloned product") using

M13–20 and the reverse primer is performed. This product is then sequenced with the same primers used for PCR.

Chromosomal Localization of Amplified Bands

The existence of standard, somatic cell hybrid template DNA permits the chromosomal localization of multiple AP-PCR bands in a single experiment. DNA panels of rodent/human somatic cell hybrids are available from PCRable panel I (Bios, New Haven, CT) and NIGMS mapping panel II (Coriell Institute for Medical Research, Camden, NJ).

AP-PCR is performed with 50 ng and 25 ng of genomic DNA from the different hybrids using the same conditions as described in Chap. VII. Human and rodent genomic DNA are used as controls. Because each cell line contains only one or a few human chromosomes, amplification of human products will indicate that this sequence is located in one of the chromosomes present in this/these cell line/s.

Some arbitrary primers produce a more complex fingerprint with rodent DNA than with human DNA, in such cases it may be difficult to get amplification of human sequences in hybrids. Under these circumstances an alternative approach is to perform the AP-PCR of genomic hybrid DNAs without radioactive label, transfer the electrophoresed products to a nitrocellulose or nylon blotting membrane (as described in "Transfer of AP-PCR gel DNA to a blotting membrane") and hybridize with the radioactively labeled product of an AP-PCR from human genomic DNA. In this case labeled human sequences will hybridize to amplified bands from hybrids, even if those are masked by the rodent DNA fingerprint.

4
Results and Comments

As shown in Fig. 1, it is possible to identify multiple polymorphisms in a single experiment. In this case normal and tumor tissue from four individuals have been analyzed and polymorphic bands appear in both samples.

Figure 2 depicts the AP-PCR amplification of NIGMS panel II hybrid DNAs. In hamster or mouse fingerprints, it is possible to distinguish bands that appear in only one of the cell hybrids and that are at the same level as the human bands. This implies that this sequence proceeds from the human chromosome(s) contained in the hybrid.

Fig. 1. Arbitrarily primed PCR analysis of four pairs of normal-tumor tissue DNA. Reaction and electrophoresis analysis were performed as described in Chap. VII, except for the primer (Blue: 5'-CCGAATTCGAAAGCTCTGA-3') and the PCR cycles. The reaction was carried out in a thermal cycler for five low stringency cycles (94 °C, 30 s; 50 °C, 1 min; 72 °C, 1 min 30 s) and 35 high stringency cycles (94 °C, 15 s; 60 °C, 15 s; 72 °C, 1 min) followed by a final extention step at 72 °C for 5 min. Case number and type of tissue (*N*, normal; *T*, tumor) is indicated at the *top*. *Arrows at left* show polymorphic bands that are characteristic of each individual

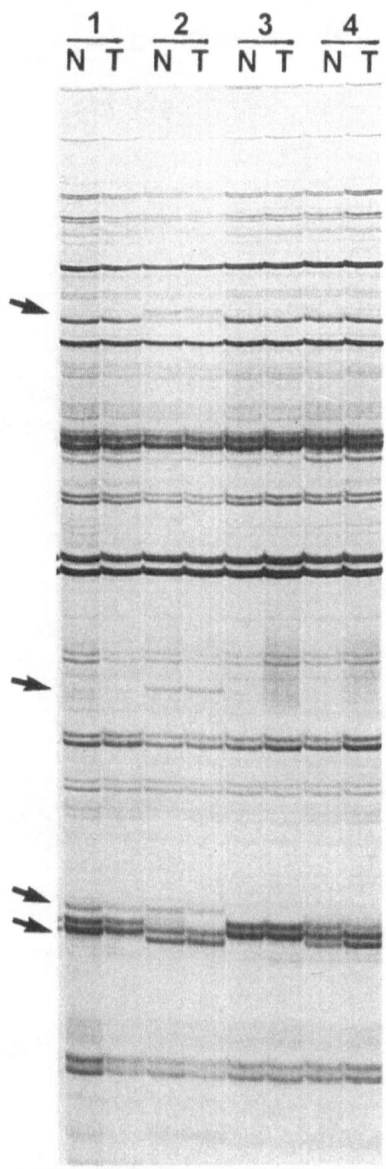

Using unbiased DNA fingerprinting by AP-PCR, we discovered recently a novel mechanism for the emergence of a profound, somatic, genomic instability that is likely to be the ultimate cause for the development of some forms of cancer (Ionov et al. 1993). This mutator pathway, characterized by the presence of ubiquitous mutations in microsatellite sequences, is the molecular symptom of a defective replication machinery.

184   R. Arribas et al.

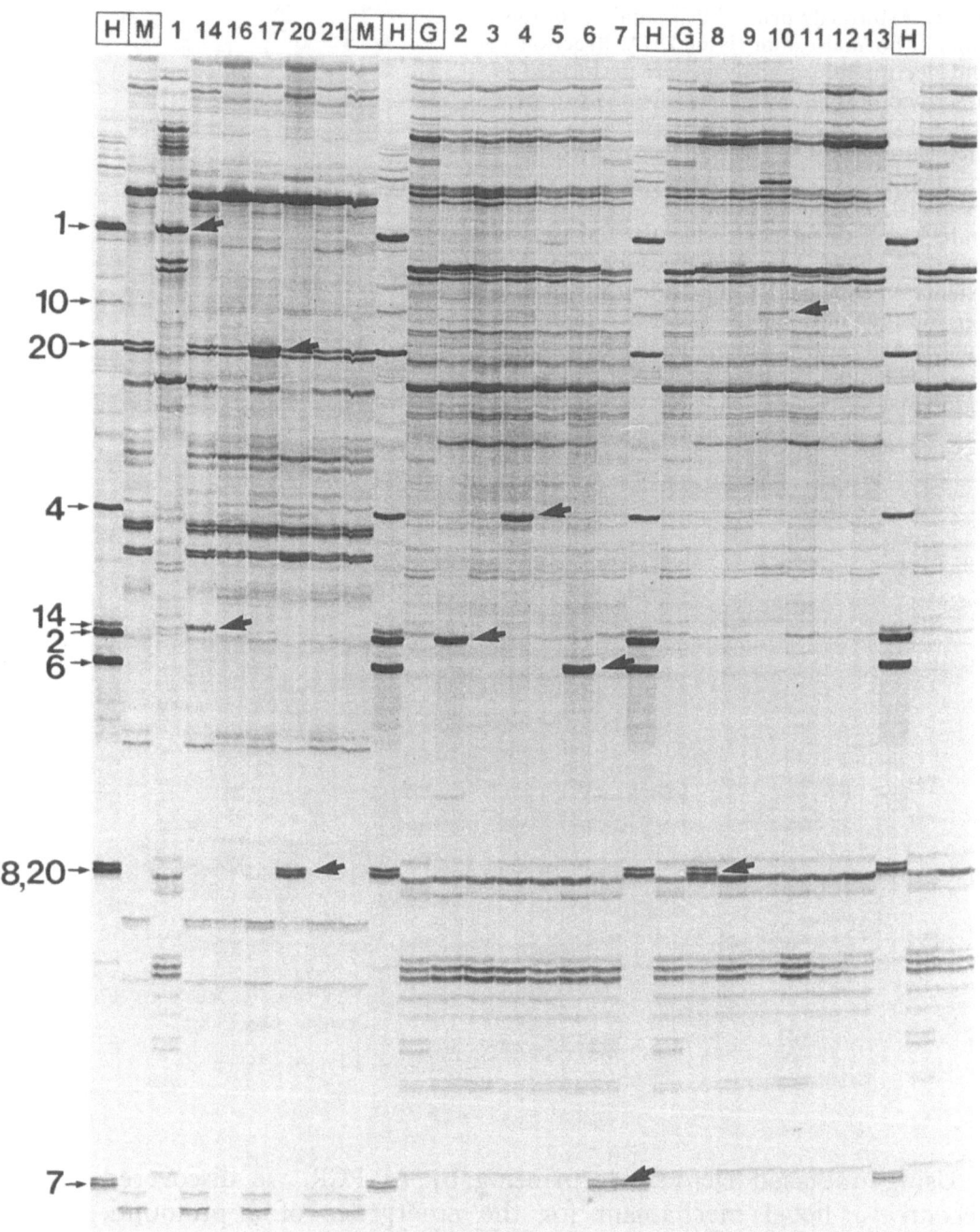

Fig. 2. Chromosomal localization of arbitrarily primed PCR bands. Reaction and electrophoresis were performed as described in Chap. VII. Lanes labeled with a *boxed letter* correspond to human (*H*), hamster (*G*), and mouse (*M*) genomic DNA. *Numbers at top* indicate the human chromosome that was present in the hybrid cell line DNA. *Arrows* show assumed human sequences in the rodent fingerprint of the hybrids; *left,* the chromosome number of each band is indicated

Acknowledgements. This work was supported in part by U.S. National Institutes of Health grants CA 68822, NS33377, AI32644 and AI34829 (to JW and MMcC) and Spanish CICYT SAF93–511 and FIS 94/37 (to MAP). RA and ST are fellows of the Spanish Ministry of Education.

References

Ionov Y, Peinado MA, Malkhosyan S, Shibata D, Perucho M (1993) Ubiquitous somatic mutations in simple repeated sequences reveal a new mechanism for colonic carcinogenesis. Nature 363:558–561

Kohno T, Morishita K, Takano H, Shapiro DN, Yokota J (1994) Homozygous deletion at chromosome 2q33 in human small-cell lung carcinoma identified by arbitrarily primed PCR genomic fingerprinting. Oncogene 9:103–108

Liang P, Pardee A (1992) Differential display of eukaryotic messenger RNA by means of the polymerase chain reaction. Science 257:967–971

Peinado MA, Malkhosyan S, Velazquez A, Perucho M (1992) Isolation and characterization of allelic losses and gains in colorectal tumors by arbitrarily primed polymerase chain reaction. Proc Natl Acad Sci USA 89:10065–10069

Sambrook J, Fitsch EF, Maniatis T (1989) Molecular cloning: a laboratory manual (2nd ed.). Cold Spring Harbor Press, Cold Spring Harbor, New York

Welsh J, Chada K, Dalal SS, Ralph D, Chang R, McClelland M (1992) Arbitrarily primed PCR fingerprinting of RNA. Nucleic Acids Res 20:4965–4970

Chapter XXI

Construction of Linkage Maps with RAPD Markers

G. J. HUNT

1
Introduction

The idea of using recombinant individuals to measure linkage distance along chromosomes comes from the research of Sturtevant, in 1913, on the fruit fly, *Drosophila melanogaster*. Since that time, the *Drosophila* map has become very detailed, consisting of about 4000 mapped genes (Kafatos et al. 1991). Molecular markers, such as restricition fragment length polymorphisms, have enabled researchers to construct detailed linkage maps. This, in turn, makes it possible to identify quantitative trait loci (QTLs) that affect characters showing polygenic inheritance and to clone genes by map-based cloning techniques. Genomic maps have been used for mendelian analysis of QTLs that affect important agronomic traits of crop plants (reviewed by Tanksley 1993). QTL analysis also has been useful for mapping and cloning genes in human genetic research (reviewed by Lander and Schork 1994), for identifying loci that affect disease transmission by mosquitoes (Severson et al. 1994, 1995) and for identifying loci that influence the behavior of mammals (e.g., Carlier et al. 1990; Crabbe et al. 1994) and insects (Hunt et al. 1995). Software has been developed both for linkage mapping and QTL analysis (Lander et al. 1987; Lander and Botstein 1989; Stam 1993) and new statistical tools are being developed to increase the power and accuracy of QTL mapping (Jansen 1994a,b; Zeng 1993, 1994; Kruglyak and Lander 1995).

New marker systems based on the polymerase chain reaction (PCR) are accelerating the rate of linkage map construction. One such system relies on specific primers that flank simple-sequence repeats (SSRs, also known as microsatellites). SSRs have been useful for making maps of mammalian genomes (e.g., Copeland et al. 1993) and are currently being developed for genetic studies of the honey bee (Estoup et al. 1993). SSRs have the advantage of being highly polymorphic. Random amplified polymorphic DNA (RAPD markers) are generated by amplification with ten nucleotide primers of arbitrary sequence and are being widely used for linkage mapping (Williams et al. 1990, 1993). Major advantages of RAPD markers are the speed, ease and relatively low cost with which they can be generated. About six to12 different RAPD markers can be generated in a single PCR from genomic DNA. Thus, several loci can be mapped with the results of one reaction. The disadvantages of RAPD markers are their sensitivity to reaction conditions and the difficulty in comparing markers from different labs. However, screening many ten nucleotide primers and choosing a reliable subset will improve the reliability of the markers generated. In addition, genetic information can be transferred between labs by cloning and sequencing RAPD marker fragments in order to design flanking primers. These primers can then be used to specifically amplify a mapped locus, thus defining a sequence-tagged site (STS; Olson et al. 1989; Hunt and Page 1994). Most polymorphic RAPD markers are scored as the presence or absence of a band, with presence being dominant over absence. But many RAPD markers appear as fragment length polymorphisms in haploid male honey bees, and some of these are codominant, and may generate heteroduplex bands in diploid, heterozygous females (Hunt and Page 1992). The protocols described in Fig. 1 are for mapping with RAPD markers in honey bees, which have haploid males (Hunt and Page 1995). The same procedures may be used on diploids with slight modifications. One researcher working with haploids should be able to construct a detailed linkage map containing 200–400 markers within a year, including DNA extraction, primer screening, generation of RAPD markers and segregation analysis. Although QTL mapping will not be included in this chapter, the processes of constructing a linkage map and of identifying QTLs can both be accomplished in the same population. For QTL mapping, it is advisable to have carefully selected lines so that the parents of the F_1 individual are likely to be fixed for alternative QTL alleles.

2
Materials

Equipment
- 1 or 2 thermal cyclers with 96-well format (PTC-100, MJ Research)
- eight-channel pipetter or a 96-prong clone transfer stamp
- gel boxes with 20×25 cm casting trays (model H4, Life Technologies, Inc.) with two combs (30-tooth, 1 mm thick)
- power supply

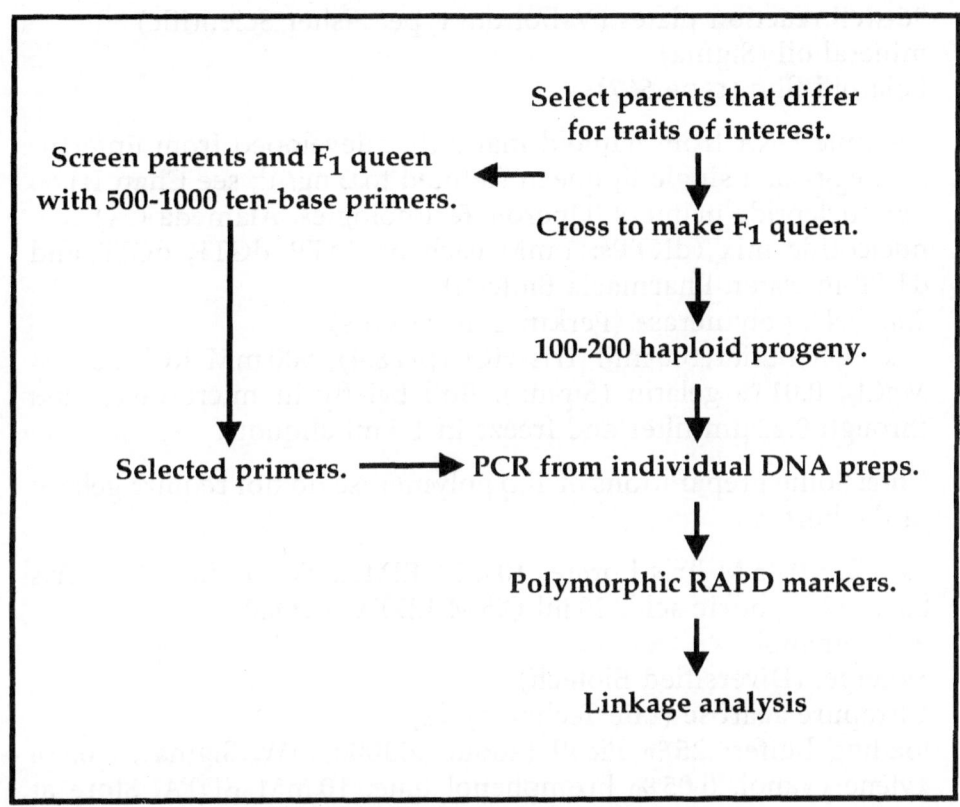

Fig. 1. Flowchart for making linkage maps of the honey bee

- UV transilluminator
- camera with orange filter (Fisher Scientific)
- Macintosh series II computer with math coprocessor (fpu)
- Excel software (Microsoft Corp)
- mapping software: MAPMAKER version 1.1 adapted for the Mac (available by writing to Dr. S. V. Tingey, P. O. Box 80402, Wilmington DE 19880–0402)

Note: A few other useful software programs for constructing primary linkage maps are listed below.

- JoinMap (IBM and Macintosh versions, Stam 1993)
- Mapmaker v3.0 and MAPMAKER/QTL1.1 (IBM and Macintosh versions, Lincoln and Lander 1992; available by anonymous FTP from genome.wi.mit.edu in the /distribution/mapmaker directory; fetch the FTP.me file first)

Supplies
- 96-well reaction plates (V-bottom type, Fisher Scientific)
- mineral oil (Sigma)
- Polaroid film (type 667)

Reagents and solutions
- genomic DNA from haploid males that developed from unfertilized eggs of a single F_1 queen (diluted to 3 ng/μl; see Chap. II)
- ten nucleotide primers (Operon Technologies, Alameda CA)
- nucleotide mix (dNTPs: 1 mM each of dATP, dGTP, dCTP and dTTP in water, Pharmacia Biotech)
- *Taq* DNA polymerase (Perkin Elmer Cetus)
- 10x PCR buffer:100 mM Tris-HCl (pH 8.4), 500 mM KCl, 20 mM $MgCl_2$, 0.01 % gelatin (Sigma). Boil briefly in microwave, pass through 0.22 μm filter and freeze in 1.5 ml aliquots.

Note: Some preparations of *Taq* polymerase do not require gelatin in the buffer.

- 5x TBE: 0.45 M Tris-borate, 10 mM EDTA. For 1 liter: 54 g Tris base, 27.5 g boric acid, 20 ml O.5 M EDTA (pH 8.0)
- 95 % ethanol
- Synergel (Diversified Biotech)
- Ultrapure agarose (Life Technologies)
- loading buffer: 25 % Ficoll (about 400000 MW, Sigma), 0.05 % xylene cyanol, 0.05 % bromphenol blue, 10 mM EDTA. Store at room temperature.
- 123 bp molecular weight ladder (Life Technologies Inc.)
- ethidium bromide stock solution: 10 mg/ml

3
Experimental Procedure

Generating RAPD Markers

1. Prepare the following mix in a 1.5 ml microcentrifuge tube for each set of 96 reactions. This provides enough mix to make 100 14 μl reactions.
   - 935 μl of water, chill on ice (use water purified to 18 mega-ohm resistivity)
   - 140 μl of 10x PCR buffer (prewarmed to 37 °C)
   - 140 μl of 1 mM dNTP nucleotide mix (store mix at –20 °C)
   - 75 μl of ten nucleotide primer (10 μM, store at –20° to –70 °C)
   - 5–6 μl of *Taq* Polymerase (5 U/μl)

Note: Some other preparations of *Taq* polymerase work equally well but may require slightly different buffers. *Taq* concentration may be optimized with a dilution series. Prepare all of the reactions for the mapping population at the same time. This will help insure uniformity of conditions. If necessary, one of the reaction plates can be kept on ice for up to 8 h.

2. After adding the *Taq*, invert the tube to mix and pipette 13 µl of reaction mix into each well of an empty 96-well plate (keep on ice).

3. Add 1 µl of dilute genomic DNA (3 ng/µl) to each well of the 96-well plate. The genomic DNA is drawn from the corresponding well of a 96-well plate that contains the mapping population of DNA with a different haploid drone in each well. The dilute genomic DNA is stored in a plate kept in a moisture chamber at 4 °C (use wet paper towels in a plastic box). This storage method will provide a stock of DNA that is stable for at least one month. While adding genomic DNA, keep the plate on ice and add the DNA with an eight-channel pipetter or a 96-prong clone transfer stamp.

Note: It may be advisable to have a more concentrated DNA stock when using the transfer stamp (10–20 ng/µl).

4. Overlay the wells of the reaction plate with 1 drop of mineral oil.

5. Place the reaction plate in the thermocycler and run the following cycle program (which takes about 3.5 h running time):
   - 94 °C for 30 s
   - 94 °C for 1 min, 35 °C for 1 min, 2 min ramp to 72 °C, 72 °C for 2 min (5 cycles)
   - 94 °C for 10 s, 35 °C for 30 s, 72 °C for 30 s (30 cycles)
   - 72 °C for 5 min

Resolving RAPD Markers in Synergel/Agarose Gels

1. Add 4.5 µl of loading buffer to amplification products in each well of 96-well plate (through oil overlay). Only one pipette tip is necessary for the entire plate.

2. Keep samples in refrigerator until electrophoresis (as long as necessary).

3. For two gels (96 samples), make up 2.4 l of 0.5x TBE.

4. Add 3–3.5 g Synergel (1.2 %–1.4 %) to an empty Erlenmeyer flask containing a stir bar and add just enough 95 % ethanol to wet the powder. Most RAPD markers will be resolved with the lower amount of Synergel.

5. Add 250 ml of 0.5x TBE and stir briefly but vigorously.

6. Add 1.75 g agarose (0.7 %) and stir.

7. Cover flask with perforated plastic wrap or small beaker and melt agarose in a microwave.

8. Pour gels with two combs (30-tooth, 1 mm thick) when the Synergel/agarose has sufficiently cooled. Cooling takes about 5 min while stirring in a room temperature water bath.

9. Load 48 samples per gel in a standard pattern with spaces between sets of 12, and including one lane that contains about 1 µg of 123 bp ladder in PCR buffer.

10. Run for 550–650 Vh (e.g., 110 V for 5–6 h; current should not exceed 60 mA when running gels at room temperature).

11. Stain with 1 µg/ml of ethidium bromide for 30 min and photograph on transilluminator with Polaroid type 667 film, using an orange filter.

Linkage Analyses

Statistical background

Most linkage mapping software packages make use of likelihood ratio testing. The Log_{10} of the odds ratio of two likelihoods (or LOD score) is the test statistic used by Mapmaker. The two likelihoods being compared when testing for linkage between two loci are the likelihood of obtaining the recombinant fraction that is estimated from the data set given linkage, vs the likelihood of obtaining this data without linkage (when the recombinant fraction is assumed to be 0.5). These likelihoods depend on the presence of marker dominance, the type of cross, and the actual data. This is a two-point analysis. Since it is a log base 10 statistic, a LOD score of 3.0 for linkage means that it is 10^3 times more likely to obtain these data if the loci are linked than if they were unlinked. LOD 3.0 is the usual threshold used for declaring linkage. In addition to the likelihoods that are compared in a two-point test, Mapmaker can also calculate the likelihood for the map of an entire linkage group (multi-point testing). The likelihoods of alternative maps are used to determine the LOD score support for the order. If the log-likelihood difference between the map with the most likely order and the next most likely order is –3.0, then the most likely order has a LOD 3.0 support (a thousand times more likely). An outline of the steps involved in linkage analyses is shown in Fig. 2.

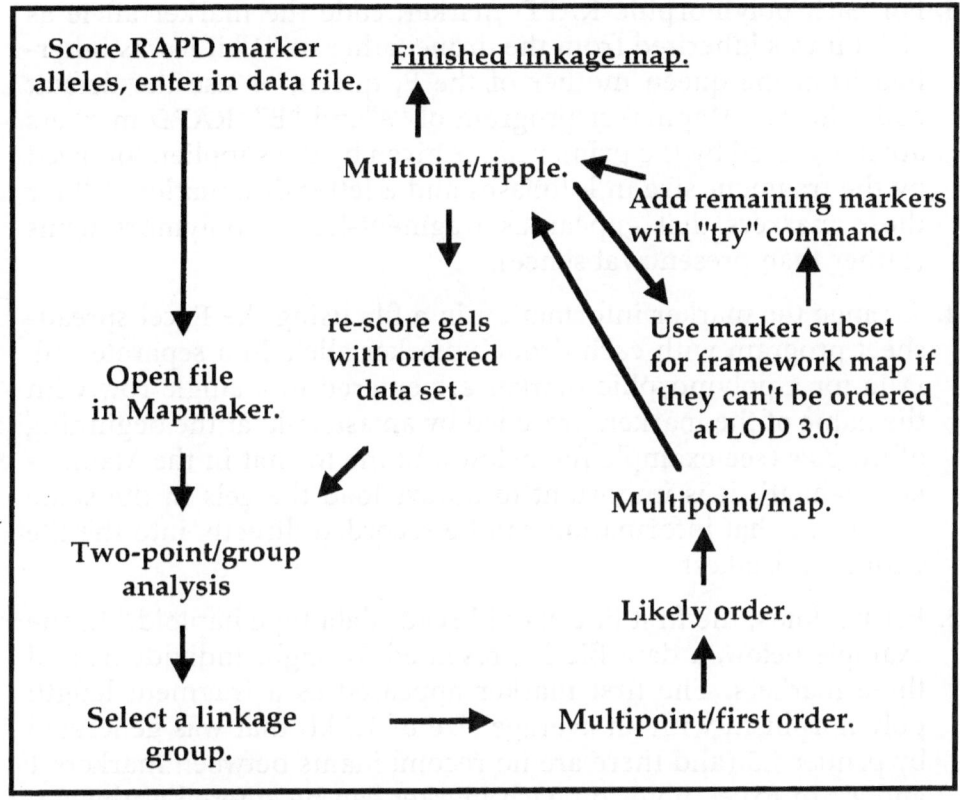

Fig. 2. Flowchart of linkage analyses

1. Many individual primers are first screened by using them in individual reactions with DNA from the parents and the F_1 queen. Primer selection is based on the generation of multiple bright bands (RAPD markers) that are polymorphic between the parents and are also inherited in the F_1 queen. In addition, other monomorphic but bright bands will also indicate a primer that works reliably.

   Note: Only a small portion of the screened primers will meet these criteria for efficient mapping (perhaps 10%–15%). The primer screen will also give information on phase (allowing you to code each allele according to which parent the allele was inherited from). Note that for diploid species, three generations are required to determine phase.

2. The selected primers are used in PCR (see above) with dilute DNA from the mapping population and the RAPD markers are resolved in gels.

Scoring and documentation of gels

3. For each polymorphic RAPD marker, code the marker allele as "0" if it was inherited from the drone father or "1" if it was inherited from the queen mother of the F_1 queen, or use the default codes for the Mapmaker program of "A" and "B". RAPD markers are designated by the primer name given by the supplier, followed by the fragment size in kilobases and a letter designation "f" for those markers that appear as fragment-length polymorphisms (rather than presence/absence).

4. Arrange the marker information in a file using the Excel spreadsheet program with each drone's marker allele in a separate cell. Data for a polymorphic marker are entered in a single row with the name of the marker, preceded by an asterisk, at the beginning of the row (see example file below and the format in the Mapmaker manual). It is important to always load the gels in the same pattern so that information can be recorded directly into the file with no mistakes!

5. For haploids, the first line should read "data type haploid." In the example below, a data file is presented for eight individuals and three markers. The first marker appeared as a fragment length polymorphism with an average size of 1.2 kb that was generated by primer C5 (and there are no recombinants between markers 1 and 2). In Excel, notes on each marker can be entered at the end of a row (e.g., "phase uncertain," or "hard to score") and files can later be organized by the order of markers as they appear in linkage groups. Save the file in text format.

Example data file

```
data type haploid
8 3 0=A 1=B
*C5 – 1.2f   0      0      1      0      0      1      1      1
*X4 – 0.6    0      0      1      0      0      1      1      1
*M5 – 0.4    1      1      1      1      0      0      0      0
```

Constructing the map

6. Open the data file in Mapmaker. Mapmaker automatically calculates all pairwise recombinant fractions and LOD scores for linkage. Use the "group" command under "two-point" in the menu ("two-point, group"). When the dialogue box appears, set an appropriate LOD score threshold and ϑ (recombinant fraction) threshold for detecting linkage. It may be desirable to use a more stringent ϑ than the default value (e.g., 0.30–0.35) in smaller data sets (such as 100) but it is generally advised to use the default LOD score of 3.0 to declare linkage.

7. Highlight and copy one of the resulting linkage groups; then open a new window and paste this sequence into the sequence box of the new window.

8. Choose "multi-point, first-order" to obtain an approximate order for the linkage group. Select the new order (by clicking 3 times with the mouse) and push the "apple" and "L" key to load the new sequence into the sequence box (or copy and paste).

   Note: The new order may be incorrect, especially if there are markers that have a lot of missing data or errors in scoring. It is also possible to determine the order based on two-point analyses using the "two-point/three-point" command (see software manual).

9. Choose "multi-point, ripple" to check the order. Mapmaker will permute all of the adjacent triplets and give the difference in the log-likelihoods for the complete, alternative maps for each possible order. The log-likelihood difference is equivalent to a LOD score and is a test statistic for the level of support for the given order. A LOD score of 3.0 is considered good support for an order. However, LOD 3 support will not be obtained for markers that are very tightly linked or have excessive errors or missing data. In this case, use a subset of markers for which there are good data to construct a framework map. Markers that are 8–20 cM apart with high LOD scores for linkage are appropriate as a framework for determining order. The "two-point, LOD table command" will display all pairwise recombination frequencies and LOD scores. Then, repeat the "ripple" command. It is also possible to rigorously compare the likelihoods of maps that represent all possible orders with the "compare" command. However, this command is only recommended if the number of markers is small (six or less) on some computers.

10. Fit markers that could not be ordered at LOD 3 onto the framework map with the "try" command. If markers were not tightly linked (e.g., ≥ 3 cM in a sample of 100 haploid drones) and cannot be ordered at $\geq$ LOD 3.0, the value of the LOD score supporting that order can be noted on the map. This value is obtained from the log-likelihood difference of the orders given by the output of the "ripple" command. Markers that are tightly linked to each other cannot be ordered at LOD 3.0 because there will be an insufficient number of informative meioses. These can be listed as a mega-locus on the map, or the most likely order can be shown.

11. Display a linkage map for that order by choosing "multi-point, map" from the menu or by pushing the "apple" and "M" key. If desired, first choose "Mapmaker, options" to switch from the default mapping function, Haldane, to the Kosambi mapping function or to change other options such as map scale.

12. In general, the "ripple" command should be repeated as a final step after a map is complete to check the order.

4
Troubleshooting

• Problems with phase: The phase may be uncertain or unknown for some markers in the mapping data set because of variations in reaction conditions during the initial primer screen or because of the difficulty of scoring the RAPD markers in the diploid queens. If no linkage is detected for these markers, a data file can be constructed with the alternative phase and again tested for linkage.

• Error checking

  - All of the gels should be scored twice in order to eliminate errors, which usually appear as spurious "double crossovers." The second scoring should be done after all data are collected, rather than immediately after the first scoring. Organize the data file so that markers are ordered by their positions in linkage groups so that the crossover positions can be seen for each individual in the data set.
  - The data should then be rechecked by scoring all of the markers a second time against the ordered data set. Data that resulted in unlikely double crossovers between tightly linked markers should be examined more closely. If a gel is ambiguous for any data point it should be entered as missing data.

  Note: It is also possible to detect such unlikely double crossovers automatically with Mapmaker v3.0 (Lincoln and Lander 1992). However, the program is more difficult to use and also flags many data points for which data are missing at adjacent markers.

  - After completing error checking, the order can be rechecked with the "ripple" command, and the final map constructed.

• Bad markers

  - A good marker is one that maps to a particular linkage group without causing much "map expansion" and also is ordered at $\geq$LOD 3.0. The results of the initial analyses may indicate markers that were difficult to order. These can be checked against the original gels to judge whether they were difficult to score and should be omitted from the data set.

- The "drop marker" command is extremely useful for detecting bad markers (markers that cause excessive errors in the data set). It gives a printout of the amount that the map contracts as each marker is dropped, along with the standard deviation. Markers that cause too much map expansion may have excessive errors in the data set. Errors tend to result in spurious recombinants.

Note: The degree of map expansion for markers at the ends of linkage groups cannot be assessed this way. The gel photos for these markers should be examined with suspicion. Some of the distal markers may be bad markers because the presence of excessive errors in the data for a particular marker that is near the end of a linkage group are likely to cause Mapmaker to place that marker at the end of the group.

- Too many linkage groups: Even with a map containing hundreds of markers there may be more linkage groups than chromosomes, indicating gaps in the map. This may be due to chance, or to "hot spots" of recombination, or to a lack of polymorphism in the region (because the region is identical in the two parents of the F_1 individual). Some strategies for confirming linkage are listed below.

  - Linkage to another group can sometimes be confirmed by increasing the sample size of progeny that are genotyped for the flanking markers. The markers with the smallest recombinant fractions with respect to the markers at the ends of linkage groups can be displayed by the "near" command.
  - If another map is constructed from a wider cross (more divergent parents), recombination may be reduced and linkage may be detected.
  - If more markers are desired to close gaps or to increase saturation near a gene of interest, methods such as bulk segregant analysis, near isogenic lines or representational difference analysis may be employed to target these regions (Giovannoni et al. 1991; Martin et al. 1991; Michelmore et al. 1991; Lisitsyn et al. 1993, 1994).

References

Carlier M, Roubertoux PL, Kottler ML, DeGrelle H (1990) Y chromosome and aggression in strains of laboratory mice. Behav Genet 20:137–156

Copeland NG, Jenkins NA, Gilbert DJ, Eppig JT, Maltais LJ, Miller JC, Dietrich WF, Weaver A, Lincoln SE, Steen RG, Stein LD, Nadeau JH, Lander ES (1993) A genetic map of the mouse: Current applications and future prospects. Science 262:57–66

Crabbe JC, Belknap JK, Buck KJ (1994) Genetic animal models of alcohol and drug abuse. Science 264:1715–1723

Estoup A, Solignac M, Harry M, Cornuet J-M (1993) Characterization of $(GT)_n$ and $(CT)_n$ microsatellites in two insect species: Apis mellifera and Bombus terrestris. Nucl Acids Res 21:1427–1431

Giovannoni JJ, Wing RA, Ganal MW, Tanksley SD (1991) Isolation of molecular markers from specific chromosomal intervals using DNA pools from existing mapping populations. Nucl Acids Res 19:6553–6558

Hunt GJ, Page RE Jr. (1992) Patterns of inheritance of RAPD molecular markers in the honey bee reveal novel types of polymorphism. Theor Appl Genet 85:15–20

Hunt, GJ, Page RE Jr. (1994) Linkage analysis of sex determination in the honey bee (Apis mellifera). Mol Gen Genet 244:512–518

Hunt, GJ, Page RE Jr. (1995) Linkage map of the honey bee, Apis mellifera, based on RAPD markers. Genetics 139:1371–1382

Hunt GJ, Fondrk MK, Dullum CJ, Page RE Jr. (1995) Major quantitative trait loci affecting honey bee foraging behavior. Genetics 141:1537–1545

Jansen, RC (1994a) High resolution of quantitative traits into multiple loci via interval mapping. Genetics 136:1447–55

Jansen, RC (1994b) Controlling the type I and type II errors in mapping quantitative trait loci. Genetics 138:871–881

Kafatos FC, Louis C, Savakis C, Glover DM, Ashburner M, Link AJ, Siden-Kiamos I, Saunders RDC (1991) Intergrated maps of the Drosophila genome: progress and prospects. Trends Genet 7:155–161

Kruglyak L, Lander ES (1995) A nonparametric approach for mapping quantitative trait loci. Genetics 139:1421–1428

Lander ES, Green P, Abrahamson J, Barlow A, Daly MJ, Lincoln SE, Newburg L (1987) MAPMAKER: An interactive computer package for constructing primary genetic linkage maps of experimental and natural populations. Genomics 1:174–181

Lander ES, Botstein D (1989) Mapping Mendelian factors underlying quantitative traits. Genetics 121:185–199

Lander ES, Schork NJ (1994) Genetic dissection of complex traits. Science 265:2037–2048

Lincoln SE, Lander ES (1992) Systematic detection of errors in genetic linkage data. Genomics 14:604–610

Lisitsyn N, Lisitsyn N, Wigler M (1993) Cloning the differences between two complex genomes. Science 259:946–951

Lisitsyn NA, Segre JA, Kusumi K, Lisitsyn NM, Nadeau JH, Frankel WN, Wigler MH, Lander ES (1994) Direct isolation of polymorphic markers linked to a trait by genetically directed representational difference analysis. Nature Genet 6:57–63

Martin GB, Williams JGK, Tanksley SD (1991) Rapid identification of markers linked to a Pseudomonas resistance gene in tomato by using random primers and near-isogenic lines. Proc Natl Acad Sci USA 88:2336–2340

Michelmore RW, Paran I, Kesseli RV (1991) Identification of markers linked to disease-resistance genes by bulked segregant analysis: A rapid method to detect markers in specific genomic regions by using segregating populations. Proc Natl Acad Sci USA 88: 9828–9832

Olson M, Hood L, Cantor C, Botstein D (1989) A common language for physical mapping the human genome. Science 245:1434–1435

Severson DW, Mori A, Zhang Y, Christensen BM (1994) Chromosomal mapping of two loci affecting filarial worm susceptibility in *Aedes aegypti*. Insect Mol Biol 3:67–73

Severson DW, Thathy V, Mori A, Zhang Y, Christensen BM (1995) Restriction fragment length polymorphism mapping of quantitative trait loci for malaria parasite susceptibility in the mosquito *Aedes aegypti*. Genetics 139:1711–1717

Stam P (1993) Construction of integrated genetic linkage maps by means of a new computer package: JoinMap. Plant J 3: 739–744

Sturtevant, AH (1913) The linear arrangement of six sex-linked factors in *Drosophila*, as shown by their mode of association. J Exp Zool 14:43–59

Tanksley SD (1993) Mapping polygenes. Ann Rev Genet 27:205–233

Williams JGK, Kubelik K, Livak J, Rafalski JA, Tingey SV, (1990) DNA polymorphisms amplified by arbitrary primers are useful as genetic markers. Nucleic Acids Res 18: 6531–6535

Williams JGK, Hanafey MK, Rafalski JA, Tingey SV (1993) Genetic analysis using random amplified polymorphic DNA markers. Methods Enzymol 218:704–740

Zeng Z (1993) Theoretical basis for separation of multiple linked gene effects in mapping quantitative trait loci. Proc Natl Acad Sci USA 90:10972–10976

Zeng Z (1994) Precision mapping of quantitative trait loci. Genetics 136:1457–1468

Chapter XXII

Pseudo-Testcross Mapping Strategy Using RAPD Markers

D. GRATTAPAGLIA

1
Introduction

Genetic linkage maps are powerful tools for detailed genetic analysis. Maps allow dissection of the architecture of polygenic traits by the investigation of the individual genomic segments underlying quantitative variation (QTLs). Mapping information can accelerate selective breeding through marker assisted selection (MAS), especially in organisms with long breeding cycles such as perennial crops and domestic animals. Comparative genome mapping provides insights into the modes and dynamics of genome evolution and patterns of variation in genetic recombination.

Linkage maps in plants and animals have been constructed using segregating populations derived from crosses between inbred lines or by exploiting the information content of three-generation pedigrees. For the majority of preferentially allogamous, highly heterozygous species with little or no history of domestication and selective breeding, such populations are not available and are frequently difficult or impossible to obtain due to a significant genetic load and time constraints. Available pedigrees for the majority of these species generally involve only two heterozygous parents and their F1. In these cases, linkage mapping can be performed for each heterozygous parental individual separately using single dose DNA polymorphisms segregating 1:1 for presence vs absence of the DNA fragment in their gametes and therefore in their F1 progeny. We called this mapping strategy "pseudo-testcross" because the testcross mating configuration of the markers is not known *a priori*, as in a conventional testcross in which the tester is homozygous recessive for the locus of interest. Rather, no prior genetic information is available thus the configuration is inferred *a posteriori* after analyzing the parental origin and segregation of the marker in the progeny. The pseudo-testcross strategy is specifically based on the selection of single-dose

DNA polymorphisms present in one parent and absent in the other. Although maps for both parents are constructed simultaneously in the same experiment, they are individual-specific and cannot be aligned or integrated as such, unless other markers common to both maps are also used (Grattapaglia and Sederoff 1994).

The pseudo-testcross mapping strategy is simple to implement and in principle can be applied with any type of molecular marker. Its potential, however, can be better exploited with the efficiency of the RAPD assay to uncover informative pseudo-testcross configurations. When two heterozygous individuals are crossed, many RAPD markers will be present in a heterozygous state in one parent and absent (i.e., null genotype) in the other, therefore segregating 1:1 in their F1 progeny. As the RAPD assay detects only one "dominant" amplified allele at a locus, (i.e., single dose polymorphism) the occurrence of such 1:1 pseudo-testcross configurations is actually enhanced. All genotypes composed by undetected (not amplified) alleles at that RAPD locus fall into the same null genotype category that is necessary to observe the 1:1 segregation of the amplified RAPD allele. Furthermore, the fact that the RAPD assay is sensitive to single base changes, contributes to a higher efficiency in scanning the genomes for polymorphisms.

The combined use of the pseudo-testcross configuration and RAPD markers is therefore a general and highly efficient strategy for the construction of genetic linkage maps in any highly heterozygous, sexually reproducing living organism. It can be immediately applied to any species without any prior genetic information. The only requirement is sexual reproduction between two individuals that results in the generation of a progeny large enough to allow the estimation of recombination frequencies between segregating markers. Its efficiency will be directly proportional to the level of genetic heterozygosity of the species under study, which is a function of its preferential mating system, and the genetic divergence between the individuals crossed.

Ritter et al. (1990) described the theoretical background for linkage analysis of markers segregating 1:1 in crosses between heterozygous parents. Wu et al. (1992) discussed the segregation of single dose RFLP in the context of mapping polyploids. Single dose RFLP or RAPD markers segregating 1:1 were used for genetic mapping in potato (Bonierbale et al. 1988) and sugar cane (Al-Janabi et al. 1993). Several labs have suggested the wide applicability of the pseudo-testcross strategy in combination with the RAPD technology for genetic mapping in perennial crops of forest and fruit trees (Carlson et al. 1991; Grattapaglia et al. 1992; Roy et al. 1992). High coverage gen-

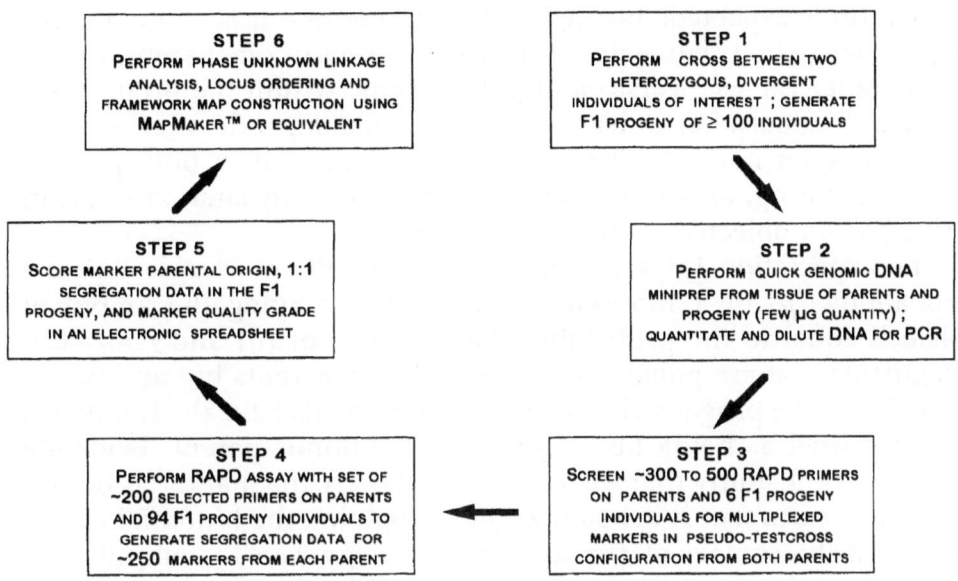

Fig. 1. Step-by-step flowchart of the pseudo-testcross mapping strategy using RAPD markers

etic maps constructed by the pseudo-testcross/RAPD strategy have been reported for several species in this group such as *Eucalyptus grandis* and *E. urophylla* (Grattapaglia and Sederoff 1994), apple (Hemmat et al. 1994), *Citrus* (Cai et al. 1994) and *Pinus* (Kubisiak et al. 1995).

This chapter details a step-by-step protocol for the construction of genetic linkage maps for individual genotypes using the pseudo-testcross mapping strategy and RAPD markers. A complete outline of this mapping strategy is shown in Fig. 1.

2
Experimental Strategy

RAPD Primer Screening

The aim of the screening step is to select for RAPD primers that amplify markers segregating for presence and absence in the F1. A RAPD marker homozygous in either parent does not result in observable segregation and is therefore discarded. Segregation requires necessarily that any parent having a marker be heterozygous. When one parent has the marker and the other does not a 1:1 ratio is expected. When both parents have the marker in heterozygosity then a 3:1 ratio is expected. Target RAPD markers in our screening strategy are those that segregate 1:1, i.e., are present in one parent, absent in the other (or vice versa) and polymorphic within the progeny set of six individuals, i.e., at least one F1 individual has the marker or at least one does not have it (Fig. 2). Markers present in both parents and segregating, although useful as it will be seen later, are not, in principle, an objective of the screening step.

The expectation for a pseudo-testcross RAPD marker is that the probability (p) of a random F1 individual carrying the amplified allele is equal to the probability that it does not (q) and $p=q=0.5$. Fragments that are polymorphic between the parents but are monomorphic in the progeny (i.e., all six F1 individuals have the fragment) are discarded as being homozygous in the donor parent. Evidently there is a probability that we mistakenly classify the marker as homozygous when in fact it is heterozygous. Applying the binomial distribution in the context of our screening strategy we can compute this probability which is analogous to the probability of getting exactly six heads in six tosses of a fair coin where the probability of getting a head (p) is equal to the probability of getting a tail (q) and $p=q=0.5$. This probability is equal to $0.5^6=0.016$ or 1.6%. Note that theoretically, fragments that are present in one parent, absent in the other and absent in all six progeny although rare (probability equal to 1.6%) should also be included as target markers. Although they are not observed in the progeny they are observed in a parent and therefore should be inherited by the progeny. However in practice these are also discarded.

PSEUDO-TESTCROSS MAPPING STRATEGY

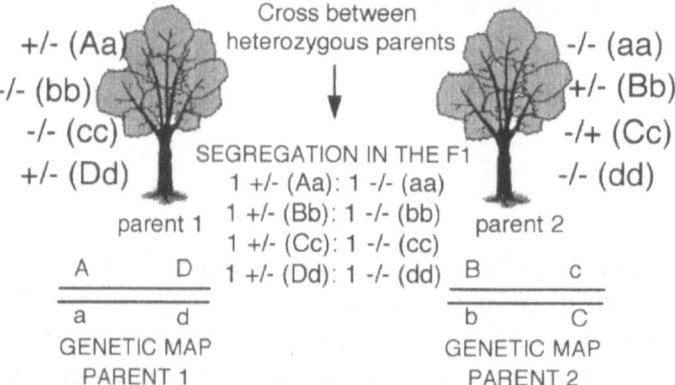

Fig. 2. Diagram of the pseudo-testcross mapping strategy

The screening setup is therefore powerful because it provides direct information on the parental origin of the RAPD marker locus with 100 % confidence and a correct determination of its genotype (homozygous or heterozygous) with a 98.4 % confidence level or a 96.8 % confidence level if we also discard fragments that are present in one parent, absent in the other and absent in all six progeny. The power of this design for distinguishing segregating markers that are present in both parents from homozygous ones is however low. The expectation for these markers is a 3:1 ratio (presence:absence), i.e., $p=0.75$ and $q=0.25$. To identify such markers we need to observe at least one F1 individual without the marker. The probability of this event is equal to 82.3 % which corresponds to the confidence level attained by the screening design adopted.

With eight DNA samples (two parents and six F1 progeny) a total of 12 arbitrary primers can be conveniently screened for pseudo-testcross polymorphic markers in a single 96-well plate. We strongly recommend screening a large number of arbitrary primers in this step. By doing so, besides allowing the selection of primers that amplify more intense, easily scorable PCR products one can screen for multiplexing ability of primers, i.e., primers that identify several pseudo-testcross marker loci simultaneously from one or both parents. The ability to apply this extra stringency will evidently depend on the level of heterozygosity and divergence of the parents and the availability of large batteries of primers. However at least two practical advantages exist that easily justify this approach: (1) it eliminates or strongly minimizes any lack of reproducibility of RAPD markers; (2) it minimizes the final cost per data point as more data points are generated with the same cost of PCR reaction and gel analysis.

Segregation Analysis and Linkage Mapping

The objective of this step is to gather segregation data on the F1 progeny for several markers and perform linkage and ordering analyses for each parent to construct both linkage maps. In order to arrive at good estimates of two-point recombination fractions and allow adequate multipoint locus ordering, a relatively large sample of meioses is necessary. A progeny of $n=100$ is generally adequate for this purpose. Note that in the testcross configuration, the linkage information content for a 1:1 segregation ratio is the same for dominant and codominant markers. A progeny size of $n=100$ allows the distinction of a 1:1 segregation ratio from a 3:1 ratio with a 99 % confidence level (Wu et al. 1992).

On a 96-well format, both parents and a progeny size of 94 individuals will be used in the segregation analysis and mapping step. Together with the six individuals analyzed in the screening step, segregation data on a total progeny size of $n=100$ individuals are obtained. Because large sample sizes need to be used for the mapping step, ideally one would aim at obtaining the largest number of markers with the least number of PCR reactions. For this reason, the efficiency of the mapping step is directly related to the effectiveness of the primer screening step in selecting multiplex primers that yield robust and easily scorable markers.

Tips In order to optimize the reproducibility of RAPD markers we have adopted four basic criteria when using this assay:

- Optimize the reaction conditions mainly concerning the quantity of DNA used and the selection of a robust set of arbitrary primers.

- Adopt an extremely systematic approach when carrying out any large scale analysis so as to maintain constant reaction conditions.

- Adopt a high stringency in the selection of which RAPD fragments to score, always looking for the more intense, easily scorable and reproducible ones tested by replicate independent assays and data scoring by different people.

- Whenever possible adopt a "hidden" replication in the genotyping work so as to come out with an estimate of repeatability for the whole experiment. The strategy we adopt is to record segregation of RAPD markers in two independent replications corresponding to two complementary sets of individuals of the same mapping population. In the first replicate, the two parents and 46 F1 progeny individuals are assayed with all the selected primers. In the second replicate, independent DNA extractions, reaction mixture preparations, gel analysis and genotype scoring are performed for another set of 48 progeny individuals. Markers that do not amplify consistently or cannot be scored reliably across the two replicates should be dropped from further analysis.

- We strongly recommend adopting an iterative mode when carrying out a mapping experiment with this protocol. While the screening step is better carried out all at once in order to rank primers more efficiently, segregation data gathering and mapping analysis should not be compartmentalized. Rather, mapping analyses should be carried out with every new set of 50–100 markers scored, until the number of markers mapped approaches the target number established for the particular application. In this manner, problems with the raw data collection and analysis can be identified and corrected earlier. Typical problems include frame shifts, inverted orientations and deletions of data arrays.

3
Materials

Equipment

- manual repeater pipette (Eppendorf)
- thermocycler for 96-well plate (U or V bottom) (MJ Research)
- large submarine gel electrophoresis unit (custom made for 30 cm wide×35 cm long gel, two 50-tooth combs, capacity of 100 lanes /15 cm run)
- power supply (500 V/250 mA)
- swinging bucket tabletop centrifuge with microtiter plate adapters
- large size UV transilluminator (30×20 cm) and gel documentation system
- Macintosh personal computer with 40 MB free HD space/16 MB RAM
- electronic spreadsheet and word processing software
- mapping software MapMaker V 2.0 for Mac

Supplies

- polypropylene or polycarbonate flexible U or V bottom 96-well plates
- 1.5 and 0.5 ml plastic tubes
- Mineral oil

Reagents and solutions

- *Taq* polymerase buffer 10x (specific composition to accompany the enzyme)
- $MgCl_2$ 25 mM
- *Taq* polymerase (5 U/µl)
- purified nonacetylated BSA (bovine serum albumin) 10 mg/ml
- dNTPs (deoxynucleoside triphosphates) 2.5 mM
- RAPD primers: dilution at 7.5 ng/µl in water (Operon Technologies Inc., Alameda, CA)
- DNA grade agarose
- 5x TBE: 54 g Tris base, 27.5 g boric acid, 20 ml 0.5 M EDTA (pH 8.0); bring to 1 l with water. **Note:** working solution is 1x TBE.
- ethidium bromide 10 mg/ml. Store in foil wrapped container; use precautions appropriate for chemical mutagen.
- DNA fragment size standard (100–4000 bp range) (e.g. 1 kb ladder–GIBCO)
- loading buffer: 0.25 % bromophenol blue, 40 % (w/v) sucrose in TE

4
Experimental Procedure

RAPD Primer Screening

RAPD assay The protocol presented is based on the original one by Williams et al. (1990) with some modifications of primer and DNA concentration, addition of purified BSA and reduction of PCR reaction volume to 13 μl, as described previously (Grattapaglia and Sederoff 1994).

When performing the RAPD assay, several RAPD reactions are assembled simultaneously both in the RAPD primer screening step and the mapping step. Common master mixes (MMs) are therefore used. In the RAPD screening step for pseudo-testcross markers, several RAPD primers are screened against the same set of two parents and six progeny. In the mapping step, several DNA samples are analyzed with the same primer. The composition of the MM will include all the components that are common to all reactions, while the variable component under scrutiny will be aliquoted separately to a 96-well plate. To allow for flexibility in performing either mapping or screening, the two potentially variable components (template DNA and primer) are diluted in a way that they are used in an equal volume (3 μl). In this way, the MM will be standard in terms of volumes no matter what step in the project is being performed.

1. With a repeater or multichannel pipette aliquot 3 μl of template DNA or 3 μl of primer to be screened to the bottom of wells. You can aliquot as many plates as you want, and store the aliquoted plates in the refrigerator for 1–2 days or for longer times at –20 °C or –70 °C wrapped in PVC wrap. Label the order of DNA distribution on the plate.

2. Thaw 10x *Taq* buffer, dNTPs, BSA and the primer(s) to be used. Make sure you have sterile distilled water ready to use.

3. Label 1.5 ml tubes with the primer code and the word MIX to denote they are MM tubes.

4. Add components into a MM according to Table 1. Several MMs can be prepared simultaneously. They can be stored at –20 °C for later use. In this case **do not** add *Taq* polymerase yet. Add only upon use. As some volume is always lost during distribution of liquid, MMs have to be calculated and prepared for approximately 5 %–10 % more than the actual number planned (e.g., calculate and prepare for 13 when 12 reactions are needed or for 100

Table 1. RAPD reactions components and quantities necessary (in μl) to prepare master mixes for a 96 reaction experiment that is either for primer screening or mapping analysis

	1 reaction	Screening[a] mix for 13	Mapping[b] mix for 50
Sterile water	2.64	34.32	132.0
10X *Tay* buffer	1.3	16.9	65.0
MgCl$_2$ (25 mM)	0.78	10.14	39.0
BSA (10 mg/ml)	1.04	13.52	52.0
dNTP's 2.5 mM	1.04	13.52	52.0
DNA or primer[c]	3.0	39.0 (DNA)	150.0 (primer)
Taq (5 U/<m>l)	0.2	2.6	10.0
total	10.0	130.0	500.0

[a] RAPD primer screening for pseudo-testcross markers with two parents and six F1 progeny (12 primers × 8 DNA samples); eight such master mixes (one for each DNA sample) are prepared.
[b] RAPD mapping assay with two primers on two parents and 46 F1 progeny (2 primers × 48 DNA samples); two such master mixes (one for each primer) are prepared.
[c] DNA at 2.0 ng/μl; primer at 7.5 ng/μl.

when 96 are needed). Although you might be tempted to prepare a large MM and then subdivide it in several single ones we recommend not doing so; instead prepare each MM separately.

5. Add *Taq* polymerase to the MMs. Mix gently by inverting the tube a few times to insure adequate homogenization of the glycerol/*Taq* solution into the MM.

6. With the repeater pipette distribute aliquots of 10 μl of MM to the wells. Add 50 μl of mineral oil to each well if the thermocycler requires so.

7. Spin the plate for 30 s in a tabletop centrifuge. Cycle RAPD reactions with the following program: 40 cycles of denaturation at 92 °C for 1 min, annealing at 35 °C for 1 min and extension at 72 °C and a final extension at 72 °C for 5 min.

8. Check if RAPD reactions worked by adding 2.0 μl of ethidium bromide 20 ng/μl to the bottom of a few randomly chosen wells in the plate and visualizing the plate on the UV transilluminator. If the PCR amplification worked accordingly, you should be able to see a distinct fluorescence in the well.

Gel electrophoresis of RAPD fragments

9. Add 3.0 μl of loading buffer to the side of each well. Spin the plate for 30 s to get the loading buffer through the mineral oil.

10. On a flat surface, prepare a large size (30 cm wide × 35 cm long × 7 mm thick) horizontal 1.5 % agarose gel in 1x TBE to load all 96 samples plus four fragment size markers, two per comb. Add ethidium bromide at 10 μg/100 ml to the cooled agarose solution. Pour gel and let harden.

11. Remove combs from gel. Load reaction products into the gel left to right starting with the first sample leaving the first and last lanes for fragment size marker. You can use the same tip to load all the samples, or one tip every 10 samples or so.

12. Load 13 μl of fragments size marker (30 ng/μl) to the first and last lanes of each comb array (top and bottom).

13. Place gel tray into electrophoresis tank and slowly add 1x TBE buffer. Make sure your gel is submarine (at least 15 mm of buffer above surface of the gel).

14. Run gel at room temperature, 10 V/cm until the dye has touched the very bottom of the gel or gel slice. You may want to check the gel after the front has migrated about 2–3 cm.

15. Look at the results on the transilluminator, cut the gel along the lower gel wells and take a separate picture or image capture for each half of the gel.

16. On the picture, annotate information on the experiment (experimental population used, primer code, segregating RAPD markers, date, operator, etc.)

17. Using a P-20 pipette and a yellow pipette tip take a sample of the segregating amplified product by touching the band into the gel visualized on the UV. Rinse the tip into a tube containing 20 μl of 20 % TE buffer. Freeze the sample at –20 °C for future use.

18. Gels can be recycled up to three times. Add 30 % of the normal amount of ethidium bromide at every recycling. TBE running buffer can be used up to six to eight times without any significant change in electrical conductivity. Polycarbonate 96-well plates can be washed, rinsed, autoclaved and reused several times.

Primer selection   All primers that amplify at least one marker segregating in the pseudo-testcross configuration should in principle be selected (Fig. 3). Depending on the genetics of the organism under study (i.e., heterozygosity and genetic divergence of the parents) a more or less stringent selection of primers can be applied, always keeping in mind that a total of ~500 markers, 250 from each parent, are typically nec-

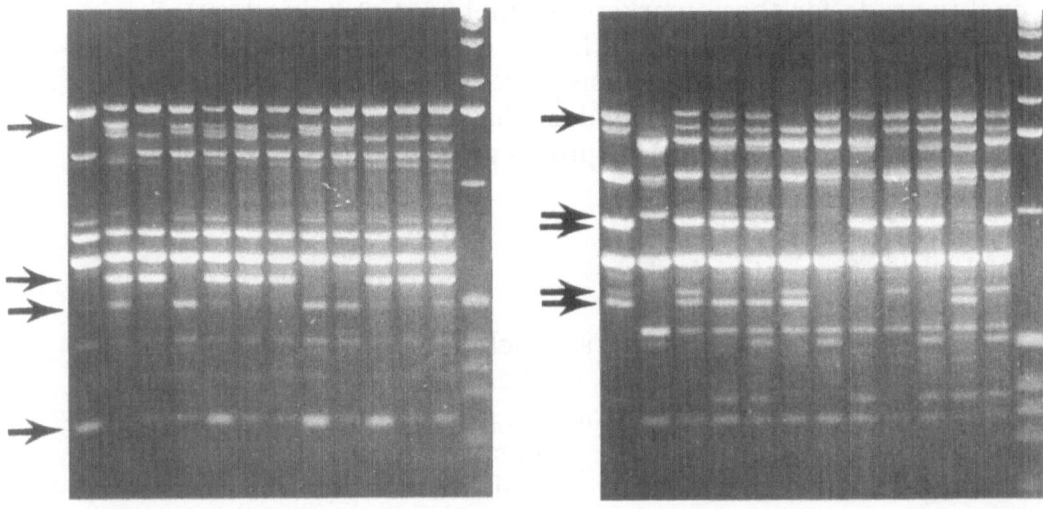

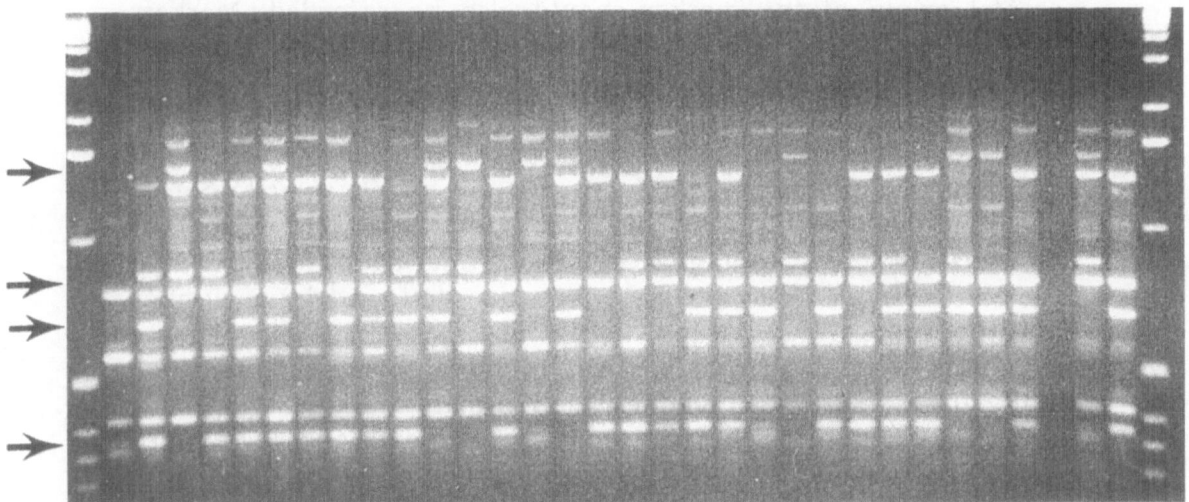

Fig. 3. *Top panels*: RAPD primer screening profiles for pseudo-testcross markers. *Left* array with primer OPU10, *right* array primer OPG12 ; *lanes left to right* in each array are RAPD assays on the two parents and 10 F1 progeny individuals and 1 kb ladder fragment size marker. *Arrows* indicate informative segregating markers. *Bottom panel*: segregation of RAPD markers; *lanes left to right* are assays on 29 F1 progeny individuals (primer OPW11), a blank lane and the two parents. *Arrows* indicate markers segregating 1:1; first and last lanes are 1 kb ladder fragment size markers

essary to construct moderate density linkage maps. We typically rank all the primers screened by a combined score that is the product of the number of pseudo-testcross markers and a subjective average quality grade for these markers (from 1 to 3, 3 being best). Sometimes, however, it is difficult to have a clear idea of the quality of the potential markers looking at only six F1 progeny, especially when only one or two progeny individuals have the marker. In these cases the overall quality of the RAPD profile generated should be taken into account.

RAPD Marker Segregation Analysis

After performing RAPD assay and gel electrophoresis as described above, pseudo-testcross RAPD markers segregating 1:1 are usually easily recognized and scored from the gel photos or captured images on the screen (Fig. 3). Markers segregating 3:1 are more difficult in this respect. We recommend scoring data directly into an electronic spreadsheet to avoid transcription errors. We use the number (1) for the presence, (2) for the absence and zero (0) for missing data. The parental origin of the marker should also be recorded. Markers should all be scored into the same spreadsheet and at the time of analysis sorted out by parent so as to obtain separate data sets for each parent. Dubious data points should be scored as missing data.

The two replicates of the mapping step should be analyzed independently in time and if possible by different people so as to obtain a measure of repeatability of scored markers. RAPD marker nomenclature should contain at least the primer code corresponding to a particular arbitrary ten-base sequence, followed by a number indicating the estimated fragment size in base pairs. Following the fragment size, separated by a slash, we have adopted a subjective quality grade from 1 to 3 denoting the fragment amplification intensity and ease of scoring, 3 being the best. For example RAPD marker G12/760/2 corresponds to a RAPD fragment amplified by Operon primer OP-G12 (corresponding to the sequence 5'-CAGCTCACGA-3'), with 760 bp and medium amplification intensity (quality grade 2). After markers have been mapped a (+) or (-) following the quality grade indicates the relative phase of linkage amongst them (Fig. 4).

Linkage Analysis and Map Construction

A detailed explanation of the statistics involved in linkage analysis is beyond the scope of this chapter. For completion, however, a brief step-by step procedure to construct the maps using the program MapMaker (Lander et al. 1987), version 2.0 for Macintosh, is given without detailed explanations of the underlying theory. More details

A

```
#
#MapMaker sample input file  of  phase-unknown pseudo-testcross RAPD marker
data/File has 6 markers duplicated and recoded, i.e. 12 markers, scored on
60 F1 progeny individuals
#
data type f2 backcross
60 12 0  symbols 1=H 2=A 0=-

*D8_1342/2   211222111211111221111122212222111222221122221121221211111112112
*D8_487/3    1222121221222211222112121111222222221222221222211111121111222121
*A10_1673/3  12122111121212121211111221112112112111111222212221111121212212
*A11_1439/1  12222212121211111211121221112111212122211222212211111111212102112
*A11_635/2   122222221121112222211222211211122221221112112221112212211201212
*D3_508/2    022222221121122222112222112222111122111112111222111212221211212
*D8_1342/2   122111222122221122222111211112221111211111221211212122222221221
*D8_487/3    2111212112111221112212122211111111121111112111122222122221112
*A10_1673/3  2121122221212121222221122212212212222211112111222221211211
*A11_1439/1  21111121212222212221211222122221211112211111211222222212120122
*A11_635/2   2111111122122111111221111221221111211122212211122211212112102121
*D3_508/2    0111111122122111111221111221221111211122212211122211212112122121
```

B

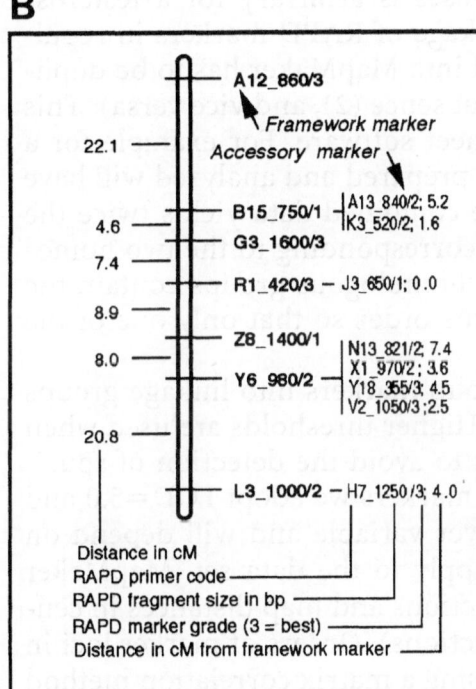

Fig. 4A,B. A MapMaker input file to analyze phase-unknown pseudo-testcross marker data. The file has six markers, duplicated and recoded, on 60 progeny individuals. B Framework linkage group of RAPD markers for a *Eucalyptus urophylla* genotype constructed using the pseudo-testcross mapping strategy can be found in several sources that discuss linkage mapping (e.g., Allard 1956; Ritter et al. 1990) and locus ordering (e.g., Buetow and Chakravarti 1987).

A single locus goodness of fit χ^2 test ($\alpha = 0.01$) is first performed on all scored marker loci in both data sets to test for departure from the null hypothesis of a 1:1 mendelian segregation ratio. Only markers that pass this test are initially used in the linkage analysis. RAPD marker loci that depart from the expected 1:1 ratio are later placed on the map by determining their most probable location in an already established gene order. Markers present in both parents and segregating 3:1 are tested for departures from this ratio and later placed on the map by testing for linkage to pseudo-testcross marker loci.

Linkage analysis is done for each parental data set separately. The genetic model for the pseudo-testcross is analogous to a testcross with parental linkage phase unknown. The MapMaker model assumes that all markers are in coupling phase and consequently it does not recognize linkages for markers in repulsion phase. The assignment of coupling and repulsion phases is arbitrary for a testcross model. To allow the detection of linkage of RAPD markers in repulsion phase the data set to be entered into MapMaker has to be duplicated and recoded, i.e. presence (1), absence (2), and vice versa). This can be done easily using a spreadsheet software. For example for a data set of six markers, the file to be prepared and analyzed will have 12 markers (Fig. 4a). Analysis of the combined data yields twice the expected number of linkage groups corresponding to the two homologs for each chromosome. The two homologous groups contain the same markers in the same exact locus order so that only one of the two is presented in the final map.

Linkage thresholds adopted to group markers into linkage groups are typically LOD$=3.0$ and $\theta = 0.40$. Higher thresholds are used when analyzing large numbers of markers to avoid the detection of spurious linkages. When analyzing ~ 200 markers we adopt LOD$=5.0$ and $\theta = 0.25$. These thresholds are however variable and will depend on the stringency that one wishes to apply to the data set. MapMaker outputs two point recombination fractions and map distances in centimorgans (Kosambi or Haldane functions). Orders of marker loci in each linkage group are established using a matrix correlation method (FIRST ORDER command) implemented by MapMaker. From this initial order, a subset of evenly spaced loci that can be ordered with a likelihood ratio support $\geq 1000:1$ establish a "framework map." Framework maps are constructed by comparing the likelihood of all permutations of all adjacent triples using the RIPPLE command. In this process, individual markers are dropped until a marker sequence is obtained that has an order at least 1000 times better than other orders, i.e., when the log likelihood difference is ≥3.0. Typically when using progeny sizes of ~100 individuals (i.e., 100 meioses) and ~250 markers, approximately 60%–70% of the markers can be ordered with this stringency. Lower stringency (log likelihood difference ≥ 2.0) can be adopted. Dropped markers are later placed on the framework map as accessory markers and located to the closest framework marker based on the two-point recombination fraction (Fig. 4b).

Linkage analysis for markers segregating 3:1 is not carried out using MapMaker. Significant associations between 3:1 and 1:1 marker loci are determined by a χ^2 test ($\alpha=0.001$). Linkage is determined if a significant departure from the 1:1 ratio is observed at the pseudo-testcross locus within the subset of individuals that are homozygous null (band absent) at a 3:1 locus. In order to have adequate statistical power to carry out this test it is critical to work with large sample sizes (≥100). Markers segregating 3:1 are tested for linkage with all markers in both parental data sets. They act as locus bridges and are therefore useful to define homologies between linkage groups so that the two separate parental maps can be aligned.

5
Results and Comments

Genetic linkage maps constructed by the pseudo-testcross strategy are in principle genotype-specific. Because the strategy is based on the selection of markers present in one parent and absent in the other, no RAPD markers are in common between the maps. Thus, it is impossible to determine homologies of linkage groups in the two maps or integrate the two maps into one. Markers present in both parents and segregating 3:1 can potentially integrate maps although a complete map merging would still be difficult to achieve, since correct locus ordering among the markers not in common between homologous linkage groups would be problematic to estimate. Overlap of RAPD marker occurrence and linkage relationships in genetic maps of different individuals will depend on the presence of the same RAPD marker loci and their allelic state. In our lab, comparative pseudo-testcross mapping studies among ten different *Eucalyptus* trees indicate that on the average 64% of the RAPD markers mapped are transferable and informative among their linkage maps. These results suggest that individual genotypes from the same population have a significant degree of genome homology for RAPD marker loci and that it will be straightforward to align linkage maps from individual trees using a set of common markers. However, because the pseudo-testcross strategy increases in efficiency with increasing divergence of the parents, the initial construction of these maps should be always carried out using the widest crosses that can be made between individuals.

6
Troubleshooting

A detailed troubleshooting guide is provided by Table 2.

Table 2. Troubleshooting guide for pseudo-testcross mapping using RAPD markers.

Step	Observation	Potential solution
RAPD assay	No RAPD bands appear on gel	1. Verify if ethidium bromide was added to gel 2. Check if some crucial PCR reaction component was missed or added in a wrong quantity (check master mix calculations)
	RAPD bands not properly separated	1. Let gel run longer or increase agarose concentration and run gel again
	Smiling RAPD bands	1. Add ethidium bromide to loading dye as directed
	RAPD bands too intense with smearing due to excessive non specific priming	1. Reduce *Taq* polymerase and/or genomic DNA in RAPD reaction
	Only one or e few RAPD fragments are amplified; fragments amplified with marginal intensity; Many DNA samples fail to amplify RAPD bands	1. Verify DNA quality (degradation) and re estimate DNA concentration 2. Do a serial dilution with 5X increments above and below concentration used to determine optimal DNA concentration window 3. Column purify genomic DNA samples or reisolate with more elaborate protocols that include extra purification steps
RAPD profiles with central smear	1. Add more buffer to the electrophoresis tank to have at least 1.5 cm of buffer above gel surface 2. Recirculate electrophoresis tank buffer	

Table 2. (continued)

Step	Observation	Potential solution
	RAPD bands are not sharply resolved	1. Load gel faster (10 wells per minute) to avoid sample diffusion in well 2. Excessive electrophoresis time; don't run overnight, rather during 5 to 6 h maximum 3. Reduce ethidium bromide into gel 4. Change electrophoresis tank buffer
Primer seceening and segregation analysis	RAPD bands do not amplify properly in parental samples but do amplify clearly in progeny samples	1. Test serial dilutions of parental DNA samples to optimize DNA concentration 2. Column purify genomic DNA samples or re isolate with more elaborate protocols that include extra purification steps
	Very few RAPD bands segregate in a pseudo-testcross	1. Use a more divergent cross 2. Use individuals that are more likely to be heterozygous (e.g. outcrossed) 3. Screen more RAPD primers
Mapping	MapMaker file does not run or MapMaker quits unexpectedly	1. Check for errors in data file preparation such as extra markers, genotypes or spaces; errors in file header or genotype data 2. Increase available RAM (random access memory) for MapMaker 3. Check compatibility of MapMaker version and processor used
	Grouping results in very large groups of markers	1. Increase LOD and/or θ thresholds to avoid spurious linkages
	Grouping results in more groups than the expected number of chromosomes	1. Increase number of markers scored to reach ~250 markers to allow for adequate expected coverage 2. Reduce LOD and/or θ thresholds to allow fusion of groups
	Ordering with log likelihood support $\geq$ 1000:1 results in very few markers in the framework map	1. Increase progeny size to increase probability of detecting double recombinants 2. Check data quality of markers by going back to gel photos 3. Reduce stringency when ordering markers

References

Al-Janabi SM, Honeycutt RJ, McClelland M, Sobral BWS (1993) A genetic linkage map of *Saccharum spontaneum* L. 'SES 208'. Genetics 134:1249–1260

Allard R (1956) Formulas and tables to facilitate the calculation of recombination values in heredity. Hilgardia 24:235–278

Bonierbale MW, Plaisted RL, Tanksley SD (1988) RFLP maps based on a common set of clones reveal modes of chromosomal evolution in potato and tomato. Genetics 120:1095–1103

Buetow, KH, Chakravarti A (1987) Multipoint mapping using seriation. I. General methods. Am J Hum Genet 41:180188

Cai Q, Guy CL, Moore GA (1994) Extension of the linkage map in *Citrus* using Random Amplified Polymorphic DNA (RAPD) markers and RFLP mapping of cold-acclimatation-responsive loci. Theor Appl Genet 89:606–614

Carlson JE, Tulsieram LK, Galubitz JC, Luk VWK, Kauffeldt C, Rutledge R (1991) Segregation of random amplified DNA markers in F1 progeny of conifers. Theor Appl Genet 83:194–200

Grattapaglia D, Sederoff RR (1994) Genetic linkage maps of *Eucalyptus grandis* and *E. urophylla* using a pseudo-testcross mapping strategy and RAPD markers. Genetics 137:1121–1137

Grattapaglia D, Chaparro JX, Wilcox P, McCord S, Werner D, Amerson H, McKeand S, Bridgwater F, Whetten R, O'Malley D, Sederoff RR (1992) Mapping in woody plants with RAPD markers: applications to breeding in forestry and horticulture. In: Proceedings of the Symposium on Applications of RAPD Technology to Plant Breeding. CSSA/ASHS/AGA, pp 37–40

Hemmat M, Weeden NF, Manganaris AG, Lawson DM (1994) Molecular-marker linkage map for apple. J Hered 85:4–11

Kubisiak TL, Nelson CD, Nance WL, Stine M (1995) RAPD linkage mapping in a longleaf pine×slash pine F1 family. Theor Appl Genet 90:1119–1127

Lander ES, Green P, Abrahamson J, Baarlow A, Daly MJ, Lincoln SE, Newburg L (1987) MapMaker: an interactive computer package for constructing primary genetic linkage maps of experimental and natural populations. Genomics 1:174–181

Ritter E, Gebhardt C, Salamini F (1990) Estimation of recombination frequencies and construction of linkage maps from crosses between heterozygous parents. Genetics 125:645–654

Roy A, Frascaria N, MacKay J, Bousquet J (1992) Segregating random amplified polymorphic DNAs (RAPDs) in *Betula alleghaniensis*. Theor Appl Genet 85:173–180

Welsh J, McClelland M (1990) Fingerprinting genomes using PCR with arbitrary primers. Nucleic Acids Res 18:7213–7218

Williams JGK, Kubelik AR, Livak KJ, Rafalski JA, Tingey SV (1990) DNA polymorphisms amplified by arbitrary primers are useful as genetic markers. Nucleic Acid Res 18:6531–6535

Wu KK, Burnquist W, Sorrells ME, Tew TL, Moore PH, Tanksley SD (1992) The detection and estimation of linkage in polyploids using single-dose restriction fragments. Theor Appl Genet 83:294–300

Chapter XXIII

Estimating Nucleotide Divergence with RAPD Data

A.G. CLARK

1
Introduction

The application of the polymerase chain reaction (PCR) with a single short oligonucleotide primer (RAPD-PCR) is an efficient and relatively inexpensive means for detecting DNA sequence polymorphisms (Williams et al. 1990; Welsh and McClelland 1990). The fact that RAPDs survey numerous loci in the genome makes the method particularly attractive for analysis of genetic distance and phylogeny reconstruction. Many applications in population biology would benefit from quantitative estimates of nucleotide diversity within species and nucleotide divergence between species based on RAPD patterns. Several authors have attempted to do this without appropriate consideration of the underlying cause for polymorphism in banding patterns, and resulting estimates can be very misleading. This chapter is intended to provide a solution to this problem.

The RAPD method is useful in genetic analysis only if variation in banding patterns represents allelic segregation. Typically this is tested by examining RAPD patterns in family-structured data. Polymorphism is detected as band presence vs absence, and in the model that has been developed to estimate divergence, we assume that loss of a band is caused by failure to prime a site in some individuals due to nucleotide sequence differences. Assuming that loss of a RAPD band only occurs when there is one or more base substitutions in a primer site, then RAPD-PCR data from haploids can be treated in a fashion very similar to restriction fragment data, with restriction sites equal in length to the primer length. RAPD patterns from diploid samples, by contrast, require consideration of the fact that presence of a band is dominant to absence of the band. A method for estimating nucleotide diversity from population samples is described along with the restrictions and criteria that must be met when RAPD data are used for estimating population genetic parameters.

2
Principles

The Model: How RAPD Patterns Change as Sequences Diverge

From the above, it should be clear that it is necessary to consider how RAPD patterns will change as nucleotide substitutions occur. The model developed by Clark and Lanigan (1993) was based on the assumption that priming of DNA synthesis occurs only when there is an exact match of the genomic DNA to the oligonucleotide primer, and a single mismatch abolishes priming. Thus a RAPD band will be lost if there is one or more nucleotide substitutions in the genome corresponding to the primer sequence. Base substitutions at other positions only matter if they produce a primer match within 4 kb of another primer site in opposite orientation (depending on PCR and gel running conditions) . The model then is based on the following assumptions:

- Presence of the primer sequence and its reverse complement within the distance of 100–4000 bp will result in amplification and appearance of a band. A single substitution in either primer site is assumed to abolish the amplification. Insertion/deletion variation will result in altered band size (or loss, in the case of a large insertion). For this reason we must assume that insertion/ deletion variation is rare compared to nucleotide substitution.
- Primer sequences are chosen at random with respect to sequence and base composition and the population is sampled at random.
- Monomorphic and polymorphic bands are scored with equal probability. One cannot bias the study toward the "more interesting" polymorphic bands.
- Polymorphic bands are assumed to behave as mendelian factors with two alleles (presence/absence of a band, with band presence dominant to band absence), and the variation is assumed to be in Hardy-Weinberg equilibrium and in linkage equilibrium with other loci.
- For analysis of interspecific divergence, it is assumed to be possible to determine which bands are truly homologous.

Even if all these assumptions are true, divergence and diversity estimates are based on a relatively small number of nucleotides (those corresponding to the primer sites), so they have relatively large standard errors. The reliability of the method was also shown by simulation to drop if the sequences have diverged greater than about

10 %. Species that have greater sequence divergence than 10 % can still be analyzed, but one must recognize that one cannot make conclusions that depend on accurate assessment of distance.

Assessment of Fit of the Model

The assumptions of the model of Clark and Lanigan (1993) are very restrictive, and one might never expect to actually satisfy them all. Because RAPDs survey whole genomes, it is not always easy to check the validity of the assumptions. One rough means of assessing the fit of the model is to compare observed and expected numbers of RAPD bands. Clark and Lanigan (1993) present theory showing that the expected interval between primer sites has a geometric distribution, and the expected size distribution of RAPD-generated fragments is also geometric. The expected number of bands has a somewhat complicated distribution (Weissing and Velterop 1993), and the distribution looks similar to a Poisson distribution. Both distributions have as parameters the genome size, the primer size, and the probability that a random run of bases matches the primer. When goodness-of-fit tests are applied to the size distribution or the number of RAPD bands, it is not unusual to reject the model. The two main reasons for rejection include non-homogeneous sequence distribution and false priming. Both departures from the model almost certainly occur, and the frequency of their occurrence remains unknown. For now, assume that the estimation method is approximate.

3
Procedure

Nucleotide Divergence Estimate

The data for calculating nucleotide divergence consist of vectors of band presence/absence for each of several RAPD primers scored for each of a group of individuals in two or more populations. From the data, the frequency of each band is calculated within each population. Let p_{xi} be the frequency of band i in population x, and let p_{yi} be the frequency of band i in population y. F, the probability of band sharing, is estimated as:

$$F = \Sigma_i \, p_{xi} \, p_{yi} \, / (\Sigma_i \, p_{xi}^2 - \Sigma_i \, p_{yi}^2)^{1/2}$$

where the sums are taken over all observed bands. $p_{xi} \, p_{yi}$ is the chance that an individual from population x and an individual from population y both have band i, and by summing over all bands, $\Sigma_i \, p_{xi} \, p_{yi}$ and normalizing by the chance of band sharing within a population (the denominator), we obtain F, the probability of band sharing between populations. The above estimate of F, described further by Nei and Takezaki (1994), differs somewhat from that presented by Clark and Lanigan (1993). The former is somewhat simpler and is presented here, although simulations have shown them to behave very similarly. For the estimation of nucleotide diversity, π, the probability of band sharing within a population, is used ($F=\Sigma_i\, p_{xi}^2\, /\Sigma_i\, p_{xi}$). Let P be the probability that no mutation has occurred at a primer site since the common ancestor of the two sequences. Following the reasoning of Nei and Li (1979) in their development of an estimate of nucleotide divergence from restriction fragment size data, the relation between F and P is:

$$F \approx P^4\,/(3-2P)$$

We can estimate P from F using the iterative approach suggested by Nei (1987):

$$P=[F(3-2P_1)]^{1/4v}$$

where P_1 is the initial trial value of P ($P_1=F^{1/4}$ is suggested), and the values of P are substituted back in as P_1 until $P=P_1$. The expected nucleotide divergence is $d=2\lambda t$, and since $P=\exp(-r\lambda t)$, this gives an estimate of nucleotide divergence,

$$d=-(2/r)ln(P)$$

Because each band is equally likely to be polymorphic, the variance in nucleotide divergence can be estimated from the variation in estimates obtained by resampling subsets of the data. A convenient numerical means to estimate the variance is to use the jackknife (Efron 1982). Clark and Lanigan (1993) suggested applying the jackknife by assessing variation in estimates across primers, but this places weight on primers that reveal more polymorphic bands. Current versions of the software jackknife over the polymorphic bands. If there are b bands observed, the jackknife provides an estimate of standard error by calculating divergence for each of the b subsets of data each having one band excluded from the analysis.

The above discussion applies to estimates of nucleotide divergence for haploid samples. Dominance of RAPDs results in the appearance of greater band sharing among diploid individuals drawn from a panmictic population than is expected among homozygous lines. Clark and Lanigan (1993) showed that by assuming Hardy-Weinberg equilibrium and using $P=1-\sqrt{(1-p)}$ for the estimate of allele frequency P based on observed band frequency p, reliable estimates of divergence

Fig. 1. RAPD data of Lanigan (1992) were used to estimate nucleotide divergence among a series of primates. Estimates were obtained with both RAPDDIP and RAPDHAP. Note that the spurious estimates obtained with RAPDHAP (which fails to correct for band dominance) are underestimates and that the degree of underestimation increases as divergence increases

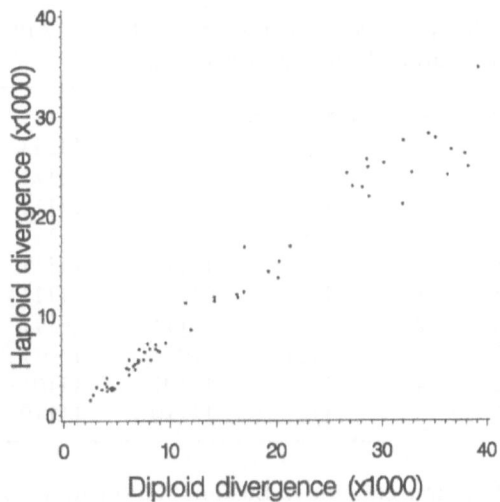

could be obtained. Failure to use this correction for dominance results in underestimating divergence (Fig. 1). Application of the estimator for F and d can then be applied as for haploid samples, substituting the corrected allele frequencies. This approach is considerably simpler than that described by Clark and Lanigan (1993).

Programs for performing estimates of nucleotide divergence from RAPD data are available through the author's page on the Penn State Biology Departments web site, "http://www.bio.psu.edu". The programs are called RAPDHAP and RAPDDIP for the haploid and diploid cases, respectively. The directory contains source and executable files, sample data files, and documentation to assist in getting the programs running. Basically one needs to create a text file providing the band presence or absence (coded as 1 and 0, respectively) for each individual in the sample. Format for this input file is provided in Table 1. Running the appropriate program (RAPDHAP or RAPDDIP) then produces an output file of the form given in Table 2. This output can be interpreted directly, or additional analysis, such as construction of phylogenetic trees, can be performed with other programs such as the MEGA package (Kumar et al. 1993).

Computer programs

Table 1. Format for the input data for RAPDDIP, a computer program that provides nucleotide divergence estimates

RAPD diploid test data.

1	1	01111	11101	11111	11111	01101	11111
1	2	10111	11111	10111	11011	01101	11011
1	2	11011	10111	11111	11111	01101	11111
1	4	11101	11111	11111	10111	01101	11111
1	5	11110	11111	11111	11111	01101	11111
2	1	11111	01111	01111	11010	10111	10110
2	2	11111	10111	10111	11010	11111	10110
2	3	10111	11011	11011	11010	11101	10110
2	4	11111	11101	11101	11010	11111	10100
2	5	11111	11110	11110	11010	11111	10110

The first column is the population number, the second column is the individual sample number, and remaining digits all specify band presence or absense.

Table 2. Sample output file generated by RAPDDIP with the input given in Table 1. Only the appropriate number of significant digits should be reported

RAPDDIP: Analysis of nucleotide divergence
RAPD diploid test data.
Nucleotide diversity within each population:

Population	pi	SE
1	0.004223	0.00131093
2	0.006528	0.00178807

Nucleotide divergence between populations:

Pop1	Pop2	d	SE
1	2	0.007694	0.00223403

4
Comments

Estimates of diversity and divergence based on RAPD data are intrinsically less reliable than estimates based on restriction site data or direct sequence data. Despite this lower reliability, RAPDs have the advantage of making genomes accessible to molecular analysis that might otherwise require far greater outlays of effort and expense. Furthermore, the accuracy of RAPD estimates of divergence may be perfectly adequate for the biological question at hand. It is important to recognize that these estimates are based on a model and that, if the assumptions of that model (listed above) are violated, the reliability of the estimators will be further compromised. In such cases, it is prudent to have additional checks of the validity of the estimates.

References

Clark AG, Lanigan EMS (1993) Prospects for estimating nucleotide divergence with RAPDs. Mol Biol Evol 10:1096–1111

Efron B (1982) The Jackknife, the Bootstrap, and other resampling plans. Soc Ind and App Math, Philadelphia.

Kumar S, Tamura K, Nei M (1993) MEGA: Molecular Evolutionary Genetics Analysis, version 1.01. Pennsylvania State University, University Park

Lanigan CMS (1992) RAPD analysis of primates: phylogenetic and genealogical considerations. PhD dissertation, Genetics Graduate Group, University of California, Davis

Nei M (1987) Molecular evolutionary genetics. Columbia University Press, New York

Nei M, Li WH (1979) Mathematical model for studying genetic variation in terms of restriction endonucleases. Proc Natl Acad Sci USA 76:5269–5273

Nei M, Takezaki N (1994) Estimation of genetic distances and phylogenetic trees from DNA analysis. Proc 5th World Cong Genet Appl Livestock Prod 21:405–412

Weissing FJ, Velterop O (1993) The expected distribution of RAPD bands. Mol Biol Evol 10:1107–1110

Welsh J, McClelland M (1990) Fingerprinting genomes using PCR with arbitrary primers. Nucleic Acids Res 18:7213–7218

Williams JGK, Kubelik AR, Livak KJ, Rafalski JA, Tingey SV (1990) DNA polymorphisms amplified by arbitrary primers are useful as genetic markers. Nucleic Acids Res 18:6531–6535

Chapter XXIV

RAPD and PAUP Analysis
for Microbial Screening Programs

F. Fujimori, H. Takaya, Y. Anzai, and T. Okuda

1
Introduction

The commercial success of pharmaceuticals derived from natural products such as mevalotin and cyclosporin has led to an increase in the number of drug companies and governmental institutions conducting natural products screening programs. During the microbiological screening phase of these programs, efficiency may be reduced by repetitive isolation/screening of similar or identical organisms. The success of this aspect of the screening program depends on the abilities of the taxonomists and microbiologists involved. We have developed a method for reducing duplication of strains in screening programs using a well-accepted molecular method (Fujimori and Okuda 1994; Anzai et al. 1994; Tanaka et al. 1994; Okuda et al. 1995).

The random amplified polymorphic DNA technique (RAPD), developed by Williams et al (1990) and Goodwin and Annis (1991), is now widely used for population studies and genetics as well as fundamental taxonomic work. We have adapted this technique to develop a dereplication tool in microbial screening programs. The reproducibility and standardization of the RAPD technique have been the subjects of several reports (Kwan 1992; Penner et al. 1993; Schierwater and Ender 1993; Molina ct al. 1995). We also evaluated every step of the method, so that a standard procedure was established to minimize the deviation of Rf values, the migration distance of electrophoretic bands. In this chapter, we describe in detail our standard methods for RAPD used in conjunction with a statistical package designed for phylogenetic analysis using parsimony, PAUP (Swofford 1991), for microbial screening programs. The entire scheme of this analysis is summarized in Fig. 1.

Cultivation in a liquid medium
↓
Template DNA preparation
↓
PCR (RAPD)
↓
Gel electrophoresis
↓
Photograph taking
↓
Rf value calculation
↓
Conversion of Rf values to binary data matrix
↓
PAUP analysis, bootstrapping with 1,000 replication

Fig. 1. Scheme of RAPD and PAUP analysis

2
Standardization of Procedure

Cultivation   Culture media and conditions are important for fluffy growth of fungi and subsequent increased DNA yield. We compared several liquid and solid media with and without glass beads or wire as a physical disruption material. Of these methods, cultivation in PYG liquid medium was the most suitable for preparing template DNAs from various fungi. For actinomycetes, we used trypticase soy broth (BBL) containing glycine.

Cell disruption and isolation of total DNA   Most moniliaceous fungi without melanoid pigments in their cell walls were processed with chitinase with the addition of glass beads, as reported by Fujimori and Okuda (1994). However, this method was sometimes unacceptable for dark colored dematiaceous fungi. Zhu et al (1993) reported that benzyl chloride readily destroyed cell walls. Their method was very effective for both actinomycetes and dematiaceous fungi (Okuda et al. 1995). The benzyl chloride method and enzymatic preparation of DNA gave identical results in terms of RAPD band patterns in actinomycetes. We also compared the effect of benzyl bromide with that of benzyl chloride in the preparation of DNA. Since benzyl bromide is more reactive, it was superior to benzyl chloride for cell wall disruption in fungi (Fujimori and Okuda 1995).

Polymerase chain reaction for RAPD   We evaluated various parameters in PCR such as the number of cycles, selection of primers, and concentrations of template, *Taq* DNA polymerase, and primers. Although Williams et al (1990) used 45 PCR cycles, we observed nonspecific bands after more than 35 cycles even without a template. With any of three primers, R2, R4, and R28, described by Goodwin and Annis (1991), more than ten bands were observed between 100 bp and 2 kbp, which were suitable for further statistical analysis. Of the three primers, however, R28 was the most suitable based on its ability to generate a larger number of DNA segments distributed over a wider range of DNA sizes. The concentration of DNA template had to be more than 1.0 at A_{260}, so that a uniform number of bands were obtained. Although all of the *Taq* DNA polymerases obtained from different manufacturers worked well, the one from Pro-Bio was most suitable for our study, probably due to its heat stability. After testing a concentration range of from 0.5 U to 8 U, we determined that the optimal concentration was 4 U/50 μl PCR solution.

Acrylamide gel (Laemmi 1970) gave clearer bands than agarose gels, as reported by Pang et al. (1992). After standardization of each step of RAPD analysis, the standard deviation of Rf values was within 0.03 even between gels. Furthermore, when a commercial gel was used, the deviation was less than 0.01. In order to calculate Rf values precisely, therefore, we recommend the use of a precast gel, such as that by Tefco Corp.

Gel electrophoresis

Photographs of DNA band patterns were analyzed either using an image analyzer or manually. The Rf value of each individual band was determined by calculating the migration distance of DNA bands from the origin. The Rf values of λ DNA (origin) and 90 bp DNA (front) were considered as 0.00 and 1.00, respectively. If the values are considered up to 10^{-2}, there should be 99 possible Rf values. Because each Rf value can be used as a variable character, there can be 99 variables. The following conversion was made using a Macro program in Microsoft Excel for Macintosh. When a DNA band was observed at a certain variable character (at a certain Rf value), the value 1 was given. The value 0 was given to any vacant position. The raw data of Rf values were then converted to a binary data matrix 1 as shown in Fig. 2. When only one gel was analyzed, or when a pre-cast gel was used, the binary data matrix was sufficient for the further analysis. In other words, standard deviations in Rf values were within 0.01. However, when more than two gels that were made in our lab were analyzed, there were errors in Rf values within 0.03. In this case, fuzzy data should be taken into account. A set of three adjacent variable characters was grouped in a new variable character. The new variable character 1 included old 0, 1, and 2; the new variable character 2 included old 1, 2, and 3; the new variable character 3 included old 2, 3, and 4; and so on. When value 1 exists within the combined variable charac-

Image acquisition and data conversion

Rf values

Rf value

	0.00	0.01	0.02	0.03	0.04	0.05	0.06	0.07	0.08	0.09	0.10	...	0.99	1.00
Strain1			I			I					I	....	....	
Strain2						I			I			....	....	
Strain3	DNA band				I			I			I	....	....	
Strain4						I			I		I	....	....	
Strain5								I				....	....	

Binary matrix 1

Variable character

	001	002	003	004	005	006	007	008	009	010	...	099	
Strain1	0	0	1	0	0	1	0	1	0	0	1	....	0
Strain2	0	0	0	0	0	1	0	0	1	0	0	....	0
Strain3	0	0	0	0	1	0	0	1	0	0	1	....	0
Strain4	0	0	0	0	0	1	0	0	1	0	1	....	0
Strain5	0	0	0	0	0	0	0	1	0	0	0	....	0

Binary matrix 2

Variable character

	001	002	003	004	005	006	007	008	009	010	...	099
Strain1	1	1	1	1	1	1	1	1	1	1	....	
Strain2	0	0	0	1	1	1	1	1	1	1	....	
Strain3	0	0	1	1	1	1	1	1	1	1	....	
Strain4	0	0	0	1	1	1	1	1	1	1	....	
Strain5	0	0	0	0	0	0	1	1	1	0	....	

Fig. 2. Conversion of Rf values to binary data matrices

ters, a new value 1 is given to the new variable character. Then, a new binary data matrix with fuzziness was completed as shown in Fig. 2. The simple conversion program is available from us in the form of a Macro file for Microsoft Excel version 5.0 for Macintosh. The request should be made via e-mail to Dr. T. Okuda <hqe01700@niftyserve. or.jp> or to Mr. H. Takaya <hiroki.takaya@roche.com>.

PAUP analysis We performed a phylogenetic analysis using parsimony, PAUP version 3.0s (Swofford 1991) for Macintosh. PAUP version 4.0 will be published soon by Swofford (1995). Requests should be made to Sinauer; e-mail <orders@sinauer.com>. The data matrix was formatted as a Nexus file for PAUP analysis as shown in Fig. 3. A simple heuristic search will also generate trees, but so many trees were obtained by this search that it was difficult to judge which tree is the most parsimonious. Therefore, if time allows, a bootstrap analysis should be performed with a heuristic search.

Although RAPD allows us to compare taxa objectively, it may give confusing results if widely separated taxa are compared. It makes no sense to compare arbitrarily generated polymorphisms between totally different genera or classes. PAUP employs "phylogenetic syste-

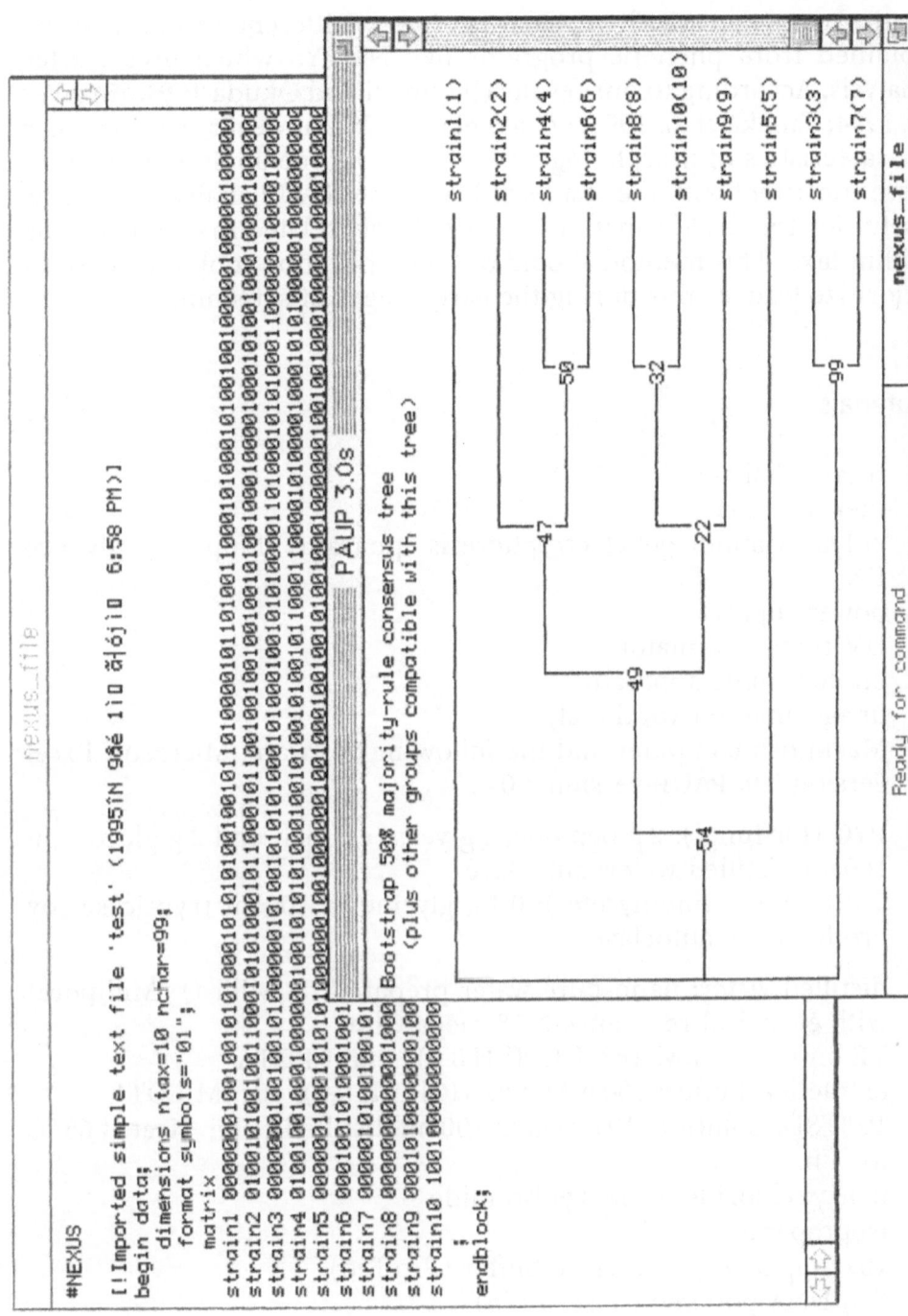

Fig. 3. A NEXUS file and the result of bootstrapping (consensus tree)

matics" or "cladistics" to make its trees. Different results may be obtained from phenetic programs like NT-SYS which uses cluster analysis. According to our results (Fujimori and Okuda 1994; Anzai et al. 1994; Tanaka et al. 1994; Okuda et al. 1995), however, various other characteristics of morphology and secondary metabolite production were consistent with the results of RAPD and PAUP analysis. RAPD is useful for the discrimination of very closely related taxa and strains within taxa. This method should be a dereplication tool to be used to prioritize lead strains during the early stages of screening.

3
Materials

Equipment
- rotary shaker
- thermocycler
- polyacrylamide gel electrophoresis apparatus (Mupid-2, Advance KK)
- power supply
- UV transilluminator
- photographic apparatus
- image analyzer (optional)
- Macintosh computer and the following software: Microsoft Excel version 5.0, PAUP version 3.0s

Media
- PYG (for fungi): 2 g peptone, 1 g yeast extract, and 2 g glucose in 100 ml distilled water; autoclave
- TSBG (for actinomycetes): 0.1 g glycine in 100 ml trypticase soy broth (BBL); autoclave

Reagents and solutions
- distilled water: nano-pure water prepared by Milli-Q (Millipore) with electrical resistance >18 mega Ohm
- TE buffer: 10 mM Tris-HCl (pH 8.0), 1 mM EDTA
- extraction buffer: 100 mM Tris-HCl (pH 9.0), 40 mM EDTA
- 10 % SDS solution: 10 g SDS in 100 ml distilled water. Heat at 65 °C for 1 h.
- benzyl chloride or benzyl bromide
- isopropanol
- 10x *Taq* DNA polymerase buffer (Pro Bio)
- *Taq* DNA polymerase (Pro Bio)
- dNTP solution: 10 μM dATP, 10 μM dTTP, 10 μM dGTP, 10 μM dCTP, in TE
- $MgCl_2$ solution: 25 mM solution provided with polymerase (Pro Bio)

- primer: 2.5 μM decamer R28, 5'-ATG GAT CCG C. R2, 5'-AGT ACA GGT C and R4, 5'-TCC TAC GCA C are also useful for analysis. These three were described by Goodwin and Annis (1991).
- mineral oil
- 50 % acrylamide solution: dissolve 29 g acrylamide and 1 g bis-acrylamide in distilled water; final volume should be 60 ml
- 50x TAE buffer: 2 M Tris-acetate (pH 8.0), 0.05 M EDTA; autoclave
- 10 % ammonium sulfate: 0.1 g ammonium sulfate in 1 ml distilled water
- sodium acetate solution: 3 M sodium acetate in distilled water (pH 5.0)
- TEMED: N,N,N',N'-tetramethylethylenediamine
- BPB-sucrose: 0.0025 % bromphenol blue, 40 % sucrose in distilled water
- ethidium bromide solution: 1 μg ethidium bromide in 1 ml distilled water
- origin marker: λ DNA
- front marker: The 90 bp DNA fragment is derived from a plasmid pUC119/118 by the PCR conditions of 92 °C, 2 min (92 °C, 45 s; 47 °C, 1 min; 72 °C, 1.5 min)×30 cycles, 72 °C, 7 min and finally 5 °C, with primers 5'-AAA ACG ACG GCC AGT and 5'-TAT GAC CAT GAT TAC.

4
Experimental Procedure

Cultivation

1. Inoculate 9 ml PYG for fungi or TSBG for actinomycetes in a 50 ml flask with a loopful of cell mass from well-grown agar culture.

2. Cultivate on a rotary shaker for 3–5 days at 27 °C with agitation at 220 rpm.

3. Pellet the mycelia at 3000 rpm for 10 min.

4. Wash the mycelia with TE buffer.

5. Pellet the mycelia at 3000 rpm for 10 min.

DNA Extraction

This method was modified from a technique using benzyl chloride reported by Zhu et al (1993). **Note:** Since both benzyl chloride and benzyl bromide are quite reactive, they should be handled carefully with adequate ventilation.

1. Suspend 200 mg wet mycelia in 500 µl extraction buffer.

2. Add 100 µl SDS solution and 600 µl benzyl chloride (or benzyl bromide) to the suspension.

3. Mix and heat the suspension at 50 °C for 30 min.

4. Add 300 µl sodium acetate solution to the suspension.

5. Cool the suspension at 0 °C for 15 min.

6. Pellet the mycelia at 12000 rpm for 10 min.

7. Add 3/5 volume isopropanol to the supernatant.

8. Centrifuge the mixture at 15000 rpm for 10 min.

9. Dry up the pellet under reduced pressure.

10. Dissolve the dried pellet in TE buffer.

Polymerase Chain Reaction Conditions for RAPD

1. Measure and adjust A_{260} of the DNA preparation to 3.0 by diluting with TE.

2. Mix 27 µl nano-pure water, 5 µl 10x buffer, 1 µl dNTP solution, 4 µl $MgCl_2$ solution, 4 U *Taq* DNA polymerase, 5 µl primer, and 5 µl template DNA prepared as above, under ice-cooled conditions. The total volume will be 50 µl except for the *Taq* DNA polymerase solution.

3. Layer 1–2 drops of mineral oil on the top of the reaction mixture.

4. Set and run a thermocycler as follows: 92 °C 1 min; 30 cycles of 92 °C 45 s, 34 °C 60 s, and 72 °C 90 s; and final cycle at 72 °C for 10 min; and finally cool down to 5 °C. Denaturation for actinomycetes should be done at 94 °C.

5. After PCR, remove the mineral oil by extraction with 50 µl chloroform.

Gel Electrophoresis

1. Mix 10 ml of 50 % acrylamide, 2 ml 50x TAE, 0.1 g ammonium persulfate in 88 ml distilled water.

2. Add 100 μl TEMED to the solute.

3. Shake the solution gently.

4. Pour 33 ml of the solution into a gel plate and insert a comb into the gel before it starts to set.

5. Store the gel at 4 °C overnight.

6. Mix 10 μl PCR product, 500 ng λ DNA as origin marker, 500 ng 90 bp DNA as front marker, and 3 μl BPB.

7. Pipette 10 μl of each DNA sample to the gel.

8. Conduct electrophoresis at 50 V for 2 h in 300 ml 1x TAE buffer.

9. Soak the gel in 30 ml ethidium bromide solution.

10. Shake the gel gently in the staining solution for 40 min.

11. Wash the gel in 50 ml distilled water by shaking gently for 20 min.

12. Take a photo of the gel using an UV transilluminator at 254 nm.

An image analyzer can measure Rf values automatically, but manual measurement is possible as follows.

Rf Value Calculation

1. Draw a straight line in the center of each lane from the origin to the front markers.

2. Measure the distance D_{of} in mm between the origin and the front marker.

3. Measure the distance D_{od} in mm between the origin marker and a target DNA band.

4. Calculate an Rf value by dividing D_{od} by D_{of}.

Conversion of Rf Values to a Binary Matrix

It is necessary to install Microsoft Excel version 5.0 with Visual Basic.

Conversion program on Microsoft Excel version 5.0

1. Open a new workbook.

2. Insert a sheet of macro module into a workbook. Select [Insert]-[Macro]-[Module] in the menu. A blank window called Module1 will appear.

3. In order to convert Rf values of 10 strains into a binary data matrix, state the following program in the window.

```
Sub Record1( )
Selection.CreateNames Top:=False, Left:=True, Bottom:=False,
Right: =False
Selection.Copy
Sheets("Sheet2").Select
Range("A4").Select
ActiveSheet.Paste
Sheets("Sheet3").Select
Range("A4").Select
ActiveSheet.Paste
Sheets("Sheet2").Select
Application.CutCopyMode=False
Range("A1").Select
ActiveCell.FormulaR1C1="binary matrix1"
Range("B3").Select
ActiveCell.FormulaR1C1="0"
Range("C3").Select
ActiveCell.FormulaR1C1="0.01"
Range("B3:C3").Select
Selection.AutoFill Destination:=Range("B3:CX3"),
Type:=xlFillDefault Range("B3:CX3").Select
Selection.NumberFormat="0.00"
Range("B4").Select
ActiveCell.FormulaR1C1="=IF(ISNA(MATCH(R[-
1]C,strain1,0)),0,1)"
Range("B4").Select
Selection.AutoFill Destination:=Range("B4:B13"),
Type:=xlFillDefault
Range("B4:B13").Select
Range("B5").Select
ActiveCell.FormulaR1C1="=IF(ISNA(MATCH
(R[-2]C,strain2,0)),0,1)"
Range("B6").Select
ActiveCell.FormulaR1C1="=IF(ISNA(MATCH
(R[-3]C,strain3,0)),0,1)"
Range("B7").Select
ActiveCell.FormulaR1C1="=IF(ISNA(MATCH
(R[-4]C,strain4,0)),0,1)"
Range("B8").Select
ActiveCell.FormulaR1C1="=IF(ISNA(MATCH
(R[-5]C,strain5,0)),0,1)"
Range("B9").Select
```

```
ActiveCell.FormulaR1C1="=IF(ISNA(MATCH
(R[-6]C,strain6,0)),0,1)"
Range("B10").Select
ActiveCell.FormulaR1C1="=IF(ISNA(MATCH
(R[-7]C,strain7,0)),0,1)"
Range("B11").Select
ActiveCell.FormulaR1C1="=IF(ISNA(MATCH
(R[-8]C,strain8,0)),0,1)"
Range("B12").Select
ActiveCell.FormulaR1C1="=IF(ISNA(MATCH
(R[-9]C,strain9,0)),0,1)"
Range("B13").Select
ActiveCell.FormulaR1C1="=IF(ISNA(MATCH
(R[-10]C,strain10,0)),0,1)"
Range("B4:B13").Select
Selection.AutoFill Destination:=Range("B4:CX13"),
Type:=xlFillDefault
Range("B4:CX13").Select
Sheets("Sheet3").Select
Range("A1").Select
ActiveCell.FormulaR1C1="binary matrix2"
Range("B3").Select
ActiveCell.FormulaR1C1="1"
Range("C3").Select
ActiveCell.FormulaR1C1="2"
Range("B3:C3").Select
Selection.AutoFill Destination:=Range("B3:CW3"),
Type:=xlFillDefault
Range("B3:CW3").Select
Selection.NumberFormat="000"
Range("B4").Select
ActiveCell.FormulaR1C1=
"=IF(OR(Sheet2!RC=1,Sheet2!RC[1]=1,Sheet2!RC[2]=1),1,0)"
Range("B4").Select
Selection.AutoFill Destination:=Range("B4:B13"),
Type:=xlFillDefault
Range("B4:B13").Select
Selection.AutoFill Destination:=Range("B4:CW13"),
Type:=xlFillDefault
Range("B4:CW13").Select
Columns("CW:CW").Select
Selection.Delete Shift:=xlToLeft
Range("A1").Select End Sub
```

4. If more than 10 strains are analyzed, the statements that begin with "Active" and "Range" will accordingly be increased as follows:

Range("B14").Select
ActiveCell.FormulaR1C1="=IF(ISNA(MATCH
(R[-11]C,strain11,0)),0,1)"

Furthermore, the statement of range in the program should also be changed. The number "13" in Range("B4:B13") and Range ("B4:CW13") should increase according to the size of the data set. This statement determines the limit of the data. If fewer than 10 strains are used, however, no change is required.

Use of this program, conversion of Rf values into binary data

A new workbook will be required each time a new conversion is considered. The sheet name and data set names, such as strain1, strain2, should not be changed.

1. Input Rf values in Sheet 1. Data set should be named strain1, strain2, and so on, in lowercase. The data will be incorporated by row.

2. Select data set so that the name such as strain1, strain2 .... will be in the first column.

3. Select [Tools]-[Macro...]- choose "Record1" in the list -[Run] in the menu.

4. Conversion will be conducted automatically, and Binary matrix 1 will be shown in Sheet 2 and a binary data matrix with fuzziness (Binary matrix 2) will be shown in Sheet 3.

5. Save the converted data matrix (either Binary matrix 1 or 2) as a simple or tab-delimited text file.

Use of PAUP

Preparation of NEXUS file by the import function

1. Drag [File]-[Import file] in the menu.

2. Select the binary data matrix file and determine "Input file format" and "Data type". The former is usually "Simple text" or "Tab-delimited text". The latter will be 'Standard'.

3. Open the file, and a NEXUS file will be generated automatically.

Execution and bootstrapping

4. Select [File]-[Execute 'File Name'] in the menu and processing will be finished within a second.

5. Drag [Options]-[Set Maxtree], and select "automatic increase by 100", because trees temporarily generated will be over 100 when more than 20–30 strains are analyzed at one time.

6. Drag [Search]-[Bootstrap], and set the number of replication to 1000 in "Bootstrap Options".

7. Bootstrap consensus tree should retain groups compatible with 50 % majority rule consensus.

8. Select "Heuristic" search; because Branch-and-bound option will generate too many trees.

9. Click "continue", and Heuristic search options will appear on the screen.

10. Instead of "General" which will be too slow, "Branch swapping" will be selected. Default values will be sufficient for the other options.

11. Click "Bootstrap".

12. One thousand replication bootstrap with ten strains, for example, will be finished within 30 min. by Macintosh LC630.

13. Drag [Trees]-[Print consensus].   **Tree generation**

14. Select "Show group frequencies". A branch with a more than 90 % consensus will be reliable.

15. Select either "Slanted" or "Rectangular".

16. Preview the consensus tree.

17. If necessary, save it as a PICT file for later display.

18. Drag [Trees]-[Print trees].

19. Click "Rooting" and set "mid point rooting" because no outgroup has been included.

20. Select "Phylogram" which is more convenient for visual differentiation of branch length.

21. Check "Show branch length".

22. Preview the phylogram. If necessary, save it as a PICT file.

23. Based on the consensus tree and phylogram, prioritize and eliminate similar or duplicate strains with a higher consensus and shorter branch length.

24. If necessary, drag [Data]-[Show distance matrix], which will show similarity index values between taxa. According to our experiences, since strains with more than 80 % similarity were almost identical in terms of morphology and secondary metabolite production, one of them can be discarded.

5
Troubleshooting

- If a sufficient amount of DNA is not obtained, use benzyl bromide instead of benzyl chloride, extend the treatment time, or scale-up the treatment amount.

- If no DNA band was observed in the gel, be sure that *Taq* DNA polymerase is intact. Nextly, confirm that correct amounts of dNTP were used. It is also useful to check temperature for denaturation whether it is suitable for the sample.

- If the shape of DNA bands are bent or curved, reduce the amount of sample to be pipetted in the gel.

Acknowledgements. We thank Dr. Maren A. Klich of USDA Southern Regional Research Center, New Orleans, and Dr. Keith A. Seifert of Agriculture Canada, Research Branch, Ottawa for their critical reading of our manuscript and valuable suggestions.

References

Anzai Y, Okuda T, Watanabe J (1994) Application of the random amplified polymorphic DNA using polymerase chain reaction for efficient elimination of duplicate strains in microbial screening. II Actinomycetes. J Antibiot 47(2):183–193

Fujimori F, Okuda T (1994) Application of the random amplified polymorphic DNA using the polymerase chain reaction for efficient elimination of duplicate strains in microbial screening. I Fungi. J Antibiot 47(2):173–182

Fujimori F, Okuda T (1995) Rapid preparation of PCR-amplifiable fungal DNA by benzyl bromide extraction. Mycoscience 36:465–467

Goodwin PH, Annis SL (1991) Rapid identification of genetic variation and pathotype of *Leptosphaeria maculans* by random amplified polymorphic DNA assay. Appl Environ Microbiol 57:2482–2486

Kwan HS (1992) Application of arbitrarily-primed polymerase chain reaction in molecular studies of mushrooms. International symposium of recent topics in genetics, physiology and technology of the basidiomycetes, pp 87–92

Laemmli UK (1970) Cleavage of structural proteins during the assembly of the head of bacteriophage T4. Nature 244:680–685

Molina F, Geletka L, Jong SC (1995) High-resolution DNA fingerprinting of microorganisms at ATCC. ATCC Quarterly Newsletter 15:1–3

Okuda T, Yanagisawa M, Fujimori F, Nishizuka Y, Takehana Y, Sugiyama M (1995) New isolation methods and polymerase chain reaction strain discrimination techniques for natural products screening programs. Can J Bot 73 (Suppl. 1):946–954

Pang JP, Chen LC, Chen LFO, Chen SCG (1992) DNA polymorphisms generated by arbitrarily primed PCR in rice. Biosci Biotech Biochem 56:1357–1358

Penner GA, Bush A, Wise R, Kim W, Domier L, Kasha K, Laroche A, Scoles G, Molnar SJ, Fedak G (1993) Reproducibility of random amplified polymorphic DNA (RAPD) analysis among laboratories. PCR Methods Applic 2:341–345

Schierwater B, Ender A (1993) Different thermostable DNA polymerases may amplify different RAPD products. Nucleic Acids Res 21:4647–4648

Swofford DL (1991) PAUP, phylogenetic analysis using parsimony, version 3.0s. Illinois Natural History Survey, Champaign Illinois, USA

Swofford DL (1995) PAUP 4.0, phylogenetic analysis using parsimony (and other methods). Laboratory of Molecular Systematics, National Museum of Natural History, Smithsonian Institution, USA.

Tanaka H, Sawairi S, Okuda T (1994) Application of the random amplified polymorphic DNA using polymerase chain reaction for efficient elimination of duplicate strains in microbial screening. III Bacteria. J Antibiot 47(2):194–200

Williams JGK, Kubelik AR, Livak KJ, Rafalski JA, Tigey SV (1990) DNA polymorphisms amplified by arbitrary primer are useful as genetic markers. Nucleic Acids Res 18:6531–6535

Zhu H, Qu F, Zhu LH (1993) Isolation of genomic DNAs from plants, fungi and bacteria using benzyl chloride. Nucleic Acids Res 21:5279–5280

Chapter XXV

Production of Specific Probes for Microorganisms

S. Fancelli and M. Bazzicalupo

1
Introduction

The problem of accurate characterization and identification of bacterial species and strains is of great relevance in many fields of microbiology. For a long time the identification of bacteria has been carried out by conventional microbiological techniques aimed at the phenotype of the cells. It is now acknowledged that these methods can fail especially for bacterial identification in environmental samples, where microorganisms often change their phenotypic characters and/or lose cultivability. Besides, phenotypic tests are not discriminating enough for identification of strains belonging to the same or closely related species. Molecular biology techniques, based on the nucleic acids rather than on the phenotype of the cells, were recently applied in bacterial identification and detection. In particular nucleic acid probes complementary to DNA sequences specific for given groups of microorganisms have been successfully designed to detect particular genotypes. Furthermore, by using such probes the problem of "viable but not culturable cells" has been overcome (Roszak and Colwell 1987), as the target DNA can be directly extracted from the samples without need for cultivation of the bacteria isolates. Among the molecular techniques, PCR represents a consolidated alternative to hybridization for the rapid and sensitive detection of bacteria, particularly when their number is very low. Specific primers, whose design is based on coding genes or on variable regions of ribosomal RNA operons, have been widely used and have been described in many reports. As a further development of PCR, the RAPD methodology has been extensively applied to determine taxonomic identity and assess genetic relationships among microbial strains; moreover it also allows identification of specific markers (Hadrys et al. 1992). The main advantages of RAPD technology include: suitability for work on anonymous genomes and applicability also when only limited quantities of DNA are available. In addition, the banding pattern is often strain specific, a feature that could be very useful in bacterial identification and monitoring. A limitation however of RAPD is that it cannot be directly used without previous isolation of the cells.

Nowadays a further development of the RAPD technique is gaining ground. It consists, after cloning and sequencing a particular RAPD fragment, of designing appropriate primers to perform a diagnostic PCR, which constitutes a rapid alternative to hybridization experiments. This technique has been recently applied to the arbuscular mycorrhizal fungus *Glomus mossae* (Lanfranco et al. 1995)

In an attempt to combine the advantages of RAPD with the specificity and power of DNA probes in monitoring a single target sequence from a mixed sample, we exploited the ability of RAPD to generate specific markers for strains belonging to the genus *Azospirillum* (Fani et al. 1993) developing the strategy outlined below.

2
Experimental Strategy

The overall strategy (Fig. 1) for detection of a particular strain in a mixed sample using the RAPD technique involves: (1) growth of the strain of interest; (2) amplification of its DNA with a random primer and (3) identification of the putative specific fragments. One particular diagnostic fragment is selected, labeled and tested as a probe for its specificity vs the amplified DNA of other strains which can occur in the sample under investigation. Finally the DNA obtained from the sample is extracted without previous isolation of the bacteria, amplified with the same primer used for generating the probe, dot-blotted and hybridized with the probe. A positive signal indicates the occurrence of the corresponding strain.

3
Materials

Equipment
- thermal cycler
- microcentrifuge for 1.5 ml Eppendorf type tubes. All microcentrifuges have approximately the same rotor radius, it is therefore more convenient to give centrifuge speed as rpm rather than *g*.
- agarose gel electrophoresis set
- UV transilluminator and photographic equipment. A video camera recorder connected to a computer is highly recommended

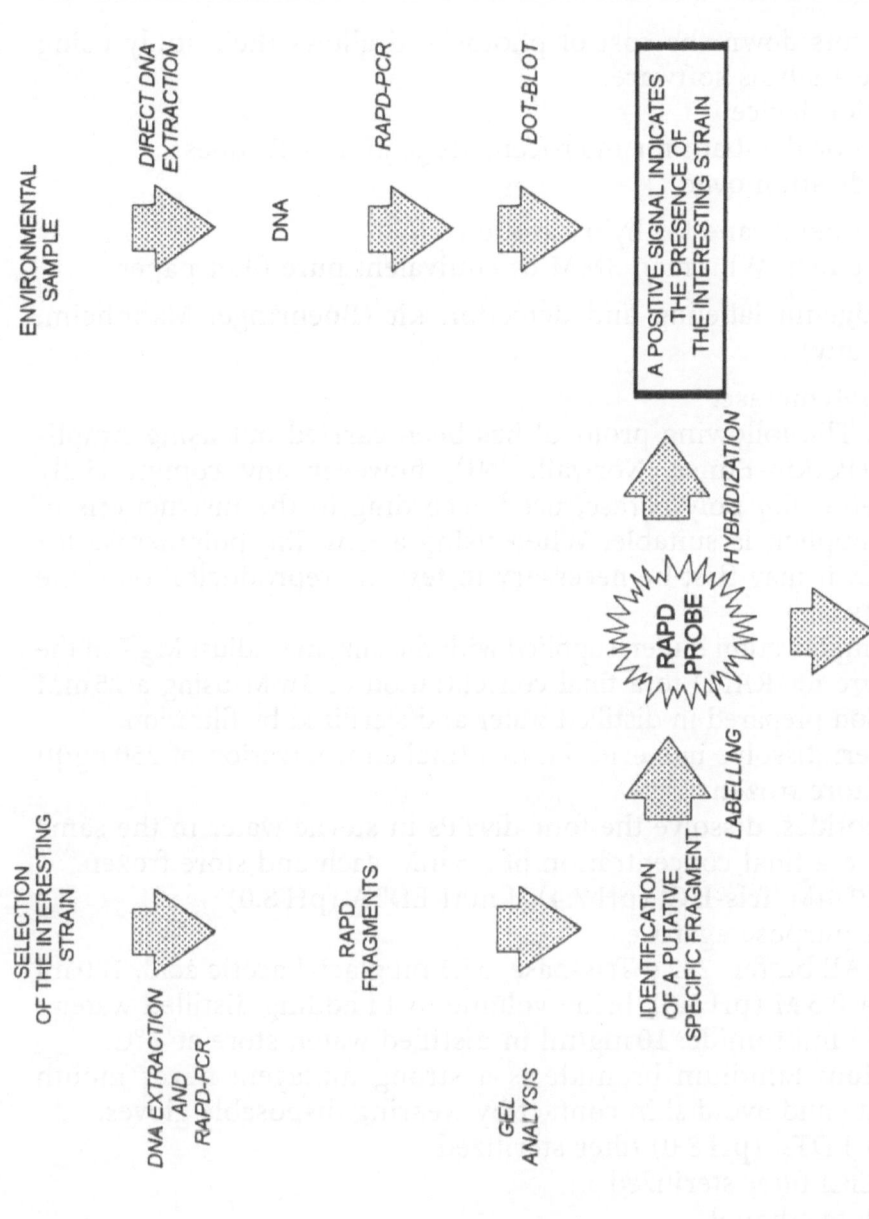

Fig. 1. Outline of the strategy for the construction of RAPD probes and detection of microorganisms

as it cuts down the cost of photos and allows their study using image analysis software.
- dot-blot device
- water- or dry-bath for microcentrifuge and PCR tubes
- hybridization oven

Supplies
- nylon membrane for hybridization
- filter paper, Whatman 3MM or equivalent pure filter paper

Kits
- digoxigenin labeling and detection kit (Boehringer-Mannheim, Germany)

Reagents and solutions
- *Taq* polymerase.
  Note: The following protocol has been carried out using Ampli-Taq (Perkin-Elmer, Norwalk, NJ), however any commercially available *Taq* polymerase, used according to the instructions of the supplier, is suitable. When using a new *Taq* polymerase for RAPD, it may first be necessary to test the reproducibility of the results.
- 10x amplification buffer, supplied with the enzyme; adjust Mg^{2+} in the mixture for RAPD to a final concentration of 3 mM using a 25 mM solution prepared in distilled water and sterilized by filtration.
- primer: dissolve in sterile TE to a final concentration of 250 ng/µl and store frozen
- nucleotides: dissolve the four dNTPs in sterile water in the same tube to a final concentration of 2.5 mM each and store frozen.
- TE: 10 mM Tris-HCl (pH 7.4), 1 mM EDTA (pH 8.0)
- multi-purpose agarose
- 50x TAE buffer: 242 g Tris-base, 57.1 ml glacial acetic acid, 100 ml EDTA 0.5 M (pH 8.0); bring volume to 1 l adding distilled water.
- ethidium bromide: 10 mg/ml in distilled water, store at 4 °C.
  Caution: Ethidium bromide is a strong mutagen; never mouth pipette and avoid skin contact by wearing disposable gloves.
- 0.2 M EDTA (pH 8.0) filter sterilized
- 4 M LiCl filter sterilized
- absolute ethanol
- denaturing solution: 1.5 M NaCl, 0.5 M NaOH
- neutralizing solution, 1.5 M NaCl, 0.5 M Tris-HCl (pH 7.2), 1 mM EDTA
- 20x SSC: 3 M NaCl, 300 mM sodium citrate. Many steps of the procedure call for dilutions of this stock solution, made with distilled water and filter sterilized.
- prehybridization solution: 5x SSC, 50 % formamide, 0.1 % N-laurylsarcosinate, 0.02 % SDS, 5 % Boehringer blocking reagent (Boehringer-Mannheim, Germany)

- hybridization solution: the same as prehybridization but with addition of the probe, previously denatured at 95 °C for 10 min. The final concentration of the labeled probe should be about 10 ng/ml of solution.
- 10 % SDS stock solution
- 0.2 M NaOH. **Caution:** Avoid contact of NaOH with skin.

4
Experimental Procedure

Identification of RAPD Marker(s)

The reaction is made up as follows:

RAPD amplification

1. Add 1 µl of DNA (10 ng/µl) extracted from bacterial cultures (see Chap. VI) to 0.1 ml PCR tubes.

2. Prepare a master mix for all the samples in order to pipette the following volumes into each tube:

   - H_2O: 15.9 µl
   - 10x buffer: 2.5 µl
   - $MgCl_2$ (25 mM): 1.5 µl
   - primer (250 ng/ml): 2.0 µl
   - dNTPs (2.5 mM): 2.0 µl

   Note: The Perkin Elmer 10x buffer contains 15 mM $MgCl_2$, so adding $MgCl_2$ is necessary to bring the final Mg^{2+} concentration to 3.0 mM.

   As an example, for 20 samples the mixture will contain: 318 µl H_2O, 50 µl 10x buffer, 30 µl $MgCl_2$ (25 mM), 40 µl primer (250 ng/ µl), and 40 µl dNTPs (2.5 mM).

3. Before adding the *Taq* polymerase to the master mix, heat the mixture at 90 °C for 2 min.

4. Cool the mixture on ice.

5. Add *Taq* polymerase (5 U/µl) 0.1 µl per tube, to the mixture (2.0 µl for 20 samples) and mix well.

6. Distribute the mixture into the tubes kept in ice (24 µl/tube).

7. Insert the tubes in the thermal cycler when the temperature has reached 90 °C.

8. Start the following thermal program:
   - 90 °C for 1 min

- 95 °C for 1 min 30 s
- 95 °C for 30 s
- 36 °C for 1 min
- 75 °C for 2 min
- Repeat steps 3–5 a further 44 times.
- 75 °C for 10 min
- 60 °C for 10 min
- Store at 4 °C.

Note: For primers longer than 17 nucleotides the annealing step must be 45 °C for 1 min

Separation of RAPD fragments by agarose gel electrophoresis

9. Load 5 ml of the amplification product onto an agarose gel (1.2 % w/v in 1x TAE buffer) containing 1 µg/ml ethidium bromide.
If the amplification profile exhibits mostly small fragments, (ranging from 100 to 500 bp or even below 100 bp), we recommend the use of a high resolution agarose, such as MetaPhor (FMC BioProducts, Rockland, ME, USA), to prepare a higher concentration gel; follow the instructions of the supplier.

10. Run the gel at 5 V/cm for about 1 h.

11. Photograph the gel under UV light for keeping record of the results and for analyzing the amplification profiles.

Caution: Protect yourself from UV radiation wearing glasses or using special protective screen.

Selection of the RAPD marker(s)

12. Select the marker to be used as probe (Fig. 2). This is one of the most critical steps in the overall strategy and deserves some caution:
- Choose the RAPD marker from among those bands with the highest intensity, as they have a higher probability of being found in amplification patterns obtained with the same DNA under different reaction conditions.
- Choose RAPD fragments ranging from 500 to 2000 bp, as they are the most reproducible ones. Furthermore, the use of very short markers as probes could result in cross-hybridization to unrelated markers, since RAPD fragments contain the same ending sequences.
- If the aim of the procedure is to obtain a highly specific probe, choose highly polymorphic bands, i.e., present in a single pattern. Alternatively, to produce a less specific probe, the choice will be directed to monomorphic bands, i.e., present in many patterns. If monomorphic fragments are not found, it is possible to label and use as a probe a mixture of two or more fragments.

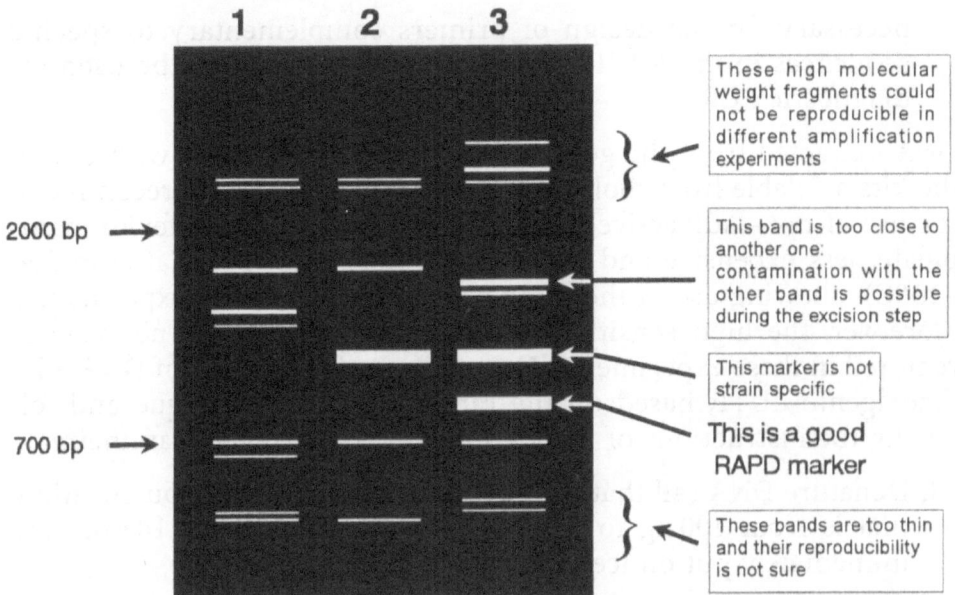

Fig. 2. Schematic representation of the electrophoretic pattern of RAPD amplification products using the DNA of three hypothetical bacterial strains with suggestions for selecting a RAPD marker. *Lane 3*, the strain for which the probe should be produced; *lanes 1* and *2*, related strains, or strains present in the same samples

Purification and labeling of the probe

The reamplification of a RAPD band by PCR results in a relatively pure DNA sequence.

Reamplification of RAPD bands

1. Proceed as described in Chap. XVIII.

2. After the reamplification, analyze the products on agarose gel (2 %), as described before, to identify pure samples containing only the desired band. In many cases it is easy to reamplify the chosen marker without contaminating bands.

The reamplified band can be labeled and used as probe without further treatment, but contaminating band(s), which might not have been revealed by electrophoretic analysis, must be absolutely avoided. After labeling, contaminating bands, if present, can give rise to false positive signals during hybridization. It is more convenient for this purpose to purify the band.

Purification of RAPD markers from agarose gel

3. Proceed as described in Chap. XVIII, Sect 4.1.

   Note: It is possible at this point to use the purified band for cloning and sequencing, as described in Chaps. XVIII and XIX. This is necessary for the design of primers complementary to specific sequences in the RAPD fragment that could in turn be used for diagnostic PCR.

Labeling of the RAPD markers

DNA extracted from the gel and purified can be labeled with one of the kits available from molecular biology companies. We recommend the use of non-radioactive labeled probes, as they are safer to manipulate, less expensive and last longer (more than 1 year if stored at –20 °C), allowing use of the same preparation for many experiments. Moreover, the high sensitivity of radioactive probes is not usually required in these experiments. Here we describe labeling of DNA with digoxigenin-dUTP, based on the random primer technique and following the instructions of the kit supplier (Boehringer Mannheim).

4. Denature DNA (all that is obtained from agarose gel purification, that is from 100 ng to 3 µg in 5–10 µl) at 95 °C for 10 min and immediately put on ice.

5. Add 2 µl of hexanucleotides and 2 µl of dNTP labeling mix (which includes digoxigenin-dUTP).

6. Bring the volume up to 19 µl with sterile water and add 1 µl of DNA polymerase (Klenow fragment).

7. Incubate at 37 °C for 3–20 h.

8. Stop the reaction by adding 2 µl of EDTA 0.2 M (pH 8).

9. Add 2 µl of 4 M LiCl and 60 µl of cold absolute ethanol to precipitate DNA; mix well and keep at –70 °C for 30 min.

10. Centrifuge at 12000 rpm for 15 min, wash the pellet with 70 % ethanol, dry the pellet and suspend in 25 µl of TE.

11. Visualize and determine the amount of labeled DNA by comparison with a standard labeled DNA included in the kit.

    Note: With digoxigenin labeled probes, detection can be performed either by a colorimetric or a chemiluminescent method. Both methods can be carried out easily by use of the appropriate kit, purchased from Boehringer and including complete instructions. Colorimetric detection is more rapid and less laborious to perform than chemiluminescent detection; the latter is however more sensitive and tends to give clearer pictures.

12. Always assess the specificity of the produced probe by hybridization with amplified DNA from a different source, as described in the next section.

Hybridization

Products of RAPD amplification of DNA from an unknown sample   **Dot blotting**
can be analyzed for the presence of one or more marker(s) by hybridization with the labeled markers. Moreover, hybridization of the
probe with amplified DNA from different bacteria, living in the environment under study, is also required in order to evaluate the specificity of the probe. These hybridizations can be accomplished by
either dot or slot blotting. These methods are carried out using a
device that allows, generally by vacuum suction, the simultaneous fil-
tering of many DNA samples onto a single membrane. Alternatively,
in the absence of the appropriate apparatus, loading of DNA samples
onto the membrane can be performed by repeated spotting of
approximately 2 µl aliquots, allowing the samples to dry between
each aliquot; this manual procedure is very time-consuming, especially when many samples must be loaded. During all of the manipulations, the membrane (we recommend the use of nylon membrane) should never be touched except by gloved hands; disposable
gloves should be worn and forceps used. Proceed as follows:

1. Pretreat the nylon membrane (appropriately cut to fit the slot of
   the device): immerse the membrane in distilled water for 5 min,
   transfer to filtered 10x SSC for 15 min and then allow to dry for
   5 min on a clean piece of Whatmann 3MM paper. Place the membrane on the dot blot apparatus according to the instructions
   included with the instrument.

2. Add TE to the DNA samples (1–10 ng) to a final volume of 30 µl
   (do not exceed a final volume of 5 µl if the loading is performed
   by manual spotting).

3. Denature DNA at 95 °C, then chill on ice for 3 min.

4. Add 1 volume (30 µl) of filtered 20x SSC; maintain the sample on
   ice until loading.

5. Load the samples onto the membrane according to the instructions of the manufacturer of the device, setting the vacuum pump
   at about 30 cm Hg and pipetting carefully, avoiding any contact of
   the tip with the membrane; avoid also the formation of air bubbles. After loading, wash each well three times with 150 µl of filtered 10x SSC.

6. After blotting, cut off the upper right corner of the membrane to
   orient samples.

7. Wet membrane in denaturing solution for 5 min.

8. Transfer membrane to neutralizing solution for 1 min.

9. Air dry the membrane on filter paper for up to an hour or dry at 80 °C for 10 min.

10. Fix the DNA to the membrane by UV cross-linking. The membrane can be fixed either by the use of a special device (for example, Stratalinker, Stratagene, USA) or by placing the membrane with the DNA side down on a UV transilluminator, whose filter has been carefully cleaned, for 3 min. After fixing, the membrane, sealed in a plastic bag, can be kept for several months before hybridization.

Hybridization   After blotting the membrane should be treated exactly as a Southern blot membrane. The protocol described here follows the procedure of Sambrook et al. (1989), washing under stringent conditions. We recommend the use of a special hybridization apparatus, e.g., those based on tubes or bottles rolling inside a temperature controlled oven, as this avoids direct handling of the membrane, necessary when using plastic bags. Hybridization is carried out as follows:

11. Insert the membrane in a hybridization bottle and add 10 ml (volumes refer to $100\,cm^2$ membrane) of prehybridization solution.

12. Incubate the membrane for 1 h at 42 °C.

13. Replace the prehybridization solution with 10 ml of hybridization solution; avoid bubble formation if hybridization is performed in a plastic bag.

14. Incubate the membrane at 42 °C overnight or for at least 6 h.

15. At the end of the hybridization, wash the membrane twice at room temperature with 10 ml of 2x SSC plus 0.1 % SDS; then wash two more times for 15 min at 68 °C with 0.2x SSC plus 0.1 % SDS. At this point the membrane is ready for the detection procedure using the colorimetric or chemiluminescent method, as described in the Boehringer instruction manual.

Stripping and reprobing of DNA blots

16. Wash the membrane twice for 10 min in TE.

17. Incubate for 10 min a 37 °C in 0.2 M NaOH plus 0.1 % SDS; replace the solution and continue the incubation for an additional 10 min.

18. Wash twice for 10 min in TE.

19. Prehybridize and hybridize with a second probe.

References

Fani R, Damiani G, Di Serio C, Gallori E, Grifoni A, Bazzicalupo M (1993) Use of random amplified polymorphic DNA (RAPD) for generating specific DNA probes for microorganisms. Mol Ecol 2:243–250

Hadrys H, Balick M, Schierwater B (1992) Application of random amplified polymorphic DNA (RAPD) in molecular ecology. Mol Ecol 1:55–63.

Lanfranco L, Wyss P, Marzachi C, Bonfante P (1995) Generation of RAPD-PCR primers for the identification of isolates of *Glomus mossae*, an arbuscular mycorrhizal fungus. Mol Ecol 4:61–68.

Roszak DB, Colwell RR (1987) Survival strategies of bacteria in natural environment. Microbiol Rev 51:365–379.

Sambrook J, Fritsch EF, Maniatis T (1989) Molecular cloning. A laboratory manual. 2nd edition. Cold Spring Harbor Laboratory Press, Cold Spring Harbor, New York

Part 2

RNA Fingerprinting

Overview

1
Introduction to RNA Fingerprinting
Through Random PCR Amplification

The identification of differentially expressed genes is an essential prerequisite in several areas of biomolecular research such as, for example, those pertaining to cell differentiation, development or neoplastic transformation. Most fundamental biological processes depend, in fact, on programs of differential gene expression, i.e., they are determined by the "choice" of which genes are expressed in a particular cell population at a given time.

Until recently, isolation of differentially expressed genes could only be achieved through approaches based on either differential screening of cDNA libraries (St. John and Davis 1979) or construction of subtracted cDNA libraries (Sargent and Dawid 1983). Although these techniques have successfully been applied, they present several disadvantages: they are rather laborious and time consuming, and require large amounts of RNA. Furthermore, differential screening detects only abundant mRNAs (0.1 % or greater); on the other hand, subtractive hybridization is quite sensitive but even more difficult to set up. Finally, a major limitation of both procedures is that only one pair of RNA samples can be compared and only a one-way comparison between them can be performed at any given time.

Most of the limitations of the hybridization-based techniques have been overcome by an innovative approach which has recently been developed by extending the random amplification strategy already established in DNA fingerprinting to the comparative analysis of RNAs derived from different sources (Liang and Pardee 1992; Welsh et al. 1992). The outcome of this polymerase chain reaction (PCR)-based methodology, referred to as RNA fingerprinting, is a pattern of amplified cDNA fragments which is displayed as a fingerprint on a polyacrylamide gel. Comparison between the patterns generated from different RNA populations allows differentially expressed species to be detected as differences between the patterns. Differentially amplified cDNAs can then be isolated, sequenced and used as probes to confirm differential expression and to screen cDNA libraries.

The great potential of this new methodology has generated much enthusiasm, evidenced by the many publications reporting its successful application as well as the ongoing efforts towards its improvement. Indeed, it is now possible to state that, 4 years after its development, RNA fingerprinting has become the method of choice for detecting and isolating differentially expressed genes.

Random Amplification of Differentially Expressed mRNAs

RNA fingerprinting methodology is based on the use of arbitrary primers to amplify partial cDNA sequences from subsets of mRNAs.

Principles The first step is the reverse transcription of a subpopulation of RNA molecules which is achieved through the use of a primer designed so that only a fraction of the RNAs is selected as template for the synthesis of first strand cDNA. Subsequently, an arbitrary oligonucleotide is used to prime second strand synthesis by a thermostable DNA polymerase in the presence of a radioactively labeled deoxynucleotide. Only those cDNAs whose sequence contains a binding site for the arbitrary primer serve as templates in this reaction and only a subset of double stranded cDNAs is, therefore, produced. PCR amplification can now take place specifically on those fragments which were tagged on both ends with the primers used in the cDNA synthesis steps. The amplification products are then resolved on a polyacrylamide gel and displayed as fingerprints by autoradiography. Amplification profiles generated from different RNA populations using the same primers are easily compared to identify possible differences, i.e., the presence of bands in one pattern only, or bands which are more intense in one pattern than in another. Any difference between the patterns reveals a variation in the abundance of the corresponding specific RNA in the original samples. In this regard, it is worth emphasizing the semiquantitative nature of the PCR amplification by arbitrary primers: the relative intensity of a given band in different fingerprints is proportional to the concentration of the corresponding template sequence in the original samples (McClelland and Welsh 1994).

Furthermore, it should be noted that RNA fingerprinting methodology potentially allows the systematic scan of all mRNA species expressed in a given cell type since every individual mRNA can be reverse transcribed and amplified by PCR provided that an appropriate number of different reactions are performed (Bauer et al. 1993).

RNA fingerprinting methodology was established by the introduction of two basic techniques referred to as differential display (Liang and Pardee 1992) and RNA arbitrarily primed PCR (RAP-PCR) (Welsh et al. 1992) both which lead to RNA fingerprint production through arbitrarily primed amplification of cDNA templates. The fundamental difference between these two procedures concerns the way in which the initial reverse transcription step is performed. In differential display, first strand cDNA synthesis is initiated by an oligo-dT based primer, whereas in RAP-PCR it is primed by an oligonucleotide of arbitrary sequence.

Methods

Other variations of the basic RNA fingerprinting strategy have also been proposed. For example Sokolov and Prockop (1994) reported a procedure in which first strand synthesis is initiated by a random 6-base primer thus generating cDNAs which represent the total population of RNA molecules.

Furthermore, RNA fingerprinting methodology has been recently enriched with a very interesting technique known as targeted RNA fingerprinting (TRF) (Stone and Wharton 1994) which allows differentially expressed members of specific gene families to be detected and cloned. In this procedure, after the arbitrarily primed reverse transcription step, cDNA amplification is performed using a degenerate primer corresponding to a coding region conserved among the members of the specific gene family of interest, thus favoring amplification of members of that family. This technique can be, therefore, considered the first choice when studying biological processes in which a specific gene family is suspected to play a role.

Although the major appeal of RNA fingerprinting by arbitrarily primed PCR lies in its speed and simplicity as well as its requirement for small amounts of RNA, the main advantages of this methodology over those previously available are that both up- and downregulated genes can be detected and multiple RNA samples can be simultaneously compared. Further substantial advantages are provided by RNA fingerprinting. It offers the possibility of estimating the proportion of transcripts differentially expressed in different cell types or different physiological, pathological or experimental conditions of a given cell type, and of inferring possible intersections between different regulatory pathways from the observation of overlaps between fingerprints corresponding to the different experimental treatments to which a given cell type is subjected (Ralph et al. 1993; McClelland et al. 1994).

Advantages

On the other hand, RNA fingerprinting has a serious limitation. It is generally acknowledged that rare transcripts are under-represented in RNA fingerprints. This is most likely due to the fact that the prob-

Problems and posssible solutions

ability of detecting a product in a fingerprint depends not only on priming efficiency but also on the relative abundance of the corresponding RNA in the sample under investigation.

The nested RAP strategy (Ralph et al. 1993), a two-round amplification by arbitrary primers, was developed in order to overcome, at least partially, this limitation. The second round amplification is primed by an oligonucleotide identical to the one used in the first round except for one or two additional arbitrary bases at its 3' end. The introduction of the additional base(s), along with the higher annealing temperature adopted, decreases the number of fragments which will be involved in the second round amplification since, obviously, there will be less templates perfectly matching the new primer. Among these amplifiable products there are fragments derived both from abundant transcripts (visible on the first gel since sufficiently amplified in the first round) and from low abundance transcripts (not visible on the gel since not sufficiently amplified). Decreasing the number of potentially amplifiable fragments causes a drastic reduction in the competition between templates and, therefore, allows detectable products to be obtained also from those fragments which correspond to rare transcripts.

Another approach aimed at detecting low abundance mRNAs was proposed by Hakvoort et al. in 1994. It relies on the combination of subtractive hybridization and differential display in order to combine the sensitivity of the former with the advantages provided by the latter. The initial subtraction step leads to an enrichment of differentially expressed transcripts so that, in the subsequent fingerprinting step, the probability of also detecting products derived from low abundance transcripts is increased.

Finally, an intrinsic problem of RNA fingerprinting must be mentioned. Different amplification products can be simultaneously present in a given electrophoretic band, so that the contamination of the isolated product with comigrating species may occur. This implies that an appropriate assay must be performed in order to confirm that the right product (i.e., the one really differentially amplified) has been isolated. As discussed below, a number of approaches are available for the identification or selection of the right clone.

2
Guide to Experimental Strategies

Experimental Steps

Figure 1 outlines the sequential steps of the whole procedure leading to the identification of differentially expressed genes. The design of a RNA fingerprinting experiment requires a pondered choice between the several alternatives available for each step. This section, which reports and briefly discusses most of these alternatives, is aimed at giving a few insights for correct experimental design.

Successful RNA fingerprinting requires undegraded RNA free of chromosomal DNA to be used as template since it is well established that all methods relying on PCR amplification by arbitrary primers are sensitive to template quality. Therefore, a rigorous purification procedure must be chosen, such as that described in Chap. XXVII, and DNA contamination must be removed by DNase treatment. It should also be noted that the isolation of poly(A)$^+$ RNA is not advisable if the method adopted is differential display (Liang et al. 1993).

Template preparation

In the original differential display protocol, synthesis of first strand cDNA is initiated by an anchored oligo-dT primer. This oligonucleotide, which consists of a stretch of 11 or 12 Ts plus 2 additional 3'

Reverse transcription and PCR amplification

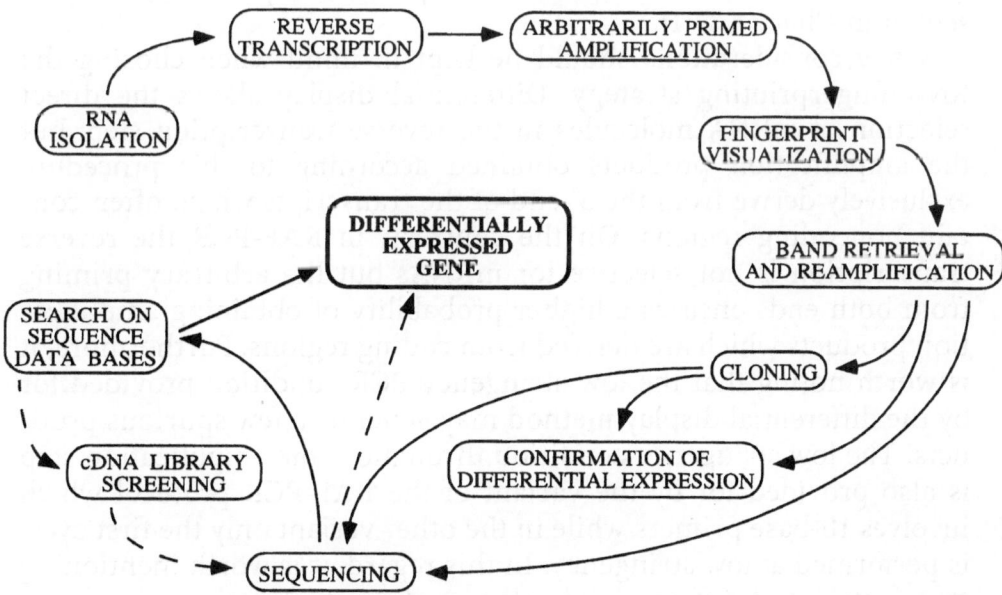

Fig. 1. Outline of RNA fingerprinting methodology

bases, primes reverse transcription only from a subpopulation of polyadenylate tailed RNAs thanks to the specificity provided by the two additional 3' bases. For example, a primer such as 5'-T_{11}GC-3' would select those mRNAs which contain the dinucleotide CG located just upstream of their poly(A) tails. Such a subpopulation of mRNAs should, by probability, represent one-twelfth of the total mRNA population because there are 12 different combinations of the two bases upstream of the poly(A) tail. The resulting cDNA is then amplified using the same primer plus an arbitrary 10-base oligonucleotide thus generating a collection of products which are derived from the 3' end of a subset of polyadenilated RNAs. More recently, 1-base-anchored oligo-dT primers in combination with arbitrary primers longer than ten bases have been successfully used (Liang et al. 1994). A further modification which involves the use of primers incorporating the universally pairing base inosine has also been proposed (Rohrwild et al. 1995). Inosine-containing primers are more expensive than conventional primers but provide the advantage of an increased fingerprint reproducibility. A detailed protocol for differential display is given in Chap. XXVI.

In RAP-PCR, reverse transcription is primed by an oligonucleotide of arbitrary sequence, thus selecting as templates only those RNA molecules which have 6–8 base matches with the 3' end of the primer, whereas second strand synthesis can be initiated by the same or a different primer. Two RAP-PCR protocols involving a pair of 10-base primers and a single 18-base primer, respectively, are presented in Chap. XXVII.

A few considerations should be kept in mind when chosing the RNA fingerprinting strategy. Differential display allows the direct selection of mRNA molecules in the reverse transcription step, but the amplification products obtained according to this procedure exclusively derive from the 3' end of the transcripts which often contain noncoding regions. On the contrary, in RAP-PCR the reverse transcription is not selective for mRNAs but the arbitrary priming from both ends ensures a higher probability of obtaining amplification products which are derived from coding regions. Furthermore, it is worth noting that the low stringency PCR condition provided for by the differential display method may generate a few spurious products. The low stringency condition throughout the amplification step is also provided for by the variant of the RAP-PCR protocol which involves 10-base primers while in the other variant only the first cycle is performed at low stringency. In this regard, it is worth mentioning that a variant of differential display involving high stringency conditions has been developed by Zhao et al. (1995).

In the aforementioned procedure by Sokolov and Prockop (1994) (not here reported), reverse transcription is primed by random 6-base oligonucleotides under conditions that allow internal sequences of most mRNAs to be reverse transcribed and, therefore, only one reaction is needed per RNA sample. In the subsequent PCR step, internal regions of the cDNA molecules are amplified using two or three longer arbitrary primers.

When performing RNA fingerprinting experiments, it is crucial to verify that the observed differences between RNA samples under investigation are not due to differences in template quality or concentration since all the described procedures are very sensitive to both these factors. As suggested by McClelland et al. (1995), the authenticity of a differentially amplified product can easily be checked if a titration of each RNA template is fingerprinted. Only those products occurring at two or more concentrations of a given template but not occurring at any concentration of the other template must be considered. When reproducible fingerprints have been obtained, only two different concentrations can be used in subsequent experiments involving the same template. On the contrary, if titration experiments do not generate reproducible fingerprints, a new RNA preparation is needed. A control performed leaving out the reverse transcriptase must be included in the experiment in order to ensure that the RNA samples are not contaminated by genomic DNA.

Fingerprint visualization and band retrieval

According to the original basic procedures (differential display and RAP-PCR), the resolution of amplification products is achieved by means of denaturing polyacrylamide gel electrophoresis which allows highly resolved fingerprint profiles containing a large numbers of products (up to ~100) to be obtained. The use of non denaturing polyacrylamide gels, described in Chap. XXVI, was introduced by Bauer et al. (1993) with the specific aim of reducing the complexity of band patterns.

The procedure developed by Sokolov and Prockop (1994), which is the extension of the random amplified polymorphic DNA (RAPD) method (Williams et al. 1990) to RNA fingerprinting, employs the easier and faster agarose gel electrophoresis to resolve the amplification products. Agarose gels are much less resolving than polyacrylamide gels, so that a number of fragments, particularly those corresponding to low abundance transcripts, can be missed. On the other hand, fragments much larger than those resolved on polyacrylamide gels can also be detected.

Radioactive labeling, which is the most commonly adopted approach for fingerprint visualization, is obtained including either

^{32}P-, ^{35}S- or ^{33}P-dNTP among the amplification reagents. ^{32}P-dNTP is not expensive and provides very high sensitivity but, due to its penetrating radiation, results in a quite low band resolution. Higher resolution can be obtained by means of dNTP labeled with ^{35}S which, however, can contaminate the PCR machine owing to its tendency to evaporate during the amplification. ^{33}P-dNTP is more expensive than the other two radiolabeled deoxynucleotides, but is a very interesting alternative since it combines the sensitivity of ^{32}P with the ability of ^{35}S to generate highly resolved fingerprints (Liang et al. 1994).

Besides radioactive labeling, a few other approaches are available to visualize amplification products.

Ethidium bromide staining, which is used in combination with agarose gel electrophoresis, is a simple way to visualize fingerprints (Sokolov and Prockop 1994). While it may be advantageous to avoid radioactive labeling, it is, on the other hand, worth noting that many differentially expressed genes might not be detected when applying this staining procedure, in view of its low sensitivity.

Silver staining of RNA fingerprints, reported in Chap. XXVII, was introduced by Lohmann et al. (1995) as a nonradioactive variant of differential display. This is a very interesting approach since it speeds up the differential display procedure without any significant loss in sensitivity and resolution compared with the standard radioactive method, and allows any band of interest to be more easily retrieved owing to its visibility on the original gel.

Fluorescence labeling, which is described in Chap. XXIX, can be used in combination either with a fluorescent image analyzer or with a simple fluorescent dye-staining method which does not require any special equipment. Due to the great potential of automatic generation of fingerprints in high throughput differential displaying, fluorescence is likely to become the most frequently used labeling approach in the future.

More recently, An et al. (1996) described an additional nonradioactive method (not here reported) which is based on chemiluminescent detection of the amplification products.

Once an RNA fingerprint displaying some differentially amplified band has been generated, confirmation of the differential expression is needed. This, as well as the characterization of differentially expressed cDNAs, always requires the band of interest to be recovered and reamplified with the same primers used in the first amplification. Several chapters in this manual report procedures for easy band retrieval from both polyacrylamide and agarose gels.

After band reamplification, several alternative strategies can be applied both to confirm differential expression and to characterize the differentially expressed cDNAs.

The reamplified fragment can be directly used as a probe in Northern blotting experiments, in order to verify whether it really corresponds to an mRNA which is differentially expressed in the samples under investigation. This is a simple strategy but does not always give the desired results: occasionally no signal is detected, presumably due to the low abundance of the mRNA corresponding to the differentially amplified fragment. An efficient procedure that may be used instead of Northern blotting, is the modified RNase protection assay described in Chap. XXXII. This technique provides several advantages: it requires small amounts of RNA, is simple, relatively fast, and sensitive enough to detect even low abundance mRNAs.

Confirmation of differential expression

Alternatively, the reamplified product can be cloned so that differential expression can be confirmed using the cloned insert as a probe. If the differential display method is adopted, an efficient screening procedure prior to cloning is advisable since false positives are never completely avoidable (Liang and Pardee 1995). The slot blot hybridization screening, described in Chap. XXX, allows the quick confirmation of differential expression thus reducing the number of fragments to be cloned and further analyzed.

Contamination of the recovered fragments by unrelated products is a frequent problem when cloning differentially amplified fragments so that some method to identify the right clone is needed. Confirmation of the correct product can be achieved by Southern blotting the original fingerprint and using the clone as a probe according to the procedure described in Chap. XX. Furthermore, it is worth mentioning that useful suggestions for selecting the correct clone can also be found in Chap. XXIX. Alternatively, the right clone can be selected through the procedure developed by Callard et al. (1994). This is a dot blot selection method in which several clones obtained from a given reamplified product are initially spotted on a nylon membrane and then probed with half of the original purified PCR product, the other half being the source from which the product was reamplified and successively cloned. Due to the prevalence in the probe of the differentially amplified fragment over the contaminating DNA, only those clones containing the former will give a strong autoradiographic signal, whereas those containing the latter will not be able to give any significant signal.

Finally, other approaches are available in the literature such as those recently described by Shoham et al. (1996), Zhao et al. (1996) and Liu and Raghothama (1996).

A fast and efficient method which avoids the laborious screening of several cloned fragments is provided in Chap. XXXI. In this procedure the right fragment to be cloned is selected through a Northern blotting experiment in which RNA samples under investigation are probed with the product recovered from the gel and reamplified. The differentially expressed mRNA (corresponding to the differentially displayed band) will bind and, therefore, "capture" the right fragment which is, then, retrieved from the membrane, amplified once again and, finally, cloned. Obviously, for the successful application of the affinity capturing approach, the reamplified fragment directly used to probe the Northern blot must be able to give a signal.

A further approach to "find" the right fragment to be cloned relies on the use of single-strand conformation polymorphism (SSCP) gels to separate it from the contaminating sequences (McClelland et al. 1995). Resolving the original differentially amplified fragment on such a gel allows the right fragment to be identified as the most prominent product detected.

Characterization of differentially amplified cDNAs

Identification of the differentially expressed gene corresponding to a given differentially amplified product requires sequencing of the isolated DNA fragment either directly or after its cloning.

Direct sequencing of arbitrarily primed PCR products can be performed according to the protocols provided by Chap. XXXIII. This chapter focuses on direct automated sequencing, however, the first protocol described can also be used for manual sequencing. In the DNA fingerprinting section of this manual, Chap. XIX reports a protocol for manual sequencing of random amplification products generated from genomic DNA templates which, of course, can also be applied to the sequencing of products derived from RNA templates.

Chapter XXXIV describes an interesting technique, referred to as ligation linked PCR (LLR), which simplifies the characterization of differential display products. The reamplification of the isolated fragment is optimized through its transient ligation to an appropriate vector and direct sequencing of the reamplified product is performed using the same highly specific primer already used, along with the oligo-dT primer, in the reamplification step. It is worth noting that a unique primer is, therefore, used in the sequencing step whatever arbitrary primer may have been used in the first amplification.

If the differentially amplified product has already been cloned, any standard sequencing procedure can be applied to determine its nucleotide sequence.

After sequencing, a search in the DNA sequence data bases will, in some cases, identify the corresponding gene. When the differentially amplified fragment does not correspond to any already known gene, cloning of the corresponding cDNA, which can be achieved using the fragment as a probe for the screening of a cDNA library (described in Chap. XXXV), becomes necessary.

Research Applications

Several successful applications of RNA fingerprinting methodology in fields such as developmental biology or neurobiology and in pathological situations such as cancer, heart disease, or diabetes have been reported. In the few years since its development, RNA fingerprinting has emerged as a highly versatile tool of great potential as demonstrated by some of these specific applications. For example it has been modified for particular use such as the identification of differentially expressed genes belonging to a specific gene family, which is described in Chap. XXXVI. Further examples are its adaptation either to a high throughput system to catalog sequences that are abnormally expressed in tumors (described in Chap. XXXVII), or to systems as peculiar as single neurons (Miyashiro et al. 1994), preimplantation mouse embryos (Chap. XXXVIII) and Trypanosomes (Chap. XXXIX).

References

An G, Luo G, Veltri RW, O'Hara SM (1996) Sensitive, nonradioactive differential display method using chemiluminescent detection. BioTechniques 20:342–346

Bauer D, Müller H, Reich J, Riedel H, Ahrenkiel V, Warthoe P, Strauss M (1993) Identification of differentially expressed mRNA species by an improved display technique (DDRT-PCR). Nucleic Acids Res 21:4272–4280

Callard D, Lescure B, Mazzolini L (1994) A method for the elimination of false positives generated by the mRNA differential display technique. BioTechniques 16:1096–1103

Hakvoort TBM, Leegwater ACJ, Michiels FAM, Chamuleau RAFM, Lamers WH (1994) Identification of enriched sequences from a cDNA subtraction-hybridization procedure. Nucleic Acids Res 22:878–879

Liang P, Pardee AB (1992) Differential display of eukaryotic messenger RNA by means of the polymerase chain reaction. Science 257:967–971

Liang P, Pardee AB (1995) Recent advances in differential display. Curr Opin Immunol 7:274–280

Liang P, Averboukh L, Pardee AB (1993) Distribution and cloning of eukaryotic mRNAs by means of differential display: refinements and optimization. Nucleic Acids Res 21:3269–3275

Liang P, Zhu W, Zhang X, Gou Z, O'Connell RP, Averboukh L, Wang F, Pardee AB (1994) Differential display using one-base-anchored oligo-dT primers. Nucleic Acids Res 22:5763–5764

Liu C, Raghothama KG (1996) Practical method for cloning cDNAs generated in an mRNA differential display. Biotechniques 20:576–580

Lohmann J, Schickle H, Bosch TCG (1995) REN display, a rapid and efficient method for nonradioactive differential display and mRNA isolation. BioTechniques 18:200–202

McClelland M, Welsh J (1994) RNA fingerprinting by arbitrarily primed PCR. PCR Methods Applic 4:s66-s81

McClelland M, Ralph D, Cheng R, Welsh J (1994) Interactions among regulators of RNA abundance characterized using RNA fingerprinting by arbitrarily primed PCR. Nucleic Acids Res 22:4419–4431

McClelland M, Mathieu-Daude F, Welsh J (1995) RNA fingerprinting and differential display using arbitrarily primed PCR. Trends Genet 11:242–246

Miyashiro K, Dichter M, Eberwine J (1994) On the nature and differential distribution of mRNAs in hippocampal neurites: implications for neuronal functioning. Proc Natl Acad Sci USA 91:10800–10804

Ralph D, Welsh J, McClelland M (1993) RNA fingerprinting using arbitrarily primed PCR identifies regulated RNase in Mink lung (Mv1Lu) cells growth arrested by TGFβ. Proc Natl Acad Sci USA 90:10710–10714

Rohrwild M, Alpan RS, Liang P, Pardee AB (1995) Inosine-containing primers for mRNA differential display. Trends Genet 11:300

Sargent TD, Dawid IB (1983) Differential gene expression in the gastrula of Xenopus laevis. Science 222:135–139

Shoham NG, Arad T, Rosin-Abersfeld R, Mashiah P, Gazit A, Yaniv A (1996) Differential display assay and analysis. BioTechniques 20:182–184

Sokolov BP, Prockop DJ (1994) A rapid and simple PCR-based method for isolation of cDNAs from differentially expressed genes. Nucleic Acids Res 22:4009–4015

St John T, Davis RW (1979) Isolation of galactose-inducible DNA sequences from Saccharomyces cerevisiae by differential plaque filter hybridization. Cell 16:443–452

Stone B, Wharton W (1994) Targeted RNA fingerprinting: the cloning of differentially-expressed cDNA fragments enriched for members of the zinc finger gene family. Nucleic Acids Res 22:2612–2618

Welsh J, Chada K, Dalal SS, Cheng R, Ralph D, McClelland M (1992) Arbitrarily primed PCR fingerprinting of RNA. Nucleic Acids Res 20:4965–4970

Williams JGK, Kubelik AR, Livak KJ, Rafalski JA, Tingey SV (1990) DNA polymorphisms amplified by arbitrary primers are useful as genetic markers. Nucleic Acids Res 18:6531–6535

Zhao S, Ooi SL, Pardee AB (1995) New primer strategy improves precision of differential display. BioTechiques 18:842–850

Zhao S, Ooi SL, Yang FC, Pardee AB (1996) Methods for identification of true positive cloned cDNA fragments in differential display. BioTechniques 20:400–404

Chapter XXVI

Differential Display of Expressed mRNAs

M. Jørgensen, M. Bévort, N. Pallisgaard, R. Hummel, C. Hansen, M. Rohde, M. Strauss, and H. Leffers

1
Introduction

Biological responses to processes including stimulation with growth factors or cell differentiation depend on changes in the expression of the 50000–100000 genes that are encoded in the genomes of higher eukaryotes. Therefore, it is important that the technology used for monitoring changes in gene expression is able to detect quantitative and qualitative differences in the expression of all of these genes, although only a subset of perhaps 8000–15000 genes are expressed at any one time in a given cell type (Fields et al. 1994). The optimal technology should be able to detect even small differences in the expression of this subset. Previously, changes in gene expression have been examined on the mRNA/cDNA level by techniques such as sub-tractive or differential hybridization and on the protein level by 2-dimensional protein gel electrophoresis. However, there are limita-tions inherent to these techniques. The hybridization technique only allows two RNA/cDNA populations to be compared at a time and will only reveal major differences in gene expression; identical 2-dimensional protein gels are technically difficult to prepare and to interpret, and only the 2000–3000 most abundant proteins can be detected by autoradiography or silver staining (Celis et al. 1991).

Recently, a new and powerful approach, called differential display of expressed mRNAs (DDRT-PCR), has been developed (Liang and Pardee 1992). This technique analyzes differences in gene expression at the mRNA level, and both qualitative and quantitative differences can be detected. The method is based on the assumption that virtu-ally all mRNAs can be reverse transcribed and amplified by the poly-merase chain reaction (PCR), if a sufficiently large set of arbitrary primers is used (Liang and Pardee 1992; Liang et al. 1995). DDRT-PCR allows for comparison of several RNA populations simulta-neously, and it is possible to detect both up- and down-regulated genes. Furthermore, DDRT-PCR only requires small amounts of RNA and is a relatively quick and simple procedure.

2
Principles and Applications

The initial step in DDRT-PCR is the preparation of RNA from the cells or tissues of interest. This can be done by acid guanidinium thiocyanate – phenol: chloroform extraction according to the method of Chomczynski and Sacchi (1987) or by similar methods (Sambrook et al. 1989). Alternatively, high yields of high quality RNA can be isolated using various commercially available RNA preparation kits (e.g., the RNeasy kit from Qiagen or the RNA extraction kit from Pharmacia). Traces of DNA must subsequently be removed from the RNA preparation by digestion with an RNase-free DNase.

The next step is reverse transcription of RNA into single stranded cDNA. At the same time, the RNA is split into either 12 or three fractions. For the latter, three different one-base-anchored oligo-dT primers ($HT_{11}V$) (V=A, C or G) (Table 1) are used. These primers will later serve as downstream primers in the PCR reactions.

Table 1. Downstream primers for DDRT-PCR

HT_{11}-V (cDNA synthesis and downstream primers for DDRT-PCR)	
HT_{11}-A	AAGCTTTTTTTTTTTA
HT_{11}-C	AAGCTTTTTTTTTTTC
HT_{11}-G	AAGCTTTTTTTTTTTG
T7-HT_{11}-VN (amplification of DDRT-PCR fragments)	
T7-HT_{11}-AA	TAATACGACTCACTATAGGGCCCAAGCTTTTTTTTTTTAA
T7-HT_{11}-AC	TAATACGACTCACTATAGGGCCCAAGCTTTTTTTTTTTAC
T7-HT_{11}-AG	TAATACGACTCACTATAGGGCCCAAGCTTTTTTTTTTTAG
T7-HT_{11}-AT	TAATACGACTCACTATAGGGCCCAAGCTTTTTTTTTTTAT
T7-HT_{11}-CA	TAATACGACTCACTATAGGGCCCAAGCTTTTTTTTTTTCA
T7-HT_{11}-CC	TAATACGACTCACTATAGGGCCCAAGCTTTTTTTTTTTCC
T7-HT_{11}-CG	TAATACGACTCACTATAGGGCCCAAGCTTTTTTTTTTTCG
T7-HT_{11}-CT	TAATACGACTCACTATAGGGCCCAAGCTTTTTTTTTTTCT
T7-HT_{11}-GA	TAATACGACTCACTATAGGGCCCAAGCTTTTTTTTTTTGA
T7-HT_{11}-GC	TAATACGACTCACTATAGGGCCCAAGCTTTTTTTTTTTGC
T7-HT_{11}-GG	TAATACGACTCACTATAGGGCCCAAGCTTTTTTTTTTTGG
T7-HT_{11}-GT	TAATACGACTCACTATAGGGCCCAAGCTTTTTTTTTTTGT
T7-SEQ (sequencing of DDRT-PCR fragments)	
	(CY5)-TAATACGACTCACTATAGGGCC

Table 2. Upstream primers for DDRT-PCR

H-AP 13-mer primers[a]

AAGCTTGATTGCC	AAGCTTAGTAGGC
AAGCTTCGACTGT	AAGCTTGCACCAT
AAGCTTTGGTCAG	AAGCTTAAGGAGG
AAGCTTCTCAACG	AAGCTTTTACCGC

other primers[a, b]

10-mer primers[c]

TACAACGAGG	GATCAAGTCC
TGGATTGGTC	GATCCAGTAC
CTTTCTACCC	GATCACGTAC
TTTTGGCTCC	GATCTGACAC
GGAACCAATC	GATCTCAGAC
AAACTCCGTC	GATCATAGCC
TCGATACAGG	GATCAATCGC
TGGTAAAGGG	GATCTAACCG
TCGGTCATAG	GATCGCATTG
GGTACTAAGG	GATCTGACTG
TACCTAAGCG	GATCATGGTC
CTGCTTGATG	GATCATAGCG
GTTTTCGCAG	GATCTAAGGC

[a] GenHunter, Nashville, TN.
[b] The primers can be designed by the user. They should include AAGCTT as the first 6 nucleotides and a G+C : A+T ratio of about 1:1. Alternatively, the primer sequences can be found in the literature or obtained by buying the kits from GenHunter Corp.
[c] Biometra, Göttingen, Germany.

As mentioned above, about 8000–15000 genes may be expressed in a human cell at any one time. DDRT-PCR should display all the transcribed mRNAs and this should be possible by the application of about 200–300 different primer combinations (Bauer et al. 1993). Therefore, we prepare up to 300 different PCR reactions using the three different one-base-anchored oligo-dT primers ($HT_{11}V$) as downstream primers, and 80–100 different 10–13-mer oligonucleotides with arbitrary sequences as upstream primers (Table 2). When establishing the DDRT-PCR technique in our laboratory, using the AmpliTaq polymerase (Perkin Elmer), we have found that the best results are obtained at a final $MgCl_2$ concentration of 1.5–2 mM and an annealing temperature of 40°–42°C throughout the PCR cycles. Under these conditions, the background is reduced without decreasing the number of bands per lane. It should be noted that the optimal $MgCl_2$ concentration may differ when other *Taq* polymerases are used. Also, we have carried out the DDRT-PCR using both [α^{35}S]-

labeled dATP and [α^{33}P]dATP and have found that both isotopes can successfully be used.

Following PCR amplification the fragments are resolved on a non-denaturing polyacrylamide gel and the result is visualized by autoradiography. Differentially displayed bands are then extracted from the dried gel and reamplified by PCR. This can pose a problem since the DDRT-PCR technique often results in the presence of more than one DNA fragment in differentially displayed bands. This has usually been solved by cloning the fragments into a plasmid vector, followed by analysis of a relatively high number of colonies to ensure that all the fragments are represented. An alternative is to perform four separate PCR reamplification reactions using the same upstream primer in the DDRT-PCR reaction and four different two-base-anchored oligo-dT primers (HT$_{11}$VN) (V=A, C or G) (N=A, C, G or T) as downstream primers. This approach is based on the assumption that the probability for two DNA fragments in one band both having the same nucleotide in the position next to the anchoring nucleotide is only 25%. Thus, in 75% of the cases in which there are two DNA fragments in one band, the two fragments can be selectively amplified by using the four different HT$_{11}$VN downstream primers. In addition, we have added a T7-RNA polymerase promoter sequence to the downstream amplification primers (T7HT$_{11}$VN) to facilitate sequencing and verification of differential expression (see below). Furthermore, the amplification reactions are performed using the Stoffel fragment polymerase (Perkin Elmer), because it is more sensitive than AmpliTaq polymerase to a mismatch at the 3'-base position. The products produced in one or more of the four reactions are then purified either from low-melting agarose gels by phenol: chloroform extractions followed by ethanol precipitation (Sambrook et al. 1989) or from normal agarose gels using a commercially available DNA purification kit, such as the QIAEX II gel extraction kit from Qiagen.

It is essential to verify the differential expression detected on the DDRT-PCR gels, since 10%–40% of the "differentially expressed" bands can be artifacts. Several different techniques are available for verification of differential expression, including northern blotting, dot or slot blots, nuclear run-on or RNase protection assays. Recently, we use only RNase protection since it is a relatively convenient and very sensitive method. Fragments that have been amplified by the T7HT$_{11}$VN primers can be used directly for the production of antisense RNA. The procedure for RNase protection is described in Chap. XXXII. Alternatively, for abundant mRNAs, analysis of slot/dot blots or northern blots may be faster and more convenient. For this we recommend labeling the fragments by random primed labeling (see Chap. XXXV). The preparation of northern and dot/slot blots is described in Sambrook et al. (1989). In addition, the antisense RNAs that are used for RNase protections can also be used as probes for in situ hybridization on, e. g., tissue sections, provided they are labeled with either $[\alpha^{33}P]$-, $[\alpha^{35}S]$-UTP or one of the substrates for nonradioactive detection instead of $[\alpha^{32}P]$UTP.

The amplified DNA fragments can be sequenced directly, without cloning, either on the ALFexpress automated sequenator from Pharmacia, using a CY-5 labeled T7 primer, or on the 373/377 sequenators from ABI, using dye terminators, as described in Chap. XXXIII. Alternatively, the fragments can be sequenced manually using $[\alpha^{35}S]$- or $[\alpha^{33}P]$dNTP (Xinkang and Feuerstein 1995) or ^{32}P-end labeled primers (Chap. XXXIII).

Finally, the amplified DNA fragments can be ^{32}P-labeled and used as probes to clone the corresponding full-length cDNAs from cDNA libraries. The procedure is described in Chap. XXXV.

Primers

The primers that are used in the different steps of the DDRT-PCR are summarized in Tables 1 and 2. For cDNA synthesis and as downstream primers for DDRT-PCR, we use three different one-base-anchored primers ($HT_{11}V$) (Table 1) (Liang et al. 1994). Alternatively, two-base-anchored primers ($T_{11}VN$) can be used (Bauer et al. 1993).

To obtain a total of 200–300 different primer combinations, DDRT-PCR is carried out using three different downstream primers in combination with 80–100 different upstream primers (H-AP primers or 10-mer primers) (Table 2). Alternatively, 12 different two-base-anchored downstream primers ($T_{11}VN$) can be combined with 25 different upstream primers as described (Bauer et al. 1993). The procedure that is described in this chapter is compatible with both 10- and 13-mer upstream primers and $T_{11}VN$ and $HT_{11}V$ downstream primers. There are alternatives to the primer design described above, e. g., the use of longer oligonucletides may improve the quality of the DDRT-PCR gels (Linskens et al. 1995); however, we have not tested the performance of those primers in our protocol. There are commercially available kits for DDRT-PCR; the kit from GenHunter (GenHunter Nashville, TN) utilises the $HT_{11}V$ downstream and 13-mer upstream primers, whereas the kit from Biometra, (Göttingen, Germany) includes the $T_{11}VN$ downstream and 10-mer upstream primers.

Table 1 also displays a list of the different downstream primers used in reamplifiction of the DDRT-PCR fragments. A T7-promoter sequence has been added to the 5'-end of the $HT_{11}V$ primer and an additional nucleotide has been added at the 3'-end, after the anchoring nucleotide. The reasons for these modifications are discussed elsewhere in the text. Also the sequence of the CY-5/fluorescein-labeled primer used for sequencing of the reamplified fragments on the ALFexpress/ALF sequenators is shown in Table 1. As discussed elsewhere, this primer also contains the sequence of the T7-RNA polymerase promoter.

3
Materials

Equipment
- thermocycler for 0.1 ml tubes and with heated lid (Perkin Elmer 9600 or similar)
- manual sequencing gel apparatus (e. g., SequiGen II from Bio-Rad)
- gel drier (e.g., BioRad model 583)
- vacuum pump
- waterbath
- tabletop centrifuge
- agarose gel apparatus

Supplies
- 0.1 ml PCR tubes
- X-ray film (e.g., Kodak BioMax MR)

Reagents and solutions
- total cellular RNA in diethylpyrocarbonate (DEPC) water
- RNase-free DNase (7.5 u/µl) (Pharmacia)
- AMV reverse transcriptase (20 u/µl) (Stratagene)
- AmpliTaq enzyme (5 u/µl) (Perkin Elmer)
- Klenow enzyme (5 u/µl)(Amersham)
- Stoffel fragment enzyme (10 u/µl) (Perkin Elmer)
- proteinase K stock solution: 20 mg/ml proteinase K in 10 mM Tris-HCl (pH 7.5); the proteinase K stock solution is autodigested at 37°C for 30 min, to remove possible traces of RNases, and stored at –20°C).
- $[\alpha^{33}P]$dATP or $[\alpha^{35}S]$-labeled dATP (10 µCi/µl) (Amersham, NEN or ICN).
- 0.1 M dATP, dCTP, dGTP and dTTP (Pharmacia)
- 50 or 100 base pair ladder (Pharmacia)
- three different $HT_{11}V$ primers (0.5 µg/µl)
- three different $HT_{11}V$ primers (10 pmol/µl)
- four different $T7-HT_{11}VN$ primers (10 pmol/µl)
- 80–100 arbitrary 13- or 10-mer primers (10 pmol/µl)
- agarose
- diethylpyrocarbonate (DEPC)

- DEPC-treated water (DEPC H$_2$O). Add 1 ml DEPC to 1 l H$_2$O, autoclave.
- 0.5 M EDTA (pH 8.0)
- 3 M sodium acetate (pH 6.0) (DEPC-treated)
- 25 mM MgCl$_2$
- ethanol (99 % and 80 %)
- phenol (molecular biology grade) saturated with TE buffer
- chloroform
- phenol:chloroform (1:1)
- 10 % ammoniumpersulfate (APS)
- TEMED (N,N,N',N'-Tetramethylethylenediamine)
- 40 % acrylamide mix (20:1): 190 g acrylamide; 10 g bis-acrylamide; add water to 500 ml. Store over mixed-bed resin (Bio-Rad: AG 501-X8) at 4°C.
- 10x TBE: 1 M Tris-borate, 10 mM EDTA
- TE buffer: 10 mM Tris-HCl (pH 7.5), 0.1 mM EDTA
- 25 mM dNTP mix: 25 mM each of dATP, dGTP, dCTP and dTTP made from 0.1 M stock solutions (Pharmacia)
- 40 μM dNTP mix: 40 μM each of dATP, dGTP, dCTP and dTTP made from 0.1 M stock solutions
- 2.5 mM dNTP mix: 2.5 mM each of dATP, dGTP, dCTP and dTTP made from 0.1 M stock solutions
- 10x labeling mix: 2.5 mM each of dGTP, dCTP and dTTP in TE buffer
- 10x DNase buffer: 200 mM Tris-HCl (pH 7.5), 75 mM MgCl$_2$
- extraction buffer: 10 mM Tris-HCl (pH 7.5), 20 mM EDTA (pH 8.0), 50 μg/ml proteinase K
- 5x cDNA synthesis buffer: 650 mM Tris-HCl (pH 8.3), 25 mM MgCl$_2$, 100 mM KCl
- 10x DDRT buffer: 100 mM Tris-HCl (pH 8.3), 500 mM KCl, 18 mM MgCl$_2$, 0.05 % gelatine, 1 % Triton-X-100
- 10x Klenow buffer: 100 mM Tris-HCl (pH 7.5), 80 mM MgCl$_2$, 10 mM DTT
- 10x Stoffel fragment buffer: 100 mM Tris-HCl (pH 8.3), 100 mM KCl
- 0.5x TBE buffer with 0.5 μg/ml ethidium bromide
- 2 % agarose (normal or low melting point) gel in 0.5x TBE with 0.5 μg/ml ethidium bromide
- loading buffer-1: 40 % Ficoll 400, 0.1225 % bromophenol blue, 0.1225 % xylene cyanol in TE buffer
- loading buffer-2: 20 % Ficoll 400, 0.1225 % bromophenol blue, 0.1225 % xylene cyanol in TE buffer
- DNA sequencing loading buffer: 10 mM EDTA (pH 8.0), 10 mM NaOH, 0.04 % bromphenol blue, 0.04 % xylene cyanol, 100 % deionized formamide

4
Experimental Procedure

RNA preparation

RNA can be isolated in various ways. If a commercially available RNA preparation kit is chosen, we refer to the detailed step-by-step protocol provided by the manufacturer. For RNA isolation by acid guanidinium thiocyanate-phenol:chloroform extraction, we refer to the protocol by Chomczynski and Sacchi (1987).

Traces of DNA can be present in the RNA preparation and must be removed. This is achieved by digestion with an RNase-free DNase:

1. Prepare a 2x DNase mix:
   10 µl 10x DNase buffer
   40 µl DEPC-treated H_2O
   0.5–1 µl RNase-free DNase
2. Mix the following:
   50 µl RNA (1–50 µg) in DEPC H_2O
   50 µl 2x DNase mix

3. Incubate at 37°C for 30 min.

4. Add 100 µl extraction buffer.

5. Incubate at 37°C for 15 min.

6. Add 200 µl phenol:chloroform (1:1); shake the sample for 5 min at room temperature and centrifuge for 4 min at full speed in a tabletop centrifuge.

7. Transfer the supernatant to a new tube and add 1/10 volume of DEPC-treated 3 M sodium acetate (pH 6.0) and 2.5 volumes of 99 % ethanol. Place at –70°C for 15 min and centrifuge in a tabletop centrifuge at 20000 rpm at 4°C for 20 min.

8. Wash the pellet carefully in 80 % ethanol and centrifuge briefly. Remove the supernatant and dry the pellet.

9. Redissolve the pellet in 20 µl DEPC H_2O.

cDNA synthesis

The protocol outlined below corresponds to one reaction. The volumes can easily be multiplied if multiple reactions are performed. The best results are obtained with the AMV reverse transcriptase, but this procedure is compatible with MMTV and MLTV (or similar) reverse transcriptases. However, if using these enzymes the incubation temperature may have to be reduced and it may be advantageous to use more units (50–100 U/reaction). Check the data sheets that accompany the enzymes.

1. Prepare 10 µl cDNA synthesis mix:
   4 µl 5x cDNA synthesis buffer
   0.5 µl 25 mM dNTP mix
   5.5 µl DEPC H$_2$O

2. Mix the following:
   1 µg total RNA in 9 µl DEPC H$_2$O
   1 µl HT$_{11}$V primer (0.5 µg/µl)

3. Incubate at 65°C for 1 min.

4. Incubate at 42°C for 1 min.

5. Add the following:
   10 µl cDNA synthesis mix
   0.5 µl AMV reverse transcriptase (10 U)

6. Incubate at 42°C for 1 h (37°C for some reverse transcriptases).

7. Add 80 µl DEPC H$_2$O ; mix and centrifuge briefly in a tabletop centrifuge.

8. Incubate at 95°C for 1 min and place on ice. The cDNA is now ready for DDRT-PCR or it can be stored in aliqouts at –80°C (repeated freezing and thawing should be avoided).

DDRT-PCR

The protocol outlined below corresponds to ten reactions. The volumes can easily be multiplied according to the number of reactions. The input cDNA should be made as described above (cDNA synthesis), otherwise it must contain 1–2 mM MgCl$_2$ and 125 µM dNTP. The final MgCl$_2$ and dNTP concentrations in the DDRT-PCR reactions are 1.72 mM and 15 µM, respectively. Both [α^{35}S]-labeled dATP and [α^{33}P]dATP can be used as an internal label in the DNA fragments; however, if ^{35}S-labeled αdATP is chosen, the exposure time should be extended to 2–3 days. Alternatively, for faster exposure, use 2 µCi ^{35}S-labeled αdATP per reaction.

1. Prepare a PCR master mix (for ten reactions) by mixing the following:
   10 µl 10x DDRT buffer
   10 µl 40 µM dNTP mixture
   10 µl HT$_{11}$V primer (10 pmol/µl)
   1.0 µl [α^{33}P]dATP (1 µCi/reaction)
   10 µl cDNA, prepared with the appropriate HT$_{11}$V primer
   59 µl DEPC H$_2$O

2. Mix and centrifuge, then add:
   2.0 µl AmpliTaq (5 U/µl)

3. Arrange and label PCR tubes as appropriate. Pipette 1.0 µl upstream primer (10 pmol/µl) into the corresponding tubes.

4. Add 10 µl PCR master mix to each tube.

5. Mix and centrifuge briefly.

6. Place the samples in the PCR thermocycler. The cycle conditions are:
   1x: 2 min at 95°C
   40x: 30 s at 94°C; 1 min at 40°C; 1 min at 72°C
   1x: 5 min at 72°C

7. Add 2 µl loading buffer-1 to each tube, mix and centrifuge briefly. The samples are now ready for electrophoresis on a nondenaturing polyacrylamide gel, or they can be stored at –80°C for up to several weeks. Alternatively, the 2 µl loading buffer-1 can be substituted by 10 µl of DNA sequencing loading buffer and the samples loaded onto a denaturing polyacrylamide gel after they have been heated to 95°C for 5 min.

Preparation of a labeled size marker for DDRT-PCR

1. Mix the following:
   1 µl marker (50 or 100 bp ladder) (Pharmacia)
   2 µl 10x Klenow buffer
   16 µl TE buffer

2. Add 1 µl Klenow enzyme (5 U/µl).

3. Incubate 5 min at room temperature.

4. Add 2 µl 10x labeling mix and 1 µl $[\alpha^{33}P]$dATP or $[\alpha^{35}S]$-labeled dATP (10 µCi).

5. Incubate 10 min at 37°C.

6. Add 80 µl TE buffer and 3 µl 0.5 M EDTA.

7. Add 100 µl phenol:chloroform (1:1); shake the sample for 5 min at room temperature and centrifuge for 4 min at full speed in a tabletop centrifuge.

8. Transfer the supernatant to a new tube and add 1/10 volume of 3 M sodium acetate (pH 6.0) and 2.5 volumes of 99 % ethanol. Place at –70°C for 15 min and centrifuge in a tabletop centrifuge at 20000 rpm at 4°C for 20 min.

9. Wash the pellet carefully in 80 % ethanol and centrifuge briefly. Remove the supernatant and dry the pellet.

10. Redissolve in 25 µl 1x DDRT-PCR buffer.

11. Add 4 µl loading buffer-1 (or equal volume of DNA sequencing loading buffer if using denaturing gels).

12. Load 1 µl per lane.

Here we describe the use of nondenaturing polyacrylamide gel electrophoresis. However, denaturing gels (DNA sequencing gels with 8 M urea) may sometimes result in better resolution of the PCR fragments.

Polyacrylamide gel electrophoresis

1. Prepare a 5 % nondenaturing polyacrylamide gel in 1x TBE buffer: 12.5 ml 40 % acrylamide mix, 10 ml 10x TBE, water to 100 ml. Add 40 µl TEMED and 400 µl 10 % APS just before use.

2. Pre-run the gel for 30 min at 800 V.

3. Load 2 µl per lane. We recommend including a DNA size marker on the gels (for preparation of marker see Sect. 4.4).

4. Run the gel in 1x TBE buffer at 1000 V for 3–4 h (or 200–300 V overnight) until the blue dye reaches the bottom of the gel.

5. Transfer the gel to 3MM paper and dry the gel.

6. Markings are made along the edges of the gel with radioactive ink to enable subsequent alignment of the autoradiogram on the gel (see below).

7. Expose the gel on X-ray film for 1–3 days.

This procedure is an alternative to the standard amplification (Liang and Pardee 1992), in which the same primers are used for both DDRT-PCR and amplification. The protocol below is compatible with amplification using either the standard downstream primers ($IIT_{11}V$ or $T_{11}VN$) or the extended primers ($T7HT_{11}VN$).

Extraction and reamplification of differentially displayed bands

The PCR reamplification protocol below corresponds to four PCR reactions, as four separate reactions are made for each extracted DNA fragment. If multiple fragments are being reamplified, the volumes can easily be multiplied.

1. Align the autoradiogram on the dried gel, and mark the bands of interest.

Extraction

2. Cut out the bands as precisely as possible and transfer to Eppendorf tubes.

3. Add 50 µl TE buffer to each tube and shake at room temperature for 10 min.

4. Incubate at 95°C for 15 min and shake for 15 min.

5. Centrifuge for 1 min at 14000 rpm.

PCR

6. Prepare a master mix with the following:
12 µl 10x Stoffel fragment buffer
16 µl 25 mM $MgCl_2$
12 µl 2.5 mM dNTP mix
48 µl H_2O
2 µl Stoffel fragment enzyme (10 U/µl)

7. Label four PCR tubes A, C, G and T, respectively. Pipette 2 µl of the appropriate downstream primer (T7-HT$_{11}$VN; 10 pmol/µl) into the corresponding tubes.

8. Add 1 µl of the appropriate upstream primer (10 pmol/µl) to each tube.

9. Add 5 µl cDNA to each tube.

10. Add 22.5 µl of the master mix to each tube.

11. Place the tubes in the PCR thermocycler. The conditions are:
1x: 2 min at 95°C
40x: 30 s at 94°C; 30 s at 42°C; 1 min at 72°C
1x: 5 min at 72°C

Purification of DNA Fragments from Agarose Gels

Reamplified DNA fragments are purified from an 1.5%–2% agarose gel. This can be done in various ways. If a commercially available DNA purification kit, such as the QIAEX II gel extraction kit (Qiagen) is chosen, we refer to the detailed step-by-step protocol from the manufacturer. Alternatively, DNA fragments can be purified from a low melting point agarose gel by phenol:chloroform extraction followed by alcohol precipitation as described below.

Extraction of DNA from low melting point agarose

1. Excise the DNA band(s) as precisely as possible.

2. Add equal volume of TE buffer.

3. Heat to 70°C for 5 min, vortex and heat for 1 min.

4. Add equal volume of phenol saturated with TE buffer; shake for 5 min at room temperature.

5. Centrifuge for 5 min at room temperature.

6. Transfer upper phase to a new tube.

7. Repeat steps 4–6.

8. Add equal volume of chloroform and shake for 2 min at room temperature.

9. Centrifuge for 2 min at room temperature.

10. Transfer upper phase to a new tube, add 1/10 volume of 3 M sodium acetate (pH 6.0) and 2.5 volumes ethanol.

11. Incubate at −70°C for 10 min, centrifuge at 4°C for 10−20 min.

12. Remove the supernatant and wash the pellet with 200 µl 80 % ethanol, dry the pellet in an exicator.

13. Redissolve pellet in 10 µl TE or DEPC H_2O. Store at −20°C.

14. Use 0.1−5 µl for sequencing as described in Chap. XXXIII.

Acknowledgements. Development of the methods was supported by The Danish Cancer Society, The Danish Biotechnology Programme, The European Commision and The Danish National University Hospital (Rigshospitalet).

References

Bauer D, Müller H, Reich J, Reidel H, Ahrenkiel V, Warthoe P, Strauss M (1993) Identification of differentially expressed mRNA species by an improved display technique (DDRT-PCR). Nucleic Acids Res 21:4272−4280

Celis JE, Rasmussen HH, Leffers H, Madsen P, Honoré B, Gesser B, Dejgaard K, Vandekerckhove J (1991) Human cellular protein patterns and their link to genome DNA sequence data: Usefulness of two-dimensional gel electrophoresis and microsequencing. FASEB J 5:2200−2208

Chomczynshi P, Sacchi N (1987) Single step method of RNA isolation by acid guanidinium thicyanate-phenol-chloroform extraction. Anal Biochem 167:157−159

Fields C, Adams MD, White O, Venter JC (1994) How many genes in the human genome. Nature Genet 7:345−346

Liang P, Pardee AB (1992) Differential display of eukaryotic messenger RNA by means of the polymerase chain reaction. Science 257:967−971

Liang P, Zhu W, Zhang X, Gou Z, O'Connell RP, Averboukh L, Wang F, Pardee AB (1994) Differential display using one-base-anchored oligo-dT primers. Nucleic Acids Res 22:5763−5764

Liang P, Bauer L, Averboukh L, Warthoe P, Rohrwild M, Müller H, Strauss M, Pardee AB (1995) Analysis of altered gene expression by differential display. Methods Enzymol 254:304−321

Linskens MHK, Feng J, Andrews WH, Enlow BE, Saati SM, Tonkin LA, Funk WD, Villeponteau B (1995) Cataloging altered gene expression in young and senescent cells using enhanced differential display. Nucleic Acids Res 23:3244−3251

Sambrook J, Fritsch EF, Maniatis T (1989) Molecular cloning: A laboratory manual. Cold Spring Harbor Laboratory Press, Cold Spring Harbor

Xinkang W, Feuerstein GZ (1995) Direct sequencing of DNA isolated from mRNA differential display. BioTechniques 18:448−452

Chapter XXVII

RNA Arbitrarily Primed PCR

M. RICOTE, J. WELSH, and M. MCCLELLAND

1
Introduction

The regulation of gene expression in higher eukaryotes continues to be a primary area of investigation in molecular biology. Delineation of the mechanisms by which deoxyribonucleic acid (DNA) coding sequences are first transcribed into ribonucleic acid (RNA) and ultimately translated into proteins is essential to an understanding of cancer and other diseases as well as to rational drug design.

A method which is becoming increasingly popular for RNA analysis takes advantage of the power of arbitrarily primed polymerase chain reaction (AP-PCR) for amplifying DNA sequences (Welsh and McClelland 1990; Williams et al. 1990). Differential gene expression between various tissues and developmental stages or between cells in vitro under different growth conditions can be rapidly and efficiently compared using the RNA arbitrarily primed polymerase chain reaction (RAP-PCR) or differential display (Welsh et al. 1992; Liang and Pardee 1992; McClelland et al. 1994).

There are several methods for detecting differential gene expression and cloning differentially expressed genes. Most of these methods fall into two general categories: subtractive hybridization and differential screening. RAP-PCR offers numerous advantages over these methods, including its simplicity and its ability to compare the fluctuations in gene expression between multiple samples simultaneously using only nanograms of RNA (McClelland et al. 1995). In addition, RAP-PCR can yield information on the overall patterns of gene expression between different cell types or between different physiological conditions of the same cell type.

RAP-PCR has many applications including the identification of developmentally regulated genes, the detection of genes that respond to various hormones or growth factors and the identification of tissue specificity of gene expression (Wong and McClelland 1994; Watson and Fleming 1994; Ralph et al. 1993). Applications of RNA finger-printing to tumors have included the identification of genes differentially expressed between normal and tumor cells in mammary epithelium and ovarian epithelium (Wong et al. 1993).

2
Principles

Strategy

In RAP-PCR, first-strand synthesis is initiated from an arbitrarily chosen primer at those sites in the RNA that best match the primer. DNA synthesis from these priming sites is performed by reverse transcription. Second-strand synthesis is achieved by adding *Taq* polymerase and the appropriate buffer to the reaction mixture. The consequence of these two enzymatic steps is the construction of a collection of molecules that are flanked at their 3' and 5' ends by the exact sequence (and complement) of the arbitrary primer. These serve as templates for PCR amplification, which can be performed in the presence of radioactive label such that the products can be displayed on a sequencing-type polyacrylamide gel (Liang and Pardee 1992; Wong et al. 1993;Welsh et al. 1992). Those PCR products representing genes that are regulated can be cloned from the gel and sequenced. Because priming is arbitrary in the first step, the lack of poly(A) tails in most bacterial RNAs is also not a problem for this method (Wong and McClelland 1994). A similar method has been developed by Liang and Pardee (1992). In this case, first-strand cDNA is made by using oligo(dT) priming from the poly(A) tail of mRNAs.

Choice of Primers

Primers are chosen with several criteria in mind. The primers should not have stable secondary structure. The sequence should be chosen such that the 3' end is not complementary to any other sequence in the primer. In particular, palindromes should be avoided. Primers from 10–20 nucleotides or more in length can be used. Primers can be obtained in kits from several companies such as Genosys (Woodlands, TX) and Operon Technologies (Alameda, CA). Longer primers have some small theoretical advantages. In particular, longer primers can be used at fairly high stringency such that any DNA contamination that persists will not contribute to the fingerprint (Welsh et al. 1992). Longer primers can contain more sequence information designed to be exploited in subsequent steps in the experiment, such as cloning and sequencing, but otherwise have no apparent advantages over 10-mers. Longer primers generate a larger number of visible fragments than do 10-mers. However, truncated forms of *Taq* polymerase such as Stoffel fragment (Perkin-Elmer, Branchburg, NJ) allows 10-mers to generate almost as many products as 20-mers produce using AmpliTaq. Also, the *Taq* Stoffel fragment gives better results with a greater fraction of arbitrarily selected 10-mers than does AmpliTaq (Sobral and Honeycutt 1993).

One could search for differentially expressed genes using motifs directed towards particular gene families of relevance to a particular biological phenomenon (see Chap. XXXVI) such as part of the Ser/Thr kinase motif (McClelland et al. 1994) or in favor of zinc fingers (Ralph et al. 1993; Stone and Wharton 1994).

Primers can be used in pairwise combination so, for example, 55 fingerprints can be generated using ten primers in all single and pairwise combinations (Welsh and McClelland 1991). Thus, using a 96-well format PCR machine and 100-well gels, it is possible to examine about 625 messages in pairwise comparison in a single experiment (25 bands/ lane×100 wells/2 concentrations/2 RNA types). One reason to use a pair of 10-mer primers is that the number of visible products in the fingerprints is increased relative to using one 10-mer primer. Furthermore, if only one of the primers is used for reverse transcription, those products that contain the first and second primer are likely to be oriented with the first primer at the 3' end and the second primer at the 5' end, thereby providing information about the sense strand of the original RNA. More than 50 % of products in reactions using pairwise combinations of primers contain both primers and are distinct from those generated using either primer individually (Welsh and McClelland 1991). In principle, when primers are used in pairwise combinations, those products that contain a different primer at each end can be sequenced directly with one of the primers (see Chap. XIX).

Some primers work better than others. It is therefore necessary to either: (1) empirically determine the optimum concentration for each primer or (2) screen primers for those that work best at a single concentration. For genetic mapping, in which a primer is to be used only once, screening many primers at a standard concentration seems to be the most efficient approach. It is usually a good idea to screen several primers and use those that give the most qualitatively distinct patterns for further work.

3
Materials

Equipment
- multichannel micropipette
- tissumizer (Tekmar)
- Beckman SW 50.17 rotors
- 96-well thermocycler (Perkin-Elmer 9600 model)
- sequencing gel electrophoresis apparatus
- gel dryer

Supplies
- Beckman polyallomer centrifuge tubes
- 5 ml syringe with 23-gauge needle
- PCR tubes (Microamp, Perkin-Elmer)
- X-ray films and exposure cassettes

Reagents and solutions
- guanidinium solution: 4 M guanidinium thiocyanate, 25 mM sodium citrate (pH 7), 0.5 % sarcosyl, 0.1 M 2-mercaptoethanol
- 5.7 M CsCl, 10 mM EDTA (pH 7.5)
- TE solution: 10 mM Tris-HCl, 1 mM EDTA (pH 7.4)
- 10 % SDS
- Phenol:chloroform:isoamyl alcohol (25:24:1)
- Chloroform:isoamyl alcohol mixture (24:1)
- 3 M sodium acetate (pH 5.2)
- 100 % ethanol
- 10 or 18-mer arbitrary primer (Genosys, Woodlands, TX or Operon Technologies, Alameda, CA)
- DNase I treatment mixture: 20 mM Tris-HCl (pH 7.5), 20 mM $MgCl_2$, 40 U/ml of Rnase-free DNase I (Boehringer Mannheim Biochemicals, Indianapolis, IN), 240 U/ml of RNase Block (Stratagene, CA).
- First-strand reaction mixture (for basic protocol, described in Sect. 4.2): 100 mM Tris (pH 8.3), 100 mM KCl, 8 mM $MgCl_2$, 20 mM DTT, 0.4 mM each dNTP, 4 µM of the first 10-mer primer, 4 U/µl Rnase block and 3.75 U/µl MuLV reverse transcriptase (RT).
- Second-strand reaction mixture (for basic protocol, see Sect. 4.2): 20 mM Tris (pH 8.3), 20 mM KCl, 8 mM $MgCl_2$, 0.4 mM each dNTP, 8 µM second 10-mer primer, 0.4 U/µl *Taq* polymerase Stoffel fragment (Perkin-Elmer, Branchburg, NJ), 0.2 Ci/µl of $[\alpha\text{-}^{32}P]dCTP$.
- first-strand reaction mixture (for alternate protocol, described in Sect. 4.3): 200 mM Tris-HCl (pH 8.3), 200 mM KCl, 16 mM $MgCl_2$, 40 mM DTT, 0.8 mM each dNTP, 40 µM primer, and 2 U/µl MuLV reverse transcriptase (RT).
- second-strand reaction mixture (for alternate protocol, see Sect. 4.3): 20 mM Tris-HCl (pH 8.3), 50 mM KCl, 4 mM $MgCl_2$, 0.2 Ci/µl

of [α-^{32}P]dCTP and 0.2 U/µl of *Taq* DNA polymerase (AmpliTaq, Perkin-Elmer Cetus). Addition of a second primer in this step is optional.
- 10x TBE: 0.89 M Tris-borate, 0.25 M EDTA (pH 8.3)
- 5 % polyacrylamide-50 % urea gel in 1x TBE buffer
- denaturing loading buffer: 95 % formamide, 0.1 % bromphenol blue, 0.1 % xylene cyanol, 10 mM EDTA

Note: It is important to control sources of contamination when performing RAP-PCR. These include both ribonuclease and nucleic acid contamination. Water used in RNA extraction solutions should be autoclaved and sterile technique should be adopted. We therefore routinely aliquot all our solutions (including small aliquots of water) and use aerosol filter tips, gloves, filter-sterilized water and sterile tubes.

4
Experimental Procedure

4.1
Extraction and Purification of RNA:
Guanidinium Thiocyanate – Cesium Chloride Centrifugation

1a. For cultured cells, to each confluent monolayer of cells, add 3 ml of guanidinium thiocyanate buffer/10^7 cells. Recover viscous lysate by scraping dishes with a policeman, remove using a 23-gauge needle and transfer to a clean tube.

1b. For tissue, add 3 ml denaturing solution/100 mg tissue and homogenize with a tissumizer at high speed until completely blended.

2. Add 1.2 g of CsCl to each 3 ml of homogenate.

3. Layer the homogenate onto a 1.2 ml cushion of 5.7 M CsCl, 10 mM EDTA (pH 7.5) in a 13×51 mm polyallomer ultracentrifuge tube.

4. Centrifuge 17 h at 32000 rpm in a Beckman SW 50.17 rotor, 20 °C (slow acceleration and deceleration).

5. Carefully remove supernatant using a Pasteur pipette, invert tube to drain, and pour remaining liquid off.

6. Resuspend the pellet in 200 µl of TE solution containing 0.1 % SDS by repeatedly drawing solution up and down in a pipet. Transfer to a clean microcentrifuge tube.

7. Add 200 µl phenol:chloroform:isoamyl alcohol (25:24:1) and shake vigorously. Centrifuge the suspension at 14000 rpm (4 °C) for 5 min.

8. Add 200 µl chloroform:isoamyl alcohol mixture (24:1) and shake vigorously. Centrifuge the suspension at 14000 rpm (4 °C) for 5 min. Transfer the upper phase to a clean microcentrifuge tube.

9. Add 1/10 vol/vol 3 M sodium acetate (pH 5.2), and 3 vol of 100 % ice-cold ethanol; precipitate overnight at –20 °C, and microcentrifuge for 10 min. Discard the supernatant.

10. Wash the pellet with 200 µl 100 % ice-cold ethanol, recentrifuge for 6 min at 14000 rpm. Discard the supernatant.

11. Resuspend RNA in 200 µl 70 % ethanol. Centrifuge for 8 min at 14000 rpm, and allow the pellet of nucleic acid to air dry.

12. Redissolve the RNA in a small volume of TE. Quantitate by diluting 10 µl to 1 ml of water and reading the A_{260} and A_{280}. Store the preparation at –80 °C.

13. Analyze equal aliquots of each RNA sample by electrophoresis on 1 % agarose gel and ethidium bromide staining in order to qualitatively compare large and small ribosomal RNAs.

4.2
RNA Fingerprinting by Arbitrarily Primed PCR Using a Pair of 10-mer Primers

RAP-PCR basic protocol

1. Prepare total RNA as described above.

2. Add 1 vol DNase I treatment mixture and incubate at 37 °C for 30 min. Phenol:chloroform extract and ethanol precipitate (steps 7–12 of Sect. 4.1).

3. Prepare treated RNA at three concentrations of about 50 ng/µl, 25 ng/µl and 12.5 ng/µl by dilution in water.

4. Add 5 µl first-strand reaction mixture to 5 µl RNA at each concentration. The reaction is ramped to 37 °C for 60 min over 5 min, held at 37 °C for 60 min, heated to 94 °C for 5 min to inactivate the reverse transcriptase, and cooled to 4 °C.

5. The reaction mixture is diluted with water to four times its original volume. Then, 10 µl of the diluted first-strand reaction is combined with 10 µl of second-strand reaction mixture, which contains the secondary primer. At this point, the reactions components are as follows: 16.25 mM Tris (pH 8.3), 16.25 mM KCl, 4.5 mM

$MgCl_2$, 0.23 mM dNTPs, 0.25 µM first primer, 4 µM second primer, and 0.2 U/µl *Taq* polymerase Stoffel fragment. Thermal cycling is performed using the following method: 94 °C for 1 min; 35 °C for 1 min; 72 °C for 2 min, for 45 cycles and 72 °C for 7 min.

6. Dilute 5 µl of each reaction in 10 µl of 95 % formamide; heat to 94 °C for 2 min and load 3 µl on a 5 % polyacrylamide-50 % urea sequencing gel Electrophorese in TBE at 55 Watts for 3 h. Dry the gel and perform autoradiography with Biomax film (Kodak).

4.3
RNA Fingerprinting by Arbitrarily Primed PCR Using Single 18-mer Primers

The protocol for RNA fingerprinting with 18 base primers is similar to that described for shorter primers; however, *Taq* polymerase is used instead of *Taq* Stoffel fragment.

Alternate protocol

1. RNA preparation was described above. Prepare treated RNA at three concentrations of 100 ng/µl, 50 ng/µl, 25 ng/µl by dilution in water.

2. Add 10 µl first-strand reaction mixture to 10 µl RNA at each concentration. The reaction is ramped to 37 °C for 5 min and held at that temperature for 60 min, followed by 94 °C for 5 min.

3. Add 20 µl of the second-strand reaction mixture to each first-strand synthesis reaction. Cycle once through 94 °C for 5 min; 40 °C for 5 min; 72 °C for 5 min; followed by 40 cycles through 94 °C for 1 min; 60 °C for 1 min; 72 °C for 2 min.

4. The reactions are analyzed on sequencing gels as described above (step 6 of basic protocol, Sect. 4.2).

Tips

- It is vital to control for the possibility that an observed difference between two RNA samples is due to slight differences in RNA quality or concentration. The simplest way to ensure the authenticity of a differentially amplified product is to fingerprint a titration of the RNA. Only products that occur at two or more concentrations in one sample and not at all in the other sample need to be considered.

- Another potential problem during RAP-PCR or differential display is genomic DNA contamination in the RNA preparation. While exhaustive purification steps may be performed, it may be impractical to remove all of the DNA, we therefore routinely treat the RNA with RNase-free DNase I (McClelland et al. 1994). A control in which the reverse transcriptase is left out must also be performed to be sure that all the RNA samples are largely DNA free. Although this control is always likely to display a few "sporadic" PCR products, these are not a problem if they are not reproducible. When reverse transcriptase is used, products from RNA will be more abundant than sporadic products from tiny amounts of contaminating DNA. Another method to detect contamination is to treat a sample of the RNA for RAP-PCR with RNase before amplification. Any bands generated must have arisen from contaminating genomic DNA present in the sample.

5
Results and Comments

An example of a RAP-PCR fingerprint generated using a pairwise sequential protocol and 10-mers is presented in Fig. 1. Here we have applied this method to identify genes differentially expressed in colorectal cancer using normal and tumorogenic tissues. Some of the genes that are differentially expressed during tumor progression in colon carcinoma may be useful as markers in the early detection of the disease. Note that some products are concentration-dependent and thus small differences in the quality or concentration of two templates can lead to spurious differences in the arbitrarily primed PCR pattern. Such problems are eliminated by the use of three concentrations, as shown in this picture. The fact that fingerprinting should be performed on at least two template concentrations differing by twofold is the single most important point that is routinely neglected by users of this technology.

When normal and tumorogenic tissues were compared, most bands were the same, but a few bands such as N1 were seen only in normal tissue, or T1 in tumor tissue. Patterns of amplified cDNAs

Fig. 1. RAP-PCR fingerprint of colorectal carcinomas using multiple primer sets. Total RNA from normal (*N*) and tumor (*T*) were prepared as described in the text, and with sequential application of two 10-mer primers. Each RNA was fingerprinted at three RNA concentrations: 500 ng (*lanes 1*), 250 ng (*lanes 2*) and 125 ng (*lanes 3*) per 10 µl reverse transcription reaction. The fingerprints were resolved on a 5% polyacrylamide-50% urea sequencing gel with electrophoresis in 1x TBE at 55 W for 3 h and visualized by autoradiography. Numbers at *left* are molecular weight markers (in nucleotides). The fingerprints were obtained using the arbitrary primer OPN24 5'-AGGGGCACCA-3' for the reverse transcription followed by addition of kinaseA1+ 5'-GAGGGTGCCTT-3' or kinaseA2+ 5'-GGTGCCTTTGG-3' before the second step

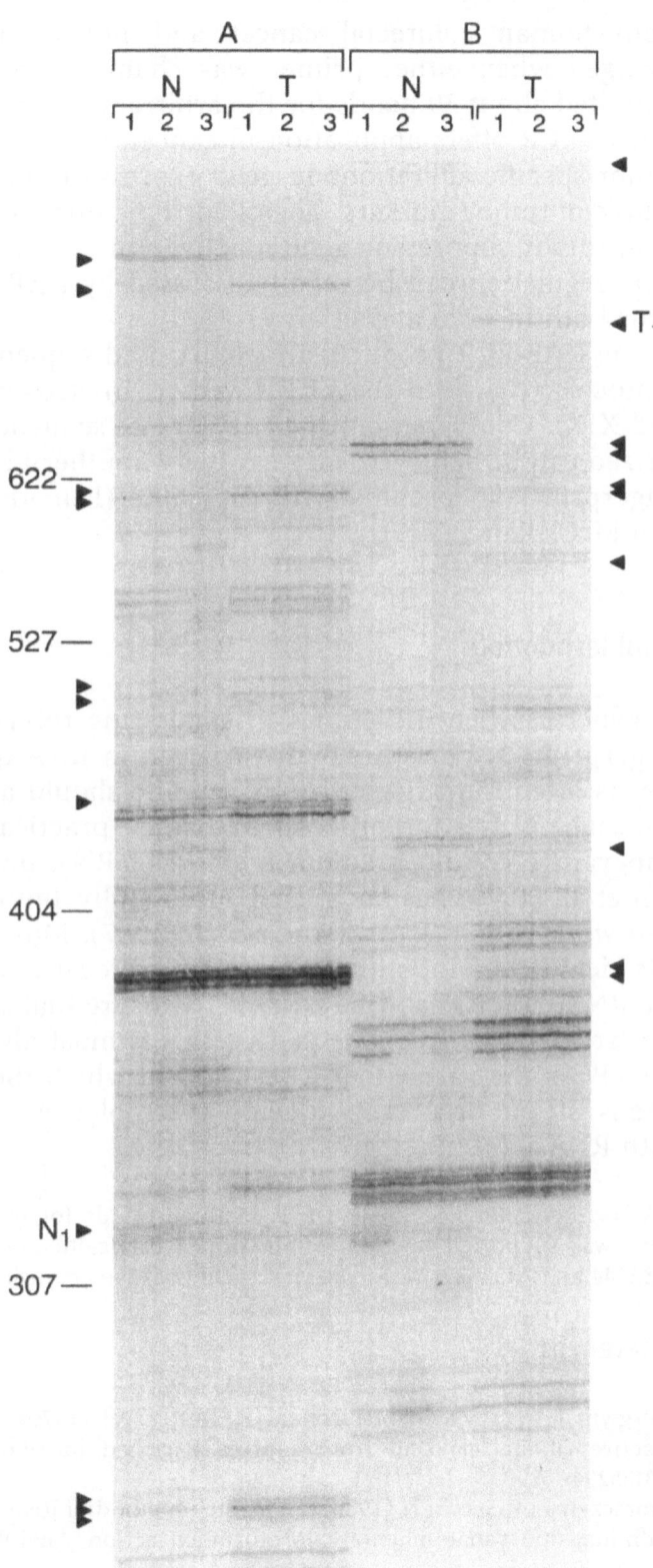

from human colorectal cancer and normal tissue were totally changed when either primer was changed. In general, each lane exhibited about 30 bands for the arbitrary 10-mer primers.

RAP-PCR offers application in cancer research in the detection of tumor-specific alterations in gene expression, providing a bountiful source of tumor markers. The pleiotropic impact of oncogene activation, tumor suppressor gene inactivation and mutator mutations in gene regulation can be readily assessed by RAP-PCR in model systems both in vitro and in vivo.

The RAP-PCR products are cloned and sequenced using standard protocols (e.g., into the pCRII vector, Invitrogen; see Chaps. XVIII and XIX). Confirmation that the correct sequence has been characterized can be achieved very easily by Southern blotting the original fingerprint and probing with the clone (Peinado et al. 1992; Wong and McClelland 1991).

6
Troubleshooting

In general, when difficulties in generating robust and reproducible fingerprints are encountered, variation in RNA quality among samples is often the culprit. Therefore, one should always use the most rigorous purification procedures that are practical. The guanidinium thiocyanate-cesium gradient method of RNA purification of Chirgwin et al. (1979) has been used successfully, but other methods may also work (Chomczynski and Sacchi 1987). Most purification methods yield RNA contaminated with genomic DNA so we routinely treat the RNA with RNase-free DNase. To ensure that all the RNA samples are largely DNA-free, adequate controls must always be included in RAP-PCR; they are either a reaction in which the reverse transcriptase is left out or a reaction using as template an RNA sample treated with RNAse (see Sect. 4.2).

Acknowledgements. We thank Dr. F. Mathieu-Daude for technical comments. This work was supported by National Institutes of Health grant CA 68822, NS33377, AI32644 and AI34829. M. Ricote was a fellow of the Spanish Ministry of Education.

References

Chirgwin J, Przybyla A, MacDonald R, Rutter WJ (1979) Isolation of biologically active ribonucleic acid from sources enriched in ribonuclease. Biochemistry 18:5294–5299
Chomczynski P, Sacchi N (1987) Single-step method of RNA isolation by acid guanidinium thiocyanate-phenol-chloroform extraction. Anal Biochem 162:156–159

Liang P, Pardee AB (1992) Differential display of eukaryotic messenger RNA by means of the polymerase chain reaction. Science 257:967–971

McClelland M, Ralph D, Cheng R, Welsh J (1994) Interactions among regulators of RNA abundance characterized using RNA fingerprinting by arbitrarily primed PCR. Nucleic Acids Res 22:4419–4431

McClelland M, Mathieu-Daude F, Welsh J (1995) RNA fingerprinting and Differential display by arbitrarily primed PCR. Trends Genet 11:242–246

Peinado MA, Malkhosyan S, Velazquez A, Perucho M (1992) Isolation and characterization of allelic losses and gains in colorectal tumors by arbitrarily primed polymerase chain reaction. Proc Natl Acad Sci USA 89:10065–10069

Ralph D, McClelland M, Welsh J (1993) RNA fingerprinting using arbitrarily PCR identifies differentially regulated RNAs in mink lung (My1Lu) cells growth arrested by transforming growth factor 1. Proc Natl Acad Sci USA 90:10710–10714

Sobral BWS, Honeycutt RJ (1993) High output genetic mapping of polyploids using PCR-generated markers. Theor Appl Genet 86:105–112

Stone B, Wharton W (1994) Targeted RNA fingerprinting: the cloning of differentially-expressed cDNA fragments enriched for members of the zinc finger gene family. Nucleic Acid Res 22:2612–2618

Watson MA, Fleming TP (1994) Isolation of Differentially Expressed Sequence Tags from Human Breast Cancer. Cancer Res 54:4598–4602

Welsh J and McClelland M (1990) Fingerprinting genomes using PCR with arbitrary primers. Nucleic Acids Res 18:7213–7218

Welsh J, McClelland M (1991) Genomic fingerprinting using arbitrarily primed PCR and matrix of pairwise combinations of primers. Nucleic Acids Res 19:5275–5279

Welsh J, Chada K, Dalal SS, Cheng R, Ralph D, McClelland M (1992) Arbitrarily primed PCR fingerprinting of RNA. Nucleic Acids Res 20:4965–4970

Williams JG, Kubelik AR, Livak KJ, Rafalski JA and Tingey SV (1990) DNA polymorphisms amplified by arbitrary primers are useful as genetic markers. Nucleic Acids Res 18:6531–6535

Wong KK, McClelland M (1992) A *BlnI* restriction map of *Salmonella typhimurium*. J Bacteriol 174:1656–1661

Wong K-K, Mok C-H., Welsh JT, McClelland M, Tsao S-W, Berkowitz RS (1993) Identification of differentially expressed RNA in human ovarian carcinoma cells by arbitrarily primed PCR fingerprinting of total RNAs. Int J Oncol 3:13–17

Wong KK, McClelland M (1994) Stress-inducible gene of *Salmonella typhymurium* identified by arbitrarily primed PCR of RNA. Proc Natl Acad Sci USA 91:639–643

Chapter XXVIII

Nonradioactive Differential Display of Messenger RNA

T.C.G. Bosch and J. Lohmann

1
Introduction

Identification and characterization of differentially expressed genes is required in many fields of modern biology and medicine. Therefore, efficient methods have been developed for identification and isolation of differentially expressed genes using molecular tools. Differential display of mRNA by PCR (DD-PCR) is a new and powerful procedure for quantitative detection of differentially expressed genes (Liang and Pardee 1992; McClelland et al. 1995). Advantages over alternative procedures include: (1) simultaneous display of all differences in gene expression between different cell fractions; (2) simultaneous detection of both up-regulation and down-regulation of genes; (3) requirement of only small amount of sample tissue; and (4) drastically reduced time of analysis. The DD-PCR procedure is based on reverse transcription of mRNA from different cell fractions using an oligo-dT-NN anchor primer followed by PCR amplification of cDNA using a set of short arbitrary primers and radiolabeled dNTP. After electrophoretic separation of the resulting fragments on sequencing gels, differential gene expression is visualized by autoradiography. Differentially expressed cDNA species can be recovered from the gel using the autoradiogram for band localization. Although straightforward, a major problem associated with the original procedure is that recovering a unique radiolabeled DNA species from the polyacrylamide gel is technically challenging and often results in failure to isolate and clone a particular cDNA (Liang et al. 1992; Bauer et al. 1993; Li et al. 1994). Here we describe a simple and nonradioactive method of differential display of mRNA that allows both rapid screening of a large number of samples for differentially expressed genes and efficient isolation of unique cDNA species. In this procedure, which we call REN-display (Lohmann et al. 1995), mRNA is isolated from two or more cell populations, reverse transcribed and PCR amplified using

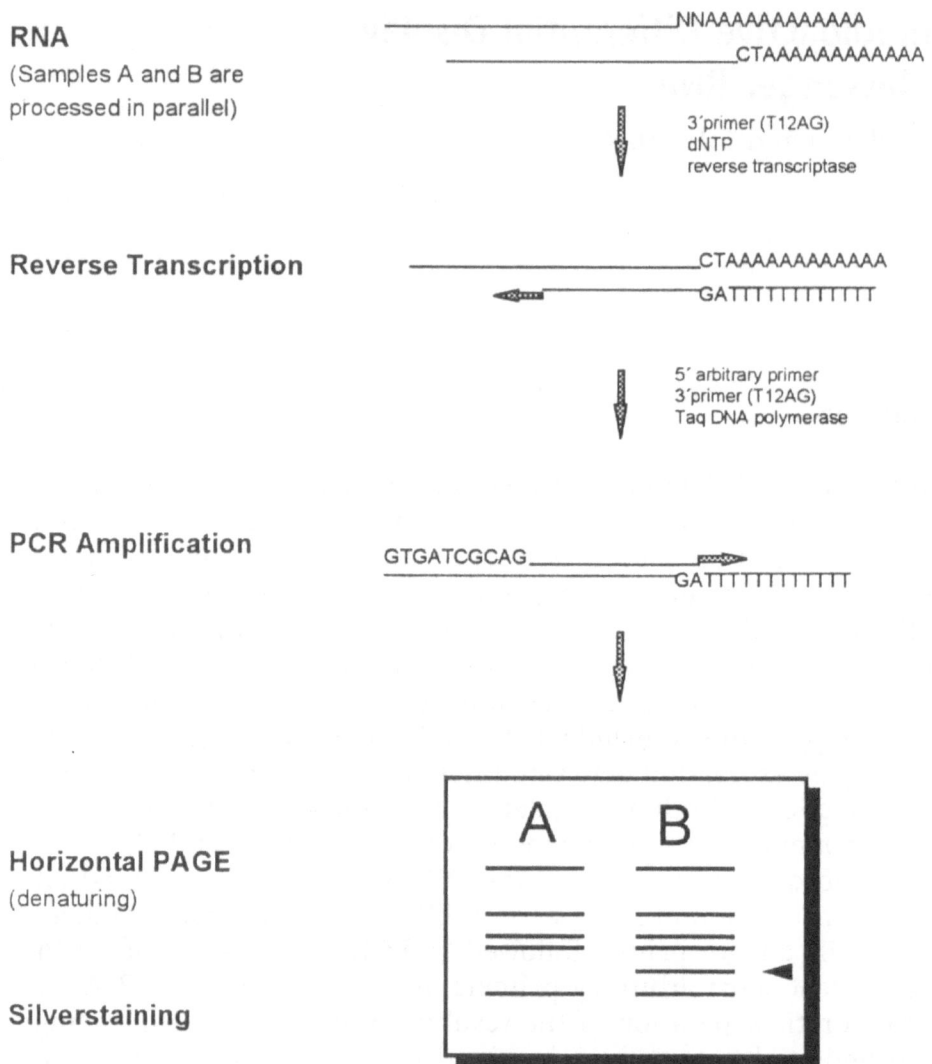

RNA
(Samples A and B are
processed in parallel)

——————————————NNAAAAAAAAAAAA
——————————————CTAAAAAAAAAAAA

3´primer (T12AG)
dNTP
reverse transcriptase

Reverse Transcription

————————————CTAAAAAAAAAAAA
—————————GATTTTTTTTTTTT

5´ arbitrary primer
3´primer (T12AG)
Taq DNA polymerase

PCR Amplification

GTGATCGCAG——————————
——————GATTTTTTTTTTTT

Horizontal PAGE
(denaturing)

A   B

Silverstaining

Fig. 1. Outline of the REN-display procedure

one 3' anchored primer and one 5' arbitrary primer. The PCR products are then resolved in adjacent lanes on horizontal polyacrylamide gels under denaturing conditions. cDNA bands are visualized by silver staining. Differentially expressed mRNA species are identified by comparing the pattern of silver stained cDNAs in adjacent lanes. Differentially expressed mRNA species then can be directly isolated from the gel and further characterized. Due to the visibility of the cDNA in the original gel, the probability of recovering a PCR product of interest is increased considerably compared to the standard radioactive procedure. The outline of the nonradioactive differential display method is shown in Fig. 1. Nonradioactive differential display is particularly valuable when applied to complex biological systems and developmental and physiological states, since it requires little biological material, allows rapid (72 h) investigation of a large number of samples, and can be carried out even in routine diagnostic laboratories.

2
Materials

- thermocycler
- Multiphor II electrophoresis unit (Pharmacia Biotech)
- Gelpool/Paperpool (PharmaciaBiotech)

Equipment

- QuickPrep mRNA purification kit (Pharmacia Biotech)
- first-strand cDNA synthesis kit (Pharmacia Biotech)

Kits

- tailing primers (e.g. 5'-$T_{(12)}$AG-3')
- arbitrary 10-mer primers (Operon, Alameda, CA)
- 100 µM dNTPs
- *Taq* DNA polymerase
- mineral oil
- sequencing stop solution: 95 % formamide, 20 mM Na_2EDTA, 0.05 % (w/v) bromphenol blue, 0.05 % (w/v) xylene cyanol
- CleanGel 15 %/48S (ETC Electrophorese Technik, Kirchentellinsfurt, Germany)
- DNA disc buffer system (ETC Electrophorese Technik, Kirchentellinsfurt, Germany): gel buffer (add 7 mol/l urea), electrode buffer
- fixing solution: 10 % acetic acid (v/v)
- staining solution: 0.1 % $AgNO_3$ (w/v), 250 µl 37 % formaldehyde
- developing solution: ice-cold 2.5 % Na_2CO_3, 250 µl 37 % formaldehyde, 250 µl 2 % sodium thiosulfate solution
- stop solution: 2 % glycine, 0.5 % EDTA-disodium
- impregnation solution: 5 % glycerol (v/v)

Reagents
and solutions

Note: All silver staining solutions (Bassam et al. 1991) are prepared in a final volume of 250 ml using ultrapure water.

3
Experimental Procedure

All solutions are prepared with water that has been treated with 0.1 % diethyl pyrocarbonate.

mRNA
preparation

1. Isolate mRNA or total RNA from about 5×10^5 cells using, for example, the Pharmacia QuickPrep mRNA purification kit and following the manufacturer's instructions. Final volume: 20 µl.

2. Calculate concentration of RNA by removing an aliquot (2 µl) of RNA suspension and determining the A_{260} in 100 µl water. Yield should be approximately 1 µg mRNA. The samples can be stored at −20 °C at this point.

Reverse transcription/ cDNA synthesis

Either 250 ng of mRNA or 3 µg of total RNA is reverse transcribed using the first-strand cDNA synthesis kit (Pharmacia) following the manufacturer's instructions:

1. Mix 5 µl bulk reaction mix, 1 µl DDT, 1.25 µl of 25 µM stock of tailing primer (e.g., 5'-T$_{(12)}$AG-3').

2. Heat 7.75 µl of RNA solution (250 ng of mRNA or 3 µg of total RNA) at 65 °C for 10 min. Thereafter place on ice.

3. Add 7.75 µl of mRNA solution to reaction mix. Final volume: 15 µl.

4. Reaction time is 1 h at 37 °C.

5. Stop reaction by heating for 5 min at 90 °C.

Polymerase chain reaction (PCR)

Dilute the resulting cDNA mixture 1:50 with water. Use 3 µl of the dilution as template for a 10 µl PCR reaction: (concentrations are of stock solutions)

1. Mix 3 µl cDNA (1:50), 1 µl 10x PCR buffer, 1 µl random 10-mer primer (5 µM), 1 µl tailing primer (25 µM), 2 µl dNTP (100 µM), 0.1 µl Taq (0.5 U) , 1.9 µl H$_2$O.

2. Cover reaction with 20 µl mineral oil.

3. PCR parameters: initial denaturing step for 5 min at 94 °C, followed by 43 cycles with cycle times of 30 s at 94 °C, 60 s at 42 °C, and 30 s at 72 °C.

4. Stop reaction by adding 7 µl sequencing stop solution to 10 µl reaction mix. The samples can be stored at −20 °C at this point.

Horizontal polyacrylamide gel electrophoresis

1. Hydrate CleanGels (ETC) in 40 ml gel buffer (containing 7 M urea) for at least 1 h on a shaker using Gelpool.

2. Cut two pieces of electrode wicks lengthwise in half and soak them in 40 ml electrode buffer using Paperpool.

3. Apply 2 ml kerosene oil to center of Multiphor II cooling plate.

4. Remove excess buffer from the gel using filter paper.

5. Lay hydrated gel on the plate.

6. Lay electrode wicks on hydrated gel.

7. Boil samples for 4 min. Chill samples on ice before loading.

8. Load 8 µl of sample solution in wells. Store remaining solution at $-20\,°C$.

9. Running conditions: Run at room temperature ($22\,°C$). Start at $400\,V_{max}$ and $11\,mA_{max}$ for 30 min. Thereafter run at $18\,mA_{max}$ for 2 h.

10. Stop electrophoresis when xylene cyanol band reaches the anode wicks.

Visualization of PCR products by silver staining

1. Fix for 30 min in 250 ml fixing solution.

2. Wash 3×2 min in 250 ml water.

3. Stain for 30 min in 250 ml staining solution.

4. Rinse gel for 30 s in water.

5. Develop for 5–15 min in 250 ml ice-cold developing solution.

6. Stop reaction by incubating for 10 min in 250 ml stop solution.

7. Impregnate by incubating for 10 min in 250 ml impregnation solution.

Elution of candidate bands

1. Transfer silver stained cDNA directly from the gel into a PCR tube using two sterile pipette tips.

2. Add 3 µl water to the isolated polyacrylamide slice.

3. Break up the gel slice and vortex briefly.

4. Centrifuge for 30 s and boil for 2 min to aid elution.

5. Vortex and centrifuge again.

6. Prepare three sterile PCR tubes containing 2 µl, 2.7 µl and 3 µl of water, respectively.

7. Add elution mix to the tubes to reach a final volume of 3 µl (in the last 3 µl tube just dip used pipette tip into the water).

8. Add to the tube containing the gel slice 1.5 µl of water and use it as your last dilution sample.

Reamplification

From the elution mixtures, use the total 3 µl of each dilution as template for a 10 µl PCR reaction:

1. Mix 3 µl eluted PCR product, 1 µl 10x PCR buffer, 1 µl random 10-mer primer (5 µM), 1 µl tailing primer (25 µM), 0.2 µl dNTP (**10 µM**), 0.1 µl Taq (0.5 U), 3.7 µl H_2O.

2. Cover reaction with 20 µl mineral oil.

3. PCR condition: initial denaturing step 5 min at 94 °C, 43 cycles of PCR with cycle times of 30 s at 94 °C, 60 s at 42 °C and 30s at 72 °C.

Confirmation of reamplification

1. For control of successful reamplification of the desired fragment, transfer 6 µl of the PCR reamplification mixture (not all!) into a new tube.

2. Freeze the remaining 4 µl of reamplification products at –20 °C.

3. Add 4 µl sequencing stop solution to the 6 µl aliquot.

4. Load reamplified fragments next to original PCR products (stored in –20 °C). Proceed as described in "Horizontal polyacrylamide gel electrophoresis" and "Visualization of PCR products by silver staining".

5. If the PCR fragment of interest was successfully amplified, use an aliquot of the frozen PCR product (see step 2) for second reamplification and further characterization (e.g., cloning and sequencing).

High yield second reamplification

To obtain large amounts of PCR product, e.g., for cloning, it is useful to perform a second reamplification step.

1. Use the same concentrations as for first reamplification in a total volume of 100 µl.

2. As template use 2 µl of first reamplification product.

3. Purify PCR product by preparative agarose gel electrophoresis.

Tips

Critical for success in REN display experiments is:

● General: Avoid both ribonuclease and nucleic acid contamination. Adopt semi-sterile technique including wearing gloves and using water treated with 0.1 % diethylpyrocarbonate. Work quickly and keep all reagents on ice.

● PCR: Concentration of cDNA-mixture is essential for successful PCR. Use dilutions of at least 1:25 to avoid interference with dNTPs and primers from the first strand cDNA synthesis mixture.

- Silver staining of cDNA: use clean glassware for all incubations, use ice-cold developer. Also, note that 2 % sodium thiosulfate stock solution is only stable for about 1 week at 4 °C.

- Elution of candidate band: The size of the gel slice used for elution of the cDNA should be as small as possible since it may contain toxic substances which inhibit the reamplification procedure.

- Cloning: After reamplification and cloning (see Chap. XVIII) of the desired PCR fragment, screen multiple clones. Any band recovered from the gel can contain more than one cDNA fragment. Thus, in order to detect the clone encoding the differentially expressed transcript, multiple cloned fragments have to be screened and analyzed by northern blotting (a protocol for northern blot analysis is described in Chap. XXXI).

4
Results and Comments

By the standard differential hybridization procedure, we previously have detected and isolated two genes expressed specifically in cells from head but not from gastric region of the freshwater polyp *Hydra* (Weinziger et al. 1994; Lopez de Haro et al. 1994). In order to detect quantitative differences in gene expression between the two cell populations, we have used REN-display. The differential pattern of silver stained PCR products from head and gastric tissue is shown in Fig. 2. Poly(A)+ RNA was isolated and reverse transcribed using the 3' anchor primer $T_{(12)}AG$. PCR amplification of the reverse transcription reaction was carried out in duplicates using the 3' anchor primer and an arbitrary 10-mer 5' primer (Fig. 2A: 5'-GGGTAACGCC-3'; Fig. 2B: 5'-AGTCAGCCAC-3'). After 43 cycles of PCR, amplified fragments were loaded onto a CleanGel 15 %/48S and separated under denaturing conditions at max 18 mA for 2.5 h at 22 °C. The cDNA fragments were then visualized by silver staining. Figure 2 shows several differentially expressed transcripts that can easily be detected by primer A (arrowheads in Fig. 2A). By using primer B (Fig. 2B) no differences in the transcript pattern can be observed.

The method not only allows rapid identification of transcripts expressed exclusively in one cell population and not in the other cell population, but even allows detection of transcripts which are up- or down-regulated in one of the two cell populations. Sensitivity and resolution are nearly identical in the nonradioactive procedure when compared to the standard radioactive technique (Lohmann et al.

A

B

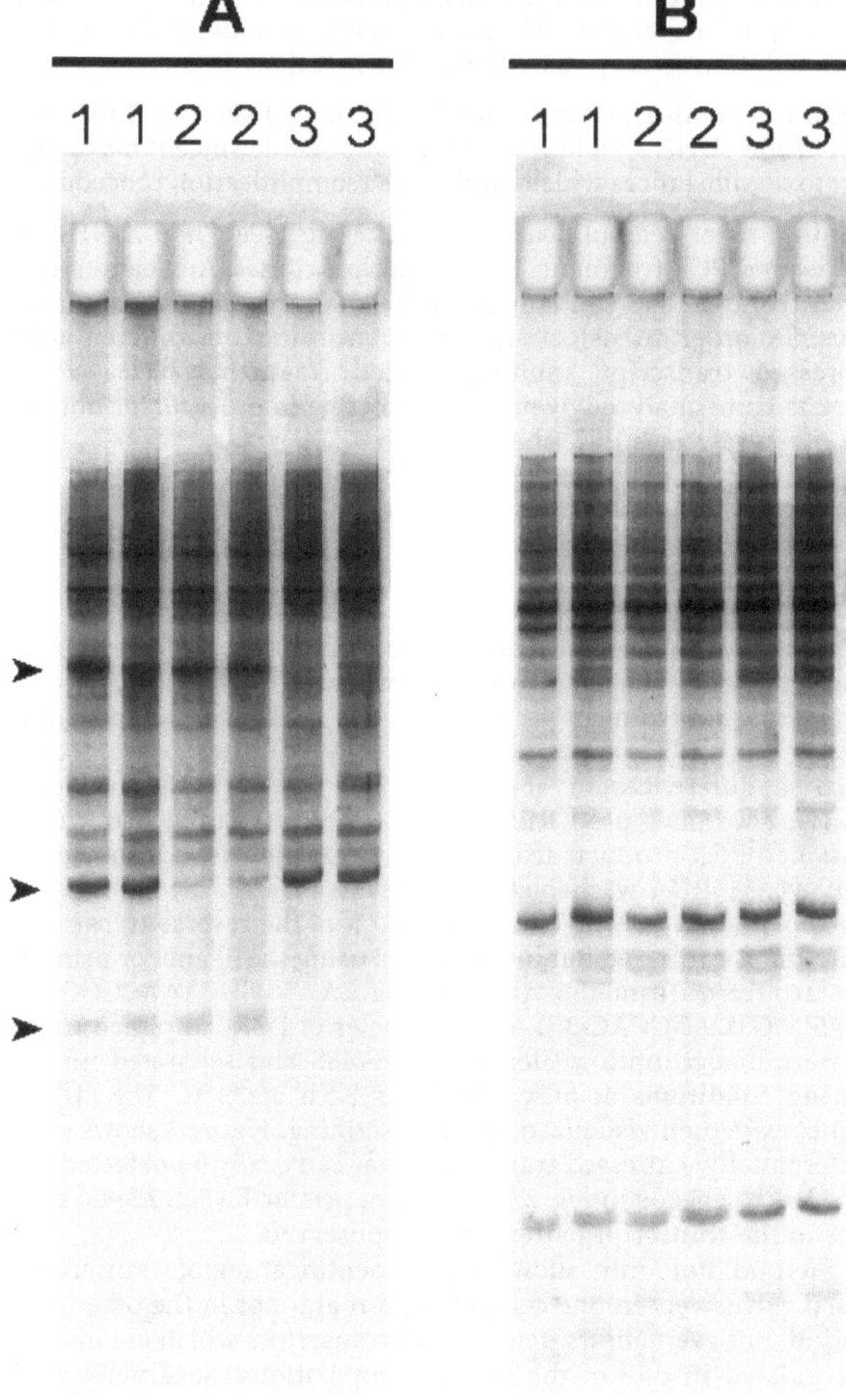

Fig. 2A,B. Differential display of hydra genes from head and gastric cells. *Lanes 1*, PCR products derived from mRNA of whole animals; *lanes 2*, PCR products derived from gastric cells; *lanes 3*, PCR products derived from head specific cells. *Arrowheads* identify differentially expressed genes. **A** Primer 5'-GGGTAACGCC-3'; **B** Primer 5'-AGTCAGCCAC-3' (see text for further details)

◄───

1995). Even low abundance PCR products (estimated to contain less than 1 ng of DNA) can be clearly detected after silver staining. Also, in contrast to alternative approaches, the complete procedure takes less than 3 days for isolation and reamplifaction of differentially expressed genes. The last and most essential step in the whole procedure is to confirm the differential expression pattern of a given PCR fragment by an independent RNA analysis technique. We and others (Bauer et al. 1993) have observed that the PCR product obtained after gel purification contains two or more independent sequences. Therefore, following reamplification and cloning (see Chap. XVIII), multiple cloned fragments have to be screened and analyzed by northern blotting (see Chap. XXXI) in order to detect the clone encoding the differentially expressed transcript. In our experience about one out of four clones contains the differentially expressed transcript of interest. In sum, the results obtained so far validate REN-display of mRNA as a method with high reproducibility and sensitivity which allows rapid identification of differences in the pattern of gene expression between two mRNA samples.

References

Bassam BJ, Caetano-Annolés G, Gresshoff PM (1991) Fast and sensitive silver staining of DNA in polyacryamide gels. Anal Biochem 196:80–83

Bauer D, Müller H, Reich J, Riedel H, Ahrenkiel V, Warthoe P, Strauss M (1993) Identification of differentially expressed mRNA species by an improved display technique (DDRT-PCR). Nucleic Acids Res 21:4272–4280

Li F, Barnathan ES, Karikó K (1994) Rapid method for screening and cloning cDNAs generated in differential mRNA display: application of Northern blot for affinity capturing of cDNAs. Nucleic Acids Res 22:1764–1765

Liang P, Pardee AB (1992) Differential display of eukaryotic messenger RNA by means of the polymerase chain reaction. Science 257:967–971

Liang P, Averboukh L, Keyomarsi K, Sager R, Pardee AB (1992) Differential display and cloning of messenger RNAs from human breast cancer versus mammary epithelial cells. Cancer Res 52:6966–6968

Lohmann J, Schickle HP, Bosch TCG (1995) REN, a rapid and efficient method for non-radioactive differential display and isolation of mRNA. BioTechniques 18(2):200–202

Lopez de Haro M, Salgado LM, David CN, Bosch TCG (1994) *Hydra* tropomyosin Trop1 is expressed in head specific epithelial cells and is a major component of the cytoskeletal structure which anchors nematocytes. J Cell Sci 107:1403–1411

McClelland M, Mathieu-Daude F, Welsh J (1995) RNA fingerprinting and differential display using arbitrarily primed PCR. Trends Genet 11:242–246

Weinziger R, Salgado LM, David CN, Bosch TCG (1994) *Ks*1, an epithelial cell specific gene, responds to early signals of head formation in *Hydra*. Development 120:2511–2517

Fluorescent Differential Display: A Fast and Safe Way for Reliable Differential Display Analysis

T. Ito and Y. Sakaki

1
Introduction

Differential display (DD), an arbitrarily primed RT-PCR fingerprinting method, provides a powerful tool to search for differentially expressed messages (Liang and Pardee 1992; see McClelland et al. 1995 for review). Unlike traditional methods based on subtractive and/or differential hybridization, DD, as well as its variants, can simultaneously compare more than two samples to reveal transcripts of various behaviors including both rare species and those with subtle changes. Although these features make DD quite attractive, it does have some drawbacks. With DD, one has to test a number of primer combinations to statistically cover the complex cellular RNA population. For instance, a simple calculation shows that ~450 primer combinations have to be tested to cover the ~15000 transcripts found in typical mammalian cells with 95 % probability, assuming each DD reaction detects 100 cDNA fragments. Therefore, it is clearly desireable to accelerate each step of DD, particularly, time-consuming and labor intensive ones such as post-run gel processing and signal detection. Furthermore, the safety problems inherent in radioactive DD methods are not negligible in a large scale experiment. Also, the high incidence of false signals reported in the literature is a serious problem.

To overcome these drawbacks, we developed two protocols (S and L) for efficient and reliable DD analysis using novel anchor primers and more standard PCR conditions (Ito et al. 1994), in contrast to the atypical ones used in the original protocol. Protocol S uses short arbitrary primers (10-mers) and modified 3'-anchored oligo-dT primers with much more improved priming efficiency than the original anchor primers. Protocol L is designed so that one can recruit longer primers, ~20 nucleotides (nt), that have been synthesized for individual purposes, as arbitrary primers for DD analysis. Both protocols ensure highly reproducible and reliable DD analysis. Moreover, they can be readily adapted to various fluorescence detection systems including automated DNA sequencers and image analyzers. These fluorescent DD (FDD) methods ensure unsurpassed throughput as well as improved operational safety with minimal operator interaction (Ito et al. 1994). Here we describe our current FDD protocols, one using a fluorescent image analyzer and the other using a simple fluorescent dye-staining method. We also refer to our procedure for correct clone selection, since this is another important step for successful DD cloning. An outline of the whole procedure is shown in Fig. 1.

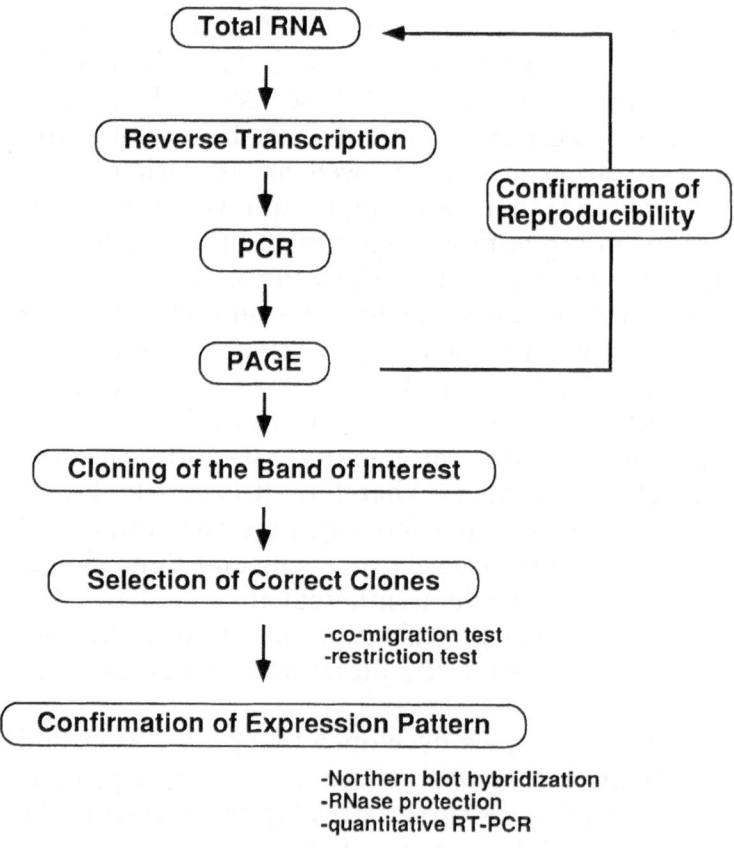

Fig. 1.   Flow-sheet of DD analysis

2
Materials

Equipment

- thermocycler
- sequencing gel apparatus
- Vistra FluorImager SI (Molecular Dynamics, CA)
- UV transilluminator
- photographic equipment with yellow gelatin filter

Kits

- preamplification kit (GIBCO-BRL, MA)
- 10-mer kit (Operon, Alameda, CA) (for protocol S): adjust to 10 µM

Reagents and solutions

- anchor primers (FITC-GT$_{15}$X (X=one of A, C or G) for protocol S, FITC-CCCGGATCCT$_{15}$X (X=one of A, C or G) for protocol L. Adjust to 50 µM)
- TE buffer: 10 mM Tris-HCl (pH 8.0), 1 mM EDTA
- GeneTaq and 10x GeneTaq buffer (Nippon Gene, Toyama, Japan)
- dNTPs mix (2.5 mM each)
- 1x TBE: 89 mM Tris, 89 mM boric acid, 2 mM EDTA
- denaturing acrylamide gel: 6 % acrylamide, 8.3 M urea, 1x TBE
- denaturing loading dye: 95 % formamide, 10 mM EDTA, 0.1 % bromphenol blue (BPB), 0.1 % xylene cyanol (XC)
- native acrylamide gel: 7 % acrylamide, 375 mM Tris-HCl (pH 8.9)
- Tris-glycine buffer: 25 mM Tris, 192 mM glycine
- native loading dye solution: 125 mM Tris-HCl (pH 6.8), 10 % glycerol, 0.1 % BPB
- SYBR green I (FMC BioProducts, Rockland, ME)
- ethidium bromide: stock solution 10 mg/ml

3
Experimental Procedure

Reverse transcription

1. Pipette 1 µl of total RNA solution (2.5 µg/µl), 1 µl of anchor primer and 8 µl of DEPC-treated water into each PCR tube. Incubate the tubes at 70 °C for 10 min.

   Note: Total RNA should be treated with RNase-free DNase to remove residual contamination of genomic DNA. The anchor primer should be synthesized and purified with special care to avoid RNase contamination.

2. During the incubation, prepare 2x reverse transcriptase (RT) solution as follows:

DEPC-treated water: 2.0×N µl
10x PCR buffer: 2.0×N µl
25 mM MgCl$_2$: 2.0×N µl
0.1 M DTT: 2.0×N µl
10 mM dNTP: 1.0×N µl
SuperScriptII: 1.0×N µl

3. Following the quenching and brief centrifugation, add 10.0 µl of 2x RT solution to each tube and mix thoroughly by gentle pipetting.

4. Place the tubes in a thermocycler programmed as follows: 25 °C for 10 min, 42 °C for 50 min, and 70 °C for 15 min.

5. Spin the tubes briefly. Add 80.0 µl of TE and mix well. Store at −20 °C as "cDNA solutions."

Polymerase chain reaction

1. Prepare the following PCR mix:

distilled water: 13.0×N µl
10x GeneTaq buffer: 2.0×N µl
dNTP mix (2.5 mM each): 1.6×N µl
anchor primer: 0.2×N µl
cDNA solution: 2.0×N µl

2. Dispense 1.0 µl of arbitrary primer solution (10 µM) to each tube or well of 96-well microtiter plate.

3. Add 0.2×N µl of GeneTaq to the PCR mix prepared as above, mix thoroughly and dispense 19.0 µl of PCR mix into each tube (or well). Overlay a drop of mineral oil, if necessary.

Note: GeneTaq is a recombinant Taq DNA polymerase largely deleted in its N terminal. It can amplify short fragments (~500 bp), but not longer ones, much more efficiently than usual full-length enzymes. We therefore routinely use 1 U of this enzyme for each FDD reaction. To improve the amplification of higher molecular weight species, use GeneTaq (0.5 U) in conjunction with full length Taq (0.5 U). Although DD can be performed with dNTPs carried over from the RT reaction, we routinely supplement dNTPs to the final concentration of 50–200 µM to ensure more efficient amplification.

4a. For Protocol S, carry out the following thermal cycling program:
1 cycle: 94 °C for 3 min, 40 °C for 5 min, 72 °C for 5 min
20–25 cycles: 95 °C for 15 s, 40 °C for 2 min, 72 °C for 1 min
72 °C for 5 min

4b. For protocol L, carry out the following thermal cycling program:
1 cycle: 94 °C for 3 min, 37 °C for 5 min, 72 °C for 5 min
20–25 cycles: 95 °C for 15 s, 55 °C for 1 min, 72 °C for 1 min
72 °C for 5 min

Gel electrophoresis protocol for FluorImager

1. Prepare a denaturing polyacrylamide gel (200×330×0.35 mm). Pre-run the gel at 1000 V for ~30 min.

2. Mix 5 µl of each PCR product with 2.5 µl of denaturing loading dye solution. Heat the tubes at 90 °C for 3 min and quench on ice. Load 6 µl of heat-denatured sample into each well.

3. Run the gel at 1500 V for ~1 h until the BPB migrates to the bottom of the gel.

4. Stop the gel and scan it without removing the glass plate. Use high-sensitivity mode and PMT voltage of 999 V. Put the gel back into the electrophoresis apparatus. Run for an additional hour until the XC reaches the bottom. Scan the gel again as above.

   Note: This two-step scanning procedure allows one to survey fragments of wider size ranges than conventional radioactive DD, in which one has to stop and dry the gel at some fixed time point, when shorter fragments might have already migrated out of the gel whereas the longer ones might be poorly separated near the top. With the wider scanning ability of FDD, no interesting bands will be missed.

5. Remove the upper glass plate, scan the gel and make an actual size print of the image.

   Note: One should confirm the reproducibility of the behavior of the target band before proceeding to excise the band.

6. Place the gel (glass plate side down) exactly onto the printed image. Excise the gel piece containing the band of interest with a blade.

7. Scan the gel again to confirm that the target band was precisely excised.

Gel electrophoresis protocol for native gel and SYBR green I staining

1. Prepare a native polyacrylamide gel (160×160×1 mm), attach the gel to the apparatus, and pour Tris-glycine running buffer. Do not pre-run the gel.

   Note: The discontinuous buffer system described here gives a much sharper pattern than does the conventional TBE buffer system.

2. Mix 5 µl of PCR product with the same volume of native loading dye solution and apply to the gel.

3. Run the gel until BPB migrates to the bottom. Current should be kept below 40 mA to avoid excess heating.

4. Remove the gel and stain with SYBR green I (1:10000 dilution in TE) for 10 min or so.

5. Place the gel on a UV (254 nm) transilluminator and take a picture using a yellow gelatin filter.

6. Excise the band of interest as quickly as possible. Rinse the gel piece with 1 µl of distilled water.

Reamplification and cloning of excised band

1. Place half of the gel piece into a PCR tube.

   Note: We have found that eluting the DNA fragments from the gel piece is unnecessary.

2. Add the following PCR mix to the tube:

   distilled water: 78.0 µl
   10x GeneTaq buffer: 10.0 µl
   dNTP mixture: 8.0 µl
   anchor primer: 0.5 µl
   arbitrary primer: 2.5 µl
   GeneTaq: 1.0 µl

3a. For Protocol S, carry out the following thermal cycling program:

   94 °C for 3 min
   20 cycles: 94 °C for 30 min, 40 °C for 2 min, 72 °C for 1 min
   72 °C for 5 min

3b. For protocol L, carry out the following thermal cycling program:

   94 °C for 3 min
   20 cycles: 94 °C for 30 min, 55 °C for 1 min, 72 °C for 1 min
   72 °C for 5 min

4. Run 5–10 µl of the reamplified product on native polyacrylamide gel. Stain the gel with ethidium bromide (0.5 µg/ml) to check the purity and amount.

   Note: Excess cycling should be avoided to suppress amplification of contaminating bands. Reduce the amount of thermal cycling if the purity of the reamplified product is poor.

5. Following ethanol precipitation, clone the fragments into a T-vector according to the conventional method (see Chap. XVIII).

1. Suspend each white colony in 40 µl of L-broth. Use 1 µl of the suspension for PCR. The condition is same as that for the reamplification step, except the volume is reduced to 20 µl.

Selection of correct clone

2. Run 0.1–1 µl of PCR product in parallel with the original FDD reaction product ("comigration test"). Select clones bearing inserts that precisely comigrate with the target band.

3. Prepare plasmid DNA from each candidate clone identified as above and determine its nucleotide sequence (see Chap. XIX).

4. Examine the nucleotide sequences of candidates to find appropriate restriction enzyme sites. Following digestion of the original FDD reaction and the amplified insert with the enzyme, run the digested products on a polyacrylamide gel to see whether the digestion pattern of the target band is similar to that of the candidate insert ("restriction test").

   Note: If one has failed to find appropriate enzyme sites, synthesize primers specific to the candidate clone and perform PCR using diluted FDD products as templates. Alternatively, this step can be done by differential screening of the transformants using cDNA or DD product as probe or by Southern hybridization between an FDD gel blot and probes generated from the candidate clones. In any case, one should prove that the target band is correctly cloned from the gel before proceeding to the next step.

5. Confirm the expression pattern of the corresponding transcripts by (semi-)quantitative RT-PCR, RNase protection (see Chap. XXXII) or northern blot hybridization (a protocol for northern blot analysis is described in Chap. XXXI).

   Note: Some short fragments, in particular, extremely AT-rich ones as well as those containing repetitive elements, are poor hybridization probes. Isolation of the upstream portions is necessary in such cases.

4
Results and Comments

Typical FDD patterns obtained by FluorImager are shown in Fig. 2. This procedure has high resolution and sensitivity, but needs special equipment. Alternatively, the simple dye-staining method described above provides a simple, albeit less sensitive, technique requiring no special equipment.

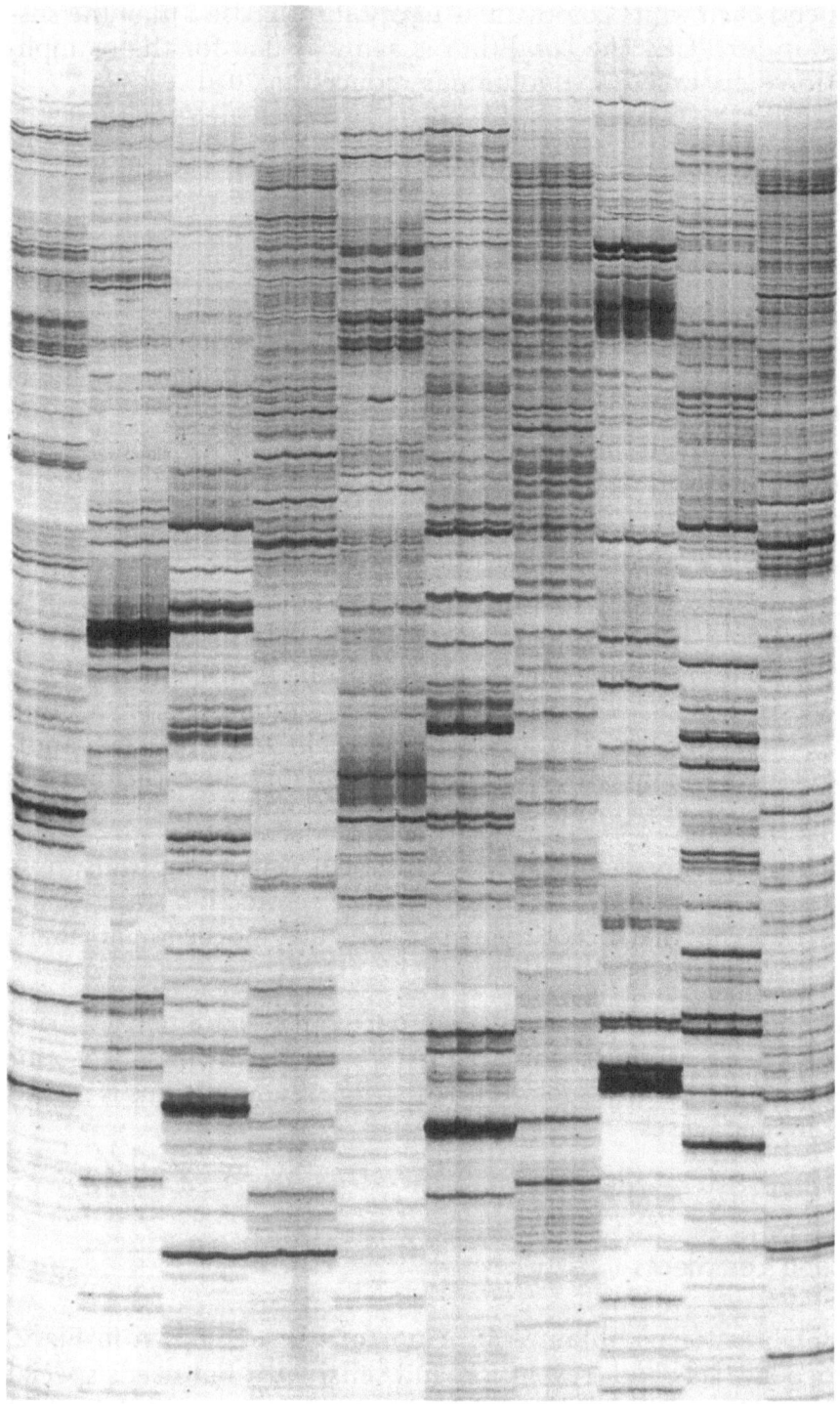

Fig. 2. An example of FDD pattern. A wild-type yeast and two gene disruptants were subjected to the FDD S protocol with ten different primer combinations. The image was obtained by FluorImager SI

For successful FDD cloning, one should pay particular attention to confirmation of the reproducibility of the target band and to selection of the correct clone. Regarding reproducibility, although reproducibility of the fingerprint with this protocol is much better than the original one, some faint bands do fluctuate from run to run. Thus we strongly recommend repeating the FDD using an independent batch of RNA samples, in particular, when pursuing those bands showing subtle changes. Regarding selection of the correct clone, one should always remember that the excised band is often contaminated with neighboring bands and sometimes contains cDNAs precisely comigrating with but distinct from the target species. Pay extra attention to avoid proceeding with incorrect clones.

We have used this protocol to clone more than 40 cDNAs from various sources. Subsequent analysis confirmed that the expression pattern of the corresponding transcripts had been faithfully reflected on FDD in all cases. We attribute the high success rate to the use of reliable protocols with the novel anchor primers under more standard PCR conditions as well as to the careful clone examination steps. We believe the procedure described here will thus provide useful hints for those performing DD with radioactive and nonradioactive detection systems.

References

Ito T, Kito K, Adati N, Mitsui Y, Hagiwara H, Sakaki Y (1994) Fluorescent differential display: Arbitrarly primed RT-PCR fingerprinting on an automated DNA sequencer. FEBS Lett 351:231–236

Liang P, Pardee AB (1992) Differential display of eukaryotic mRNA by means of polymerase chain reaction. Science 257:967–971

McClleland M, Mathieu-Daude F, Welsh J (1995) RNA fingerprinting and differential display using arbitrarily primed PCR. Trends Genet 11:242–246

Chapter XXX

Slot Blot Hybridization Screening

L. E. CHALIFOUR

1
Introduction

The differential display technique (Liang and Pardee 1992; Liang et al. 1994) is gaining acceptance as a useful method to identify the 3' ends of discordantly expressed mRNAs in comparisons between samples. Identification of differentially expressed amplified DNA bands in the initial experiment is only the first step in determining discordant expression. Differential expression must be confirmed by northern blots, RNase A protection or some other method. In general, we found that a large number of fragments were identified in the initial differential display analysis. When RNA from tissue or cells is difficult to acquire it is sometimes impractical to perform northern blots in large numbers. To circumvent this problem we developed a rapid method to confirm discordant expression (Chalifour et al. 1994; Mou et al. 1994). This method requires no more total RNA than that required in the original differential display; therefore, confirmation of discordant expression need not consume large amounts of precious sample. In this report we detail a slot blot hybridization method that uses RNA samples to generate radioactive probes followed by hybridization to reamplified gene fragments immobilized on duplicate membranes using a slot blot manifold. We found fewer gene fragments identified by the initial differential display were similarly discordantly expressed in the screening method. Thus, this rapid screening method significantly decreased the number of gene fragments that required the more laborious and expensive northern analyses.

An outline of the slot blot hybridization method is shown in Fig. 1.

1. Prepare radioactive DNA from RNA samples

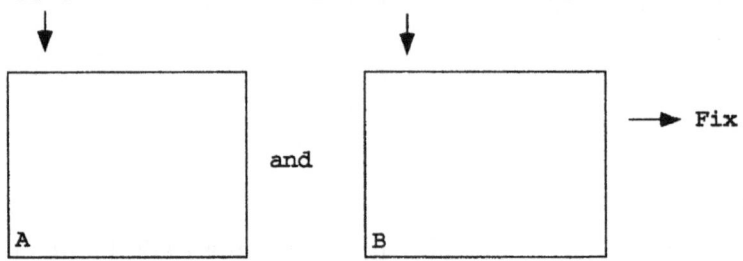

2. Prepare duplicate membranes

   Apply denatured re-amplified DNA to duplicate membranes

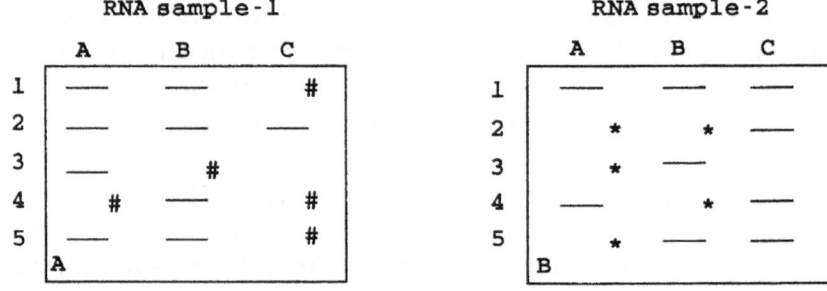

3. Hybridize the [^{32}P]-dsDNAs to the duplicate membranes

   Pre-hybridize ⟶ Hybridize ⟶ Wash and expose to X-ray film

4. Compare hybridization pattern

Fig. 1. Flow-sheet demonstrating the generation of radioactive probes from RNA, hybridization to duplicate membranes and comparison

2
Materials

Equipment
- slot blot manifold (Tyler Research Instruments, Edmonton, Alberta, Canada)
- water bath or hybridization oven (Bellco Glass Co)

Supplies
- Sephadex G50 spin column or any method designed to remove unincorporated radioactive nucleotides
- Gene Screen Plus (NEN/DuPont) or similar nylon membrane
- X-ray film and cassettes

- purified DNA-free total RNAs prepared from a separate experiment than that used for the initial differential display
- Superscript II reverse transcriptase (RNase H minus, GIBCO/ BRL)
- oligo(dT) primers (Pharmacia)
- random primers (pdN)$_6$ (Pharmacia)
- Klenow DNA polymerase
- [α^{32}P]dCTP (ICN, 3000 Ci/mM specific activity)
- reamplified gene fragments identified by differential screening
- TE: add 1 ml of 1 M Tris-HCl (pH 7.4) and 0.2 ml of 0.5 M EDTA (pH 8) and dilute to 100 ml with distilled water for a final concentration of 10 mM Tris and 1 mM EDTA. Autoclave and store at room temperature.
- 0.5 µg (100 nM) oligo(dT) per µl TE: dissolve vial contents in the appropriate amount of TE, aliquot in small amounts and store at −20 °C. Thaw quickly and keep on ice prior to use.
- 5x reverse transcriptase buffer: supplied by the manufacturer, thaw quickly and keep on ice.
- 0.1 M DTT: dissolve 0.772 g DTT in 5 ml distilled water and filter to sterilize. Do not autoclave.
- 10 mM dNTPs: purchase 100 mM stock solutions of all four dNTPs (Pharmacia). Add 10 µl each of dCTP, dATP, dGTP and dTTP to 60 µl of TE. Store frozen at −20 °C. Thaw quickly and keep on ice prior to use.
- 0.5 mM dNTPs: dilute 5 µl of 10 mM dNTPs to 100 µl with TE. Store frozen at −20 °C. Thaw quickly and keep on ice prior to use.
- [α^{32}P] dCTP: purchase radioactive label containing indicator dye.
- 4.2 µg random primers (pdN)$_6$ per µl TE (Pharmacia): dissolve vial contents in appropriate amount of TE, aliquot into Eppendorf tubes and store frozen at −20 °C. Thaw quickly and keep on ice prior to use.
- 10x Klenow buffer: supplied by the manufacturer. Store frozen at −20 °C. Thaw quickly and keep on ice prior to use.
- 0.4 M Tris-HCl (pH 7.5): dissolve 48.44 g Tris base in 900 ml water, adjust to pH 7.5; adjust to 1 l and autoclave. Store at room temperature.
- 0.5 M NaOH/1 M NaCl: dissolve 2 g NaOH and 5.86 g NaCl in 100 ml distilled water and autoclave. Store at room temperature.
- 20x SSC: 3 M NaCl/0.3 M sodium citrate (pH 7.0): dissolve 351 g NaCl and 176.4 g sodium citrate in 1750 ml distilled water, adjust to pH 7.0; adjust to 2 l and autoclave. Store at room temperature.
- 2x SSC: dilute 100 ml of 20x SSC to 1 l and autoclave. Store at room temperature.

- 2x SSC/1 % SDS: dilute 100 ml of 20x SSC to 800 ml water, add 10 g SDS, stir to dissolve and adjust to 1 l. Make the day of use.
- 0.1x SSC/0.125 N NaOH: add 0.5 ml of 20x SSC to 100 ml sterile water and add 0.5 g NaOH. Make the day of use.
- 0.5 M Tris HCl (pH 7.5)/0.5 M NaCl: Dissolve 60.55 g Tris base and 29.25 g NaCl in 900 ml water, adjust to pH 7.5, adjust to 1 l and autoclave. Store at room temp.
- prehybridization solution, 10 % dextran sulfate/1 M NaCl/1 % SDS: Calculate the amount of solution to make by measuring the length and width of the membrane in centimeters and inserting these measurements into the formula L×W×0.2 ml×number of blots. For 100 ml of solution add 10 g of dextran sulfate and 1 g SDS to 90 ml of water and put in a 62°–65 °C water bath until dissolved. This will take some time so start dissolving the buffer early in the morning if you want to use it in the afternoon. When the dextran sulfate is dissolved add 5.85 g NaCl, adjust so the total volume is 100 ml and put it back in the water bath. Keep it in the water bath until use. A membrane cut for a square 48 slot blot manifold requires about 10 ml of buffer.
- hybridization solution, 10 % dextran sulfate/1 M NaCl/1 % SDS/ 50 µg sheared herring sperm DNA/$1-2\times10^6$ cpm probe/ml solution. Calculate the amount of buffer to prepare by measuring the length and width of the membrane and inserting the numbers into the formula, L×W×0.05 ml×number of blots. For 25 ml of buffer start with 25 ml of prehybridization buffer and add 0.1 ml of 5 mg/ml denatured sheared herring sperm DNA. Add the appropriate volume of denatured radioactive probe so that there is $1-2\times10^6$ cpm/ml buffer. For a membrane cut for use in a square 48 slot blot manifold about 2.5 ml of buffer are sufficient.
- 5 mg/ml denatured, sheared herring or salmon sperm DNA: Add 200 mg salmon or herring sperm DNA (Sigma) to 40 ml TE in a 50 ml conical tube. Add 200 µl proteinase K (10 mg/ml) and incubate at 60 °C for 1–2 h. Shear through repeated passages, about 5–10 times, through a syringe and 18-gauge needle. Extract the proteins using phenol/chloroform (1:1) and then chloroform alone; ethanol precipitate the DNA. Collect the DNA and resuspend it in 40 ml TE. Place in a boiling water bath for 20–30 min to denature, then immediately put on ice. Store in small aliquots at –20 °C.

3
Experimental Procedure

Preparation of Radioactive DNA from RNA Samples

The method is based on production of a double stranded DNA probe using random primers and Klenow DNA polymerase with the first-strand cDNA as the substrate DNA.

1. Pipet 1–5 µg DNA-free total RNA from each sample to be compared into separate 0.5 ml Eppendorf tubes, add 1 µl oligo(dT) primer and bring the total volume to 11 µl with sterile water.

Prepare first-strand cDNA

2. Heat the RNA/oligo(dT) mixtures to 70 °C for 10 min then put them on ice for 5 min.

3. To each tube add:
   - 4 µl 5x reverse transcriptase buffer
   - 2 µl 0.1 M DTT
   - 1 µl 10 mM dNTP mix
   - 0.5 µl [α^{32}P]dCTP, 50 µCi

Mix by tapping, then put the tubes at 37 °C for 2–3 min.

4. Add 0.5 µl Superscript reverse transcriptase (diluted to 50 U/µl in the dilution buffer supplied by the manufacturer prior to use). Mix by tapping gently and put at 37 °C for 2 h.

5. After 2 h either use immediately or freeze at –70 °C for use later.

6. Remove 1 µl and dilute to 10 µl with TE. Add 1 µl of 10x gel loading buffer and electrophorese through a 1 % agarose gel with TBE as the running buffer. Use DNA markers, for example λ DNA cut with HindIII (Pharmacia), to act as size markers in a neighboring lane.

   Caution: Set aside a gel apparatus specifically for the electrophoresis of radioactive samples. Dispose of the radioactive running buffer and gel in the radioactive waste according to the rules and regulations of your institution. Monitor the area, your gloved hands, the gel box and everything that might have come in contact with the gel with a Geiger counter. Clean any contaminated equipment or the bench top immediately.

7. Wrap the gel in plastic wrap and take a photo of the gel.

8. Put the wrapped gel in a cassette and expose it to X-ray film. After 2 h or overnight, develop the film. A smear of radioactivity, labeled first-strand cDNAs, should be evident from 0.5 kb to 2 kb.

Random prime the first-strand reaction

9. Add 1 μl of the (pdN)₆ random primers to the first-strand reaction tube, heat to 90 °C for 10 min then immediately put on ice for 5 min.

10. Add:
    - 2.5 μl 0.5 mM dNTPs (dATP, dGTP, dTTP)
    - 2.5 μl Klenow DNA polymerase buffer (supplied by the manufacturer)
    - 5 μl [α³²P]dCTP
    - 1 μl Klenow DNA polymerase (5 U/μl)
    and incubate at 37 °C for 2–4 h or overnight at room temp.

11. Stop the reaction by the addition of 75 μl TE.

12. Separate the incorporated radioactivity from the unincorporated nucleotides using a Sephadex G50 spin column or some other method, or ethanol precipitate the DNA.

13. Denature the probe by either boiling or alkali.
    - Denaturation by boiling: place the tube in a floater, put the floater in boiling water for 5–10 min then immediately put the tubes on ice for 5 min.

    Caution: A drawback to the boiling method is that unless the top of the tube is tightly closed the tube can sometimes open, spill the contents and spray radioactivity. Therefore, when using the boiling method be sure to close the tubes tightly and boil the samples in a fume hood dedicated to radioactive usage.

    - Denaturation by brief treatment with alkali: add one-tenth volume of 1 N NaOH, mix gently by tapping and let sit at room temperature for 5 min. Neutralize by adding one-tenth volume of 1 N HCl and mix by tapping.

The denatured DNA can now be used immediately or frozen at –20 °C.

Preparation of Duplicate Membranes

The method involves the reamplification of the DNA fragments isolated from the differential display experiment and their immobilization onto membranes to be prepared during the PCR step.

Prepare membranes and assemble the slot blot manifold

The method described uses Gene Screen Plus membrane, but other nylon membranes are also suitable. Make a schematic diagram of the slot blot manifold and indicate where you intend to load the different DNA samples. Make certain that the orientation of the membrane is unambiguous.

1. Cut two membranes to a size slightly larger than the size of the slot blot manifold. The slightly larger size allows easier positioning and handling, and permits the use of small identification markers to be placed on the margins.
   - When cutting any membrane wear gloves and use a clean scalpel or a pair of scissors dedicated to this use.
   - Use a lead pencil to put letters or some other identification marks on the margins of the membranes so that the orientation of the membranes can be easily identified. Do not use marker pen as the ink often comes off in the pre-hybridization/hybridization and washing steps.

2. Cut a piece of filter paper, for example Whatman 3 M, and put it on the bottom of a plastic tray. Add enough 0.4 M Tris, pH 7.5, to wet the filter paper thoroughly then pour off any excess.
   - Put the membrane on top of the filter paper, using forceps to manipulate the membrane, and let it sit for 5 min. Do not allow any air bubbles underneath to remain.
   - Add enough 0.4 M Tris, pH 7.5, so the membrane can be submerged in the buffer and let it soak for 5 min. Make sure the entire membrane is wet and no air bubbles are underneath the membrane.

3. Make sure the manifold is clean and dry before assembly. Assemble the slot blot manifold according to the manufacturer's instructions. In general, this means placing a wet (wet the filter paper in 0.4 M Tris, pH 7.5) piece of Whatman 3 M paper on the slotted bottom, putting the membrane on top and placing the top of the manifold on the membrane. Clamp the assembly together tightly. Do not let the membrane dry.
   - Place the membrane in the manifold so the pencil identification marks are uppermost. The marks make it easier to later identify which side of the membrane harbors the immobilized DNA, and orient the membranes in the same direction prior to film exposure.

Prepare the reamplified DNA

1. Reamplify the DNA isolated from the differential display experiment using the identical parameters and primers to those in the original experiment.

2. Denature the DNA. To 50 µl of amplification mix, add 50 µl of 0.5 M NaOH/1 M NaCl, mix gently by tapping and leave at room temperature for 10 min.

3. Dilute the DNA. Mix the denatured DNA with 100 µl of 0.1x SSC/ 0.125 N NaOH.

Immobilize DNA onto membranes

4. Divide the diluted and denatured DNA samples into two equal halves, one-half for each of the two duplicate membranes. If more than two samples are to be compared then divide the reamplified DNA into the necessary number of tubes.
   - Consult the schematic diagram of the slot blot manifold and order the DNA samples in two racks, one for each membrane, to correspond exactly to your plan.
   - Load all the half-volume DNA samples onto the slots of the manifold, without suction, and let the DNAs sit on the membrane for 30 min.
   - Apply the suction so the DNA solution is drawn through the slots slowly. Once all the DNA solution is gone from the slots release the vacuum slowly.

5. Disassemble the manifold carefully, remove the membrane using forceps and rinse it in 0.5 M NaCl/0.5 M Tris, pH 7.5, briefly.

6. Fix the DNA by either air drying or baking at 80 °C for 1 h or UV cross-linking. All the procedures will adequately fix the DNA and the choice of protocol solely depends on time constraints and equipment on hand. However, the duplicate membranes must both be fixed by the same method.
   - When air drying, hang the membrane up by a corner using small bulldog clips dedicated to this purpose and choose a location away from any direct heat or noxious fumes, i.e. not in the fume hood where phenol might be stored or in use.
   - When baking, place the membrane between a larger piece of Whatman 3 M filter paper. Close the envelope using paper clips making sure the clips clamp the filter paper edges only and not the membrane too.
   - Commercial cross-linkers are available and are preferable to using an UV gel box. Follow the manufacturer's instructions for fixing DNA. In general, fixation of a wet membrane is preferable to fixation of a dry membrane and the DNA side of the membrane should be uppermost.
   The dried, fixed membranes can be placed in between pieces of Whatman 3 M filter paper and stored for later use.

7. Reassemble the manifold with wet Whatman paper and membrane, clamp it together and load the second set of half-volume DNA samples in the same pattern and manner as was used for the first membrane. Fix the second membrane at the same time or follow the same procedure with the second membrane as was used for the first.

Hybridization of the [α^{32}P] Double-Stranded DNA to the Membranes

The hybridization solutions described here were designed with the Gene Screen or Gene Screen Plus membranes in mind. Other membranes might require different buffers so consult the manufacturer's instructions. The method describes hybridization in plastic bags, but hybridization in bottles also works well. The choice depends on the availability of the hybridization oven. Follow the instructions for assembly of the membrane and any screens into the bottles. The volumes of prehybridization and hybridization to add can be maintained. Washing can be accomplished in the bottles or the membranes can be removed and washed as described.

The prehybridization and hybridization mixes will take some time to prepare so start dissolving the mixes early in the morning if you want to use them in the afternoon. Keep them in the water bath until use.

Prepare prehybridization and hybridization mixes

Calculate the amount of solution to make by measuring the length and width of the membrane in centimeters and inserting these measurements into the formula L×W×0.2 ml×number of blots for the prehybridization and L×W×0.05 ml×number of blots for the hybridization mix. For a membrane cut for a square 48 slot blot manifold about 10 ml of mix is required for the prehybridization and 2.5 ml mix for the hybridization.

1. Place a piece of Whatman 3 M filter paper in the bottom of a plastic tray and add 2x SSC. Using forceps and wearing gloves lower the membrane onto the 2x SSC so it floats: once it is clear that the membrane is wet submerse it and let it soak for about 1 min. Do not allow any air bubbles underneath the membrane or obvious dry spots to remain.

Prehybridize

2. Place each membrane in a separate plastic bag; heat seal three sides and leave one side, the tallest, open.

3. Add 10 ml of prewarmed prehybridization mix to each membrane slowly to minimize the number of bubbles.

4. Remove any bubbles using the side of the lab bench, your gloved fingers or a pencil and heat seal the remaining side. Push down slightly to be sure the sides are well sealed.

5. Place the membranes in a plastic tray and fill the tray with water from the water bath.

6. Incubate at 60 °C for 2–4 h.

Hybridize

7. Add $1-2\times 10^6$ cpm denatured radioactive DNA to each ml of warmed hybridization mix.

8. Cut off a small corner of the plastic bag. Squeeze out the pre-hybridization mix completely.

9. Using a blade or scissors cut off a good portion of the tip of a 1 ml pipet tip and insert it in the corner hole. The tip acts like a funnel keeping the hole open and acting as a port to load the hybridization mix.

10. Slowly load the hybridization mix through the pipet tip using a Pasteur pipet and bulb.

11. Remove the pipet tip, carefully squeeze out any bubbles, reseal the corner, check to be sure the seal is secure, place the membrane back into the tray in the water bath and incubate overnight.

   Caution: Monitor your gloved hands, pipets, heat sealer, bulbs and anything else that might have come in contact with the radioactive label with a Geiger counter. Dispose of contaminated material according to the rules and regulations of your institution.

Wash and expose the membranes.

12. Cut open the plastic bag and use tweezers to remove the membrane. Dispose of the hybridization mix and bag in the radioactive waste.

13. Put the membrane in a plastic tray with about 100 ml of 2x SSC for 5 min with agitation. Remove the wash and dispose in the radioactive waste. Repeat.

14. Add 100 ml of 2x SSC/1 % SDS and wash for 30 min with agitation. Repeat.

15. Monitor the membrane with a Geiger counter after each wash and when the background reaches about 500 cpm wrap the wet membranes in plastic wrap and expose them to X-ray film for 2–3 days. If the film is too dark the wet membranes can be rewashed at higher temperatures and reexposed.

Compare the hybridization pattern

16. Align the X-ray film so that the exposures from the two membranes are in the same orientation.

17. Mark on the schematic diagram (not on the film you may want it later for photographic purposes) those gene fragments present in one sample and absent in another with different colored highlighter pens or some other marks. As in the first differential display do **NOT** score as differentially expressed those gene

fragments in which there is simply a difference in amount of hybridization.

18. Mark on the schematic diagram those fragments which no longer appear to be differentially expressed or in which no hybridization occurred in either sample. These gene fragments likely do not represent discordantly expressed mRNAs.

Clone, sequence and perform northern analyses or some other confirmatory test on only those gene fragments with a clear hybridization signal in one sample and no detectable signal in the other sample.

Choose gene fragments

4
Results and Comments

The technique uses similar methods to those used to generate cDNA libraries. We used it with known genes and plasmids containing β-actin and GAPDH as control genes to identify test genes particularly expressed in one tissue vs another (Chalifour et al. 1994; Holder et al. 1995). When combined with the differential display technique it permits quick confirmation of discordant expression. It also permits the rescreening of fragments originally isolated from one experiment with another experiment entirely. For example, fragments identified as differentially expressed in 3T3 cells with or without serum can be rescreened to determine if any of those fragments also are potentially discordantly expressed in 3T3 cells grown in the absence or presence of a specific growth factor.

When comparing the hybridization signal from two or more membranes it is important to score as differentially expressed only those fragments which are present in one sample and absent in the other. Because no attempt is made to adjust the concentration of DNA loaded onto the membranes to any constant amount, a difference in hybridization intensity between one fragment, for example A, and another fragment, for example B, does not indicate a difference in expression levels of the two genes. Thus, while it is possible to conclude that an all or none difference in hybridization intensity of gene A between tissue 1 and tissue 2 indicates that gene A is expressed in tissue 1 and not tissue 2, it is not possible to conclude that tissue 1 expressed more fragment A than fragment B.

5
Troubleshooting

- The purity of the input RNA is of paramount importance. The RNA must be DNA-free. If genomic DNA contaminates the RNA preparation all gene fragments may be identified as not discordantly expressed because both samples will have hybridized to all gene fragments. We found that RNA methods designed to minimize genomic DNA contamination still often contained small amounts of genomic DNA. Thus, we found it preferable to routinely incubate all total RNA preparations with RQ1 DNase, or similar RNase-free DNase, to completely remove all remaining genomic DNA. We routinely treat RNA with RQ1 DNase, followed by incubation with proteinase K (to remove the DNase), phenol extraction, chloroform extraction and collection of the RNA by ethanol precipitation.

- In general, if low cpms are obtained in the generation of the double-stranded DNA radioactive probes it is because the first-strand reaction did not proceed well. Gel electrophoresis and exposure of a sample of the first-strand reaction will help identify if this problem occurred. If few or no cpms are obtained, the first-strand reaction should be repeated before going on to generation of the double-stranded DNA probe. The input RNA should demonstrate clear 28S and 18S bands on electrophoresis. A new RNA sample should be prepared and the experiment repeated if extensive smearing is observed upon electrophoresis.

- If different amounts, for example a 20-fold difference in the nucleotide incorporation into sample A vs sample B double-stranded DNA probes, are obtained the number of cpms added to the membranes should be adjusted to about the same level. Differences of two- to five-fold are not adjusted because an absolute difference in hybridization intensity is important and not a quantitative difference.

References

Chalifour LE, Fahmy R, Holder EL, Hutchinson EW, Osterland CK, Schipper HM, Wang E (1994) A method for analysis of gene expression patterns. Anal Biochem 216:299–304
Holder EL, Al Moustafa AE, Chalifour LE (1995) Molecular remodelling in hypertrophied hearts from polyomavirus large T-antigen transgenic mice. Mol Cell Biochem, in press

Mou L, Miller HM, Li J, Wang E, Chalifour L (1994) Improvements to the differential display method for gene analysis. Biochem Biophys Res Comm 199:564–569

Liang P, Pardee AB (1992) Differential display of eukaryotic messenger RNA by means of the polymerase chain reaction. Science 257:967–971

Liang P, Zhu W, Zhang X, Guo Z, O'Connell RP, Averboukh L, Wang F, Pardee AB (1994) Differential display using one-base anchored oligo-dT primers. Nucleic Acids Res 22:5763–5764

How To Find and Clone the Appropriate cDNA Fragments Generated in Differential mRNA Display by Using Northern Blot for cDNA Capture

K. Karikò

1
Introduction

Differential Display of mRNA

A new technique called differential mRNA display has been recently established and used to identify differences in subsets of mRNA samples (Liang and Pardee 1992; Liang et al. 1992). This technique enables one to analyze any mechanism which involves changes in specific mRNA levels. By utilizing this technique, differentially expressed genes with known or previously unreported sequences have already been identified (Sager et al. 1993; Aiello et al. 1994; Zou et al. 1994). The key element of this technique is the use of sets of anchored and arbitrary primers to generate cDNA fragments in reverse transcription followed by polymerase chain reactions (RT-PCRs). The cDNA fragments are resolved and compared on sequencing gels. The resulting cDNA patterns reflect differences in the mRNA levels and composition. Differentially displayed cDNAs are isolated, cloned, sequenced and used as probes on northern blots to confirm differences in the particular mRNA level. The general strategy seems to be simple and straightforward, although increasing number of reports on methodical refinements (Liang et al. 1993; Li et al. 1994; Reeves et al. 1995) suggest that the involved procedures are technically very challenging.

As part of our studies investigating the effect of basic fibroblast growth factor (bFGF) on human endothelial cells we used the differential display technique. We also encountered several technical difficulties. One problem, experienced by others as well (Liang et al. 1993; Sager et al. 1993), was the failure to clone the cDNAs that would reconfirm differences of mRNA levels demonstrated on the display. When northern blots were probed with cloned cDNA fragments, frequently, no signal or no differences in message levels were detectable.

The most probable reason for the failure was the fact that DNA recovered from what appeared to be a unique band contained additional undetectable, overlapping, or unresolved cDNAs. Contaminant cDNAs might have been copurified with the differentially displayed cDNAs. In fact, our studies revealed that each band recovered from the sequencing gel contained three or more different cDNA fragments. To avoid the laborious screening of multiple cloned fragments, we applied a different approach for cloning the appropriate differentially displayed cDNAs. We used northern blots to affinity capture cDNA fragments that were radiolabeled in PCR prior to hybridization. Fragments of cDNAs which demonstrated differences in mRNA levels of the bFGF-treated and untreated cells were recovered from the membrane after hybridization and cloned. Repeated northern analyses, now using the cloned inserts as probes, confirmed that affinity capturing is an effective and rapid method for selecting the appropriate cDNA fragments from differential display isolates.

In this chapter, application of the northern blot method for affinity capturing of cDNA is described. This method provides a rapid and effective approach for screening, selecting and cloning cDNAs obtained by a wide variety of techniques including differential mRNA display and RNA fingerprinting (Welsh et al. 1992).

Utilization of Nitrocellulose Membrane as Solid Support To Immobilize RNA and To Capture cDNA

There are several molecular biotechniques which require immobilization of RNA on solid support. The most well known of these techniques is northern analysis (Sambrook et al. 1989), in which nitrocellulose filter is used as support material to immobilize the RNA. The principle of the northern analysis relies on the fact that membrane-bound denatured RNA can efficiently anneal to a DNA probe with complementary sequences under conditions of salt concentration and temperature that stabilize hydrogen bonding between the complementary RNA-DNA sequences. If required, specifically annealed single-stranded cDNA probe can also be removed under conditions which destabilize hydrogen bonding. Such conditions can be achieved using low salt solution and high temperature for incubation (Sambrook et al. 1989).

This chapter focuses on protocols describing: (1) preparation and utilization of differentially displayed cDNA fragments as radiolabeled probes for northern analysis; (2) recovery of the appropriate cDNA probe from the membrane of the northern blot; (3) cloning of the cDNA probe and confirming the proficiency of the screening method.

2
Materials

<table>
<tr><td>

- thermal cycler
- agarose gel electrophoresis apparatus
- UV transilluminator
- scintillation counter

</td><td>Equipment</td></tr>
<tr><td>

- agarose
- scintillation vials
- Nytran membrane (Schleicher and Schuell)
- XAR film (Kodak) and exposure cassette with intensifier screen

</td><td>Supplies</td></tr>
<tr><td>

- PCR kit (Perkin-Elmer)
- pCR-Script SK(+) cloning kit (Stratagene)
- Random Prime labeling kit (Boehringer Mannheim)

</td><td>Kits</td></tr>
<tr><td>

- 10 mg/ml ethidium bromide stock solution
- DNA size markers
- $[\alpha^{32}P]dCTP$
- formaldehyde
- deionized formamide (Clonetech)
- 10 % SDS stock solution
- 3 M sodium acetate (pH 5.0)
- ethanol
- 10 mg/ml glycogen
- Pfu DNA polymerase (Stratagene)
- TE: 10 mM Tris-HCl (pH 7.4), 1 mM EDTA (pH 8.0)
- TBE: 89 mM Tris, 89 mM boric acid, 0.2 mM EDTA
- MESA buffer (Sigma): 40 mM MOPS (pH 7.0), 10 mM sodium acetate, 1 mM EDTA
- gel loading solution (Sigma): 0.05 % (w/v) bromphenol blue, 40 % (w/v) sucrose, 100 mM EDTA (pH 8.0)
- 10x SSC buffer: 150 mM sodium citrate (pH 7.0), 1.5 M NaCl
- prehybridization buffer: 5x SSC, 50 % (v/v) deionized formamide, 5x Denhardt's solution (1x Denhardt's solution contains 0.02 % BSA, 0.02 % Ficoll 400, 0.02 % Polyvinyl pyrrolidone), 0.1 % SDS, 10 mM Tris-HCl (pH 7.5) and 0.1 mg/ml sonicated salmon sperm DNA (Sigma). Store at −20 °C.
- 10x Pfu buffer: 200 mM Tris-HCl (pH 8.5), 100 mM ammonium sulfate, 20 mM $MgCl_2$, 1 mg/ml BSA, 1 % Triton X-100. Store at −20 °C.

</td><td>Reagents
and solutions</td></tr>
</table>

3
Experimental Procedure

Elution of differentially displayed band from the sequencing gel

First, select the differentially displayed band that you wish to further characterize. To avoid chasing artifacts, it is critical to confirm that the pattern of the selected band is reproducible. Next, elute the selected DNA fragment from the sequencing gel.

1. Use the autoradiogram to localize the selected band on the sequencing gel.

2. Cut out band from the gel using a razor blade.

3. Put the acrylamide gel slice into a microcentrifuge tube, add 50 µl TE. Incubate 2 h at 65 °C. Vortex intermittently.

4. Pellet the gel debris in a microcentrifuge using top speed for 2 min.

5. Collect the supernatant, which contains the cDNA fragment, and store at −20 °C

Preparation of radiolabeled cDNA probe

In this step, the cDNA fragment will be reamplified in two rounds of PCR. In the first round of PCR the template is the DNA sample that was eluted from the sequencing gel. In the second round of PCR the gel-purified product of the first round of PCR is reamplified. The PCR conditions and primers should be the same as were used for generation of cDNA fragments that have been displayed. The only exception is that in the first round of PCR no radioactive nucleotide should be included. In the second round of PCR [α^{32}P]dCTP should be added to the reaction mix.

1. Amplify the DNA sample in a final volume of 25 µl using PCR condition (Perkin-Elmer kit). Measure 2 µl cDNA (sequencing gel eluent) into a 0.5 ml Eppendorf tube. Add 12.2 µl water, 2.5 µl 10x PCR buffer, 1.2 µl anchored oligo-dT primer (0.1 mM), 1.2 µl arbitrary 10-mer primer (0.1 mM), 1.7 µl MgCl$_2$ (1.5 mM), 1.0 µl dATP (1.25 mM), 1.0 µl dGTP (1.25 mM), 1.0 µl dCTP (1.25 mM), 1.0 µl dTTP (1.25 mM) and 0.2 µl AmpliTaq (5 U/µl).

2. The PCR should be performed according to the following program:
   1 cycle: 94 °C for 2 min
   35 cycles: 94 °C for 30 s, 42 °C for 60 s, 72 °C for 40 s,
   final elongation cycle: 5 min at 72 °C

3. Resolve 15 μl of PCR reaction products on 2% agarose/TBE gel.

4. Estimate the size of the ethidium bromide stained and UV visualized PCR product by comparing it to appropriate molecular weight markers (note that the size of the PCR product should be the same as it was estimated on the differential display).

5. Cut out agarose gel slice containing the PCR product using a razor blade.

6. Remove DNA from the dissected slice. (You might use the simplest high speed filter spinning to recover DNA from gel.)

7. Save gel eluent (~20–30 μl) containing PCR amplified differential displayed band. Store sample at –20 °C.

8. Reamplify 2 μl aliquot of cDNA sample (agarose gel eluent) in a final volume of 25 μl using PCR conditions described above with the exceptions that the dCTP concentration should be 0.125 mM and 1 μl [α^{32}P]dCTP (3000 Ci/mmol) should be also included.

9. Separate 15 μl aliquot on 2% agarose/TBE gel.

   Caution: Be aware that the sample is radioactive. Use protective shields while loading the sample or handling the gel.

10. Visualize ethidium bromide stained DNA by UV.

11. Cut out gel slice containing the desired band.

12. Elute radiolabeled DNA (~30–50 μl) from gel slice using high speed filter spinning.

13. Measure radioactivity in eluent by Cherenkow effect. First, place Eppendorf tube containing ^{32}P-labeled PCR product into a scintillation vial. Measure counts in the aqueous solution (without adding scintillation cocktail) on the ^{3}H channel of a scintillation counter. Adjust values, considering that based on the Cherenkow effect, ^{32}P is measured only by 50% efficiency at the ^{3}H channel.

14. Keep aqueous solution of radiolabeled DNA at –20 °C until it is used for probing the northern.

Northern analysis

First, according to standard procedures (Sambrook et al. 1989), prepare a northern blot which contains aliquots of RNA samples that have been used to generate the differential display. Second, the northern blot should be hybridized with radiolabeled probe to detect differences in mRNA levels.

1. Prepare denaturing gel containing 1.4 % agarose, 0.22 M formaldehyde and 100 ng/ml ethidium bromide in 1x MESA buffer.

2. Incubate 2 µl RNA sample (~10 µg) after addition of 1 µl 5x MESA buffer, 5 µl formamide (deionized) and 2 µl formaldehyde for 15 min at 55 °C.

3. Cool denatured RNA samples on ice, then add 2 µl gel loading solution.

4. Load and resolve samples in gel submerged into 1x MESA buffer also containing 0.22 M formaldehyde. Gel should run at 5 V/cm for about 2–3 h, or when the bromphenol blue dye front reaches two thirds of the gel length. Test even loading of the samples by illuminating the gel on UV box.

5. Transfer RNA samples to Nytran membrane by capillary action using 10x SSC buffer. This process requires 10–16 h.

6. Use UV irradiation to covalently cross-link RNA samples to membrane.

7. Place air-dried filter into prehybridization solution and incubate at 42 °C for 2 h. Then supplement the solution with the probe. As probe, use the differentially displayed cDNA fragment that was labeled in PCR (see above). The probe should be denatured by boiling for 10 min. Extend filter hybridization in the presence of the probe at 42 °C for 16–22 h.

8. Wash blot three times with 2x SSC and 0.1 % SDS at room temperature for 15 min, then wash once with 0.1x SSC and 0.1 % SDS at 52 °C for 10 min.

9. Expose blot to Kodak XAR film using an intensifier screen at –70 °C for 4–48 h.

10. Analyze signals on the autoradiogram. Signals should have different intensities in the different lanes of the filter which contain the compared RNA samples. Additional signals with same intensities in the different samples might also be detected.

Recovering the probe from the northern blot

First, select an autoradiogram and the corresponding northern blot which demonstrate that the differentially displayed mRNA is indeed expressed differently in the tested set of RNA samples. Second, elute the captured cDNA probe from the filter and save it for cloning and further characterization. Although there is increasing evidence suggesting that even membrane bound DNA can serve as template for DNA polymerases (Kainz and Kainz 1989; Maskell et al. 1993), we could not amplify the captured cDNA unless it was removed from the filter. Control experiments excluded the possibility that the membrane would interfere with the PCR.

1. Cut out a piece of filter ~10 square mm in area containing the captured complementary DNA probe. Use a sharp scalpel to avoid stretching of the elastic nylon membrane. Autoradiogram can aid band localization.

2. Strip probe from filter by boiling in 100 µl water for 5 min. The efficiency of stripping can be measured by detecting the radiolabeled probe in the water phase by Cherenkow effect (see above).

   Note: filter interferes with detection and must be removed before measurement.

3. Recover the eluted probe by precipitation in 0.3 M sodium acetate (pH 5.0) and 75% ethanol for 16 h at –20 °C, using glycogen (0.5 µg/µl) as carrier.

4. Collect pellet in microcentrifuge using top speed for 2 min. Wash pellet with 80% ethanol at –20 °C, then reconstitute it in water and store at –20 °C.

Cloning the recovered probe

1. Amplify an aliquot of the recovered probe using PCR conditions described above with the exception that no radioactive dNTP should be included.

2. Resolve PCR product in 2% agarose/TBE gel and remove DNA from gel by high speed filter spinning. Purify by ethanol precipitation, then reconstitute in 25 µl water.

3. Use Pfu DNA polymerase to flush the ends of the AmpliTaq amplified PCR product (Costa and Weiner 1994). The following procedure is recommended. First incubate 5.0 µl purified PCR product (10–500 ng) in a final volume of 10.0 µl containing 1x Pfu buffer, 1 mM dNTP mix (0.25 mM for each) and 0.25 U/µl Pfu DNA polymerase for 30 min at 72 °C.

4. Clone blunt-ended product into *Srf* site of the pCR-Script SK(+) vector using the pCR-Script cloning kit.

5. Release cloned insert, label it by random priming (Random Prime labeling kit) and use it as a probe on a northern blot which contains aliquots of RNA samples that have been used to generate the differential display as well as to capture the cDNA probe.

6. Northern analysis should result in an autoradiogram that is almost identical to those obtained previously with the labeled PCR probe.

Such a result would confirm that the plasmid carries the insert which generated the positive signal on the first northern blot.

References

Aiello LL, Robinson GS, Lin Y-W, Nishio Y, King GL (1994) Identification of multiple genes in bovine retinal pericytes altered by exposure to elevated levels of glucose by using mRNA differential display. Proc Natl Acad Sci USA 91:6231–6235

Costa GL, Weiner MP (1994) Protocols for cloning and analysis of blunt-ended PCR-generated DNA fragments. PCR Methods Applic 4:s95-s106

Kainz P, Kainz V (1989) Nick translation of DNA immobilized on nylon membranes. Anal Biochem 178:260–262

Li F, Barnathan ES, Kariko K (1994) Rapid method for screening and cloning cDNAs generated in differential mRNA display: application of Northern blot for affinity capturing of cDNAs. Nucleic Acids Res 22:1764–1765

Liang P, Averboukh L, Keyomarsi K, Sager R, Pardee AB (1992) Differential display and cloning of messenger RNAs from human breast cancer versus mammary epithelial cells. Cancer Res 52:6966–6968

Liang P, Averboukh L, Pardee AB (1993) Distribution and cloning of eukaryotic mRNAs by means of differential display: refinements and optimization. Nucleic Acids Res 21:3269–3275

Liang P, Pardee AB (1992) Differential display of eukaryotic messenger RNA by means of the polymerase chain reaction. Science 257:967–971

Maskell D, Szabo M, High N (1993) PCR amplification of DNA sequences from nitrocellulose-bound, immunostained bacterial colonies. Nucleic Acids Res 21:171–172

Reeves SA, Rubio M-P, Louis DN (1995) General method for PCR amplification and direct sequencing of mRNA differential display products. BioTechniques 18:18–20

Sager R, Anisowicz A, Neveu M, Liang P, Sotiropoulou G (1993) Identification by differential display of alpha 6 integrin as a candidate tumor suppressor gene. FASEB J 7:964–970

Sambrook J, Fritsch EF, Maniatis T (1989) Molecular cloning: A laboratory manual, 2nd edition. Cold Spring Harbor Laboratory Press, Cold Spring Harbor

Welsh J, Chada K, Dalal SS, Cheng R, Ralph D, McClelland M (1992) Arbitrarily primed PCR fingerprinting of RNA. Nucleic Acids Res 20:4965–4970

Zou Z, Anisowicz A, Hendrix MJC, Thor A, Neveu M, Sheng S, Rafidi K, Seftor E, Sager R (1994) Maspin, a serpin with tumor-suppressing activity in human mammary epithelial cells. Science 263:526–529

Chapter XXXII

Verification of Differential Display Results by RNase Protection

R. Hummel, M. Jørgensen, M. Bévort, M. Grønborg, and H. Leffers

1
Introduction

The differential display-polymerase chain reaction (DDRT-PCR) technique is a powerful method for detecting differentially expressed genes (Liang and Pardee 1992). Nonetheless, DDRT-PCR, with about 300 different primer combinations, typically results in about 50–100 differentially displayed bands, of which about 10 %–40 % may be artifacts. The number of artifacts can be reduced by including several identical RNA samples; however, because of the high number of artifacts it is essential to verify the differential expression. This can be done by several different quantitative or semi-quantitative approaches, e.g., northern blotting, nuclear run-on assay, quantitative PCR and RNase protection.

Since the limiting factor often is the amount of RNA that is available, the method that should be used must require a small amount of RNA and, at the same time, it must be sufficiently sensitive to detect even very rare mRNAs. Furthermore, the method of choice must be rapid, since verification of 50–100 differentially expressed genes is a very time consuming part of the DDRT-PCR analysis. Northern blot analysis (Ausubel et al. 1995) is easy and reliable but not very sensitive and, thus, may not detect rare mRNAs. Furthermore, northern blot analysis of a large number of differentially expressed genes is difficult to handle and requires several northern blots and therefore a large amount of RNA. The nuclear run-on assay (Marzluff and Huang 1985) is very time consuming and not easy to apply to small amounts of RNA. While quantitative RT-PCR (Wang et al. 1989) is easy, very sensitive and allows for the handling of many samples simultaneously, it nonetheless requires specific primers for every sample.

We have developed a modified RNase protection protocol for the verification of the differential gene expression detected on DDRT-PCR gels. RNase protection assay (Saccomanno et al. 1992) is a very sensitive method to detect and quantitate the relative levels of a mRNA in different RNA samples. The detection limit of this assay for a specific mRNA is less than 0.001 % of the total mRNA population, using only 0.2 µg total RNA per reaction (Ferré 1994). This protocol eliminates the need for cloning the PCR fragments prior to performing the antisense in vitro transcription, thus reducing the assay time considerably. The modified RNase protection assay is a simple and relatively quick procedure (3 days) for verification of differential gene expression displayed by the DDRT-PCR method.

2
Principles and Applications

Normally an amplified fragment has to be purified and cloned into a plasmid vector that includes a T7, T3 or SP6 RNA polymerase promoter. Several colonies then have to be analyzed for the presence of the fragment in the right orientation, the plasmids have to be purified, linearized and then subjected to in vitro transcription. We have circumvented this by using downstream primers that are composed of a T7 RNA polymerase promotor sequence attached to the 5'-end of the antisense $HT_{11}VN$ downstream primer ($T7HT_{11}VN$) (V=A, C or G; N=A, C, G or T), for the reamplification of the DDRT-PCR bands (see Chap. XXVI). Thus, reamplification of the DDRT-PCR fragments generates a linear template that includes a T7 RNA polymerase promotor sequence in the antisense direction (Davanloo et al. 1984; Tabor and Richardson 1985). The amplified fragment is purified from a low melting agarose gel and in vitro transcription is initiated using approximately 0.2 pmol of the purified fragment, with the incorporated T7 RNA polymerase promotor sequence, in the presence of T7 RNA polymerase, $[\alpha^{32}P]UTP$ and NTPs. This reaction produces radioactively labeled run-off transcripts for use in the RNAse protection assay. The in vitro transcription products are resolved on a small denaturing polyacrylamide gel and full-length transcription products are excised and eluted overnight.

The next step is the annealing of the purified full-length transcripts to total RNA from the RNA samples that should be analyzed. As a negative control, the antisense in vitro transcript is hybridized to *E. coli* tRNA. The next day, the hybridization mixtures are digested with an RNase (RNAse ONE), followed by a phenol:chloroform extraction and ethanol precipitation. We recommend using RNAse ONE instead of the commonly used RNases A and T1, since RNAse ONE does not have any nucleotide preferences (RNase A cuts preferentially at C residues and T1 at G residues). The RNase protection samples and approximately 100 cpm of each in vitro transcript (undigested control) are then dissolved in a formamide loading buffer and resolved on a denaturing polyacrylamide gel. The dried gel is subsequently exposed to an X-ray film. Using the exposed film, the respective protected bands (10 nucleotides shorter than the full length probes) can be quantitated either manually or using a densitometric scanner and the relative expression of the gene in the respective RNA samples can be compared.

To reduce the number of protection samples it is possible to perform multiple RNase protection. Up to five different in vitro transcription probes with different sizes, i.e., 100, 150, 200, 250 and 300 nucleotides can be mixed and hybridized to the RNA samples. This significantly reduces the number of RNase protection samples.

3
Materials

– small polyacrylamide gel apparatus	**Equipment**
– sequencing gel apparatus (e.g., SequiGen II from BioRad)	
– tabletop centrifuge	
– β-counter (mini-monitor)	
– waterbath	

- X-ray film   **Supplies**
- exposure cassette and intensifying screen

- diethylpyrocarbonate (DEPC)   **Reagents and solutions**
- DEPC treated water (DEPC H_2O) (Add 1 ml DEPC to 1 l water, autoclave)
- reamplified DDRT-PCR cDNA fragment resuspended in DEPC H_2O or TE buffer
- 10x T7 transcription buffer: 400 mM Tris-HCl (pH 8.0), 60 mM $MgCl_2$, 50 mM DTT, 10 mM spermidine
- 10x UTP labeling mix: 2.5 mM each of ATP, GTP and CTP, 0.2 mM UTP in TE buffer
- $[\alpha^{32}P]$UTP (800 Ci/mM) (Amersham, NEN or ICN)
- RNA-Guard (32400 U/ml; Pharmacia Biotech) or similar
- T7-RNA polymerase (40 U/μl; Pharmacia Biotech)
- DNAse I (RNase-free) (10 U/μl; Pharmacia Biotech)
- TE buffer: 10 mM Tris-HCl (pH 7.5), 0.1 mM EDTA
- phenol (molecular biology grade) saturated with TE buffer
- chloroform

- 3 M sodium acetate (pH 6.0)
- 99 % and 80 % ethanol
- deionized formamide
- formamide loading buffer: 10 ml deionized formamide, 200 µl 0.5 M EDTA, 0.05 % xylene cyanol, 0.05 % bromophenol blue
- 10x TBE: 1 M Tris-borate, 10 mM EDTA
- 40 % acrylamide mix (20:1): 190 g acrylamide, 10 g bis-acrylamide, water to 500 ml. Store over mixed-bed resin (Bio-Rad: AG501-X8) at 4 °C.
- urea (Sigma)
- TEMED (N,N,N',N'-tetramethylethylenediamine)
- 10 % ammonium persulfate
- 20 % sodium dodecyl sulphate (SDS)
- extraction buffer: 10 mM Tris-HCl (pH 7.5), 20 mM EDTA (pH 8.0) 0.5 % SDS, 50 µg/ml Proteinase K
- 10 mg/ml RNase-free tRNA (e.g., from wheat germ or *E. coli*) (Boehringer Mannheim)
- 10x hybridization buffer: 400 mM PIPES (pH 6.4), 10 mM EDTA, 4 M NaCl
- RNase ONE (1 U/µl) (Promega)
- RNase digest mix: 40 µl 10x RNase ONE buffer (Promega), 360 µl water, 1 µl RNase ONE (Promega), prepare just before use
- proteinase K (10 mg/ml) (Sigma)
- RNase stop mix: 20 µl 10 % N-lauryl sarcosine, 10 µl proteinase K (10 mg/ml), 5 µl tRNA (1 µg/µl)
- 10 % acetic acid

4
Experimental Procedure

T7 RNA polymerase transcription. The reamplified fragment can be isolated in various ways, however, it is very important that the samples are RNase-free before the T7 RNA polymerase transcription assay is initiated. We normally purify our PCR reamplified fragment as described in Chap. XXVI. If other procedures are used, it is highly recommended to include a phenol: chloroform extraction.

The protocol outlined below corresponds to one reaction which will produce enough labeled probe to perform 10–20 RNase protection assays.

1. Mix in an Eppendorf tube:
   1 µl 10x T7 transcription buffer
   1 µl 10x UTP labeling mix

2.5 μl (25 μCi) [α^{32}P]UTP (800 Ci/mmol)
0.75 μl RNA-Guard (Pharmacia)
1–4 μl DDRT-PCR DNA fragment (20–40 ng)
0.5 μl T7-RNA polymerase (20 U)
DEPC H$_2$O to 10 μl

2. Incubate at 37 °C for 1 h. Add 1 μl (10 U) DNAse I (RNase-free). Incubate at 37 °C for 15 min. Add 80 μl TE.

3. Extract once with one volume phenol:chloroform.

4. Transfer upper aqueous phase to a new tube. Add 10 μl 3 M sodium acetate (pH 6.0). Add 250 μl 99 % ethanol.

5. Precipitate at –20 °C for 30 min, wash with one volume 80 % ethanol, dry the pellet.

6. Dissolve in 6 μl formamide loading buffer.

7. Denature at 95 °C for 3 min, place on ice.

1. Prepare an 8 % denaturing (8 M urea) polyacrylamide gel in 1x TBE buffer. For 60 ml: 12 ml 40 % acrylamide mix, 6 ml 10x TBE, 25.2 g urea, 21 ml water. Add 40 μl TEMED and 400 μl 10 % APS just before making the gel. **Polyacrylamide gel electrophoresis**

2. Pre-run the gel for 30–60 min at 50°–55 °C.

3. Load the denatured probe (6 μl).

4. Run the gel until the xylene cyanol dye is 5–10 cm from the bottom of the gel (depending on the length of the probe).

5. The gel is covered with Saran wrap and exposed to X-ray film for 1–5 min.

1. Excise the band corresponding to the full-length labeled probe. **Purification of full-length probe from the polyacrylamide gel**

2. Extract overnight in 200 μl extraction buffer.

3. Transfer buffer to a new tube.

4. Re-extract the gel piece with 100 μl extraction buffer.

5. Combine the samples and extract once with one volume phenol:chloroform (1:1). Add 30 μl 3 M sodium acetate (pH 6.0). Add 2.5 volumes 99 % ethanol (780 μl)

6. Precipitate at –20 °C for 30 min, wash with one volume 80 % ethanol, dry the pellet.

7. Dissolve in 20 μl DEPC H$_2$O.

Hybridization and RNase digestion

The protocol outlined below corresponds to one reaction with the appropriate controls.

1. Prepare tubes with:

   0.2–1 µg of each of the appropriate total RNAs in 1 µl DEPC H₂O

   0.8 µg tRNA in 1 µl (control sample)

   Add 500 cps (mini-monitor) to each tube (1–2 µl of the labeled RNA fragment). Add DEPC H₂O to 3 µl (0–1 µl). Add 24 µl deionized formamide. Add 3 µl 10x hybridization buffer and mix.

2. Incubate at 85 °C for 10 min.

3. Quickly transfer tubes to hybridization temperature: Tm–(20°–25 °C). We recommend using 45 °C as the standard.

   $T = 79.8 + 18.5 \times \log[Na^+] - 0.35 \times (\% \text{ formamide}) + 58.4 \times (G+C) + 11.8 \times (G+C)^2$

4. Incubate at hybridization temperature for 14–24 h. Proceed the next day with the following steps:

5. Place tubes on ice for 5 min. Add 450 µl ice-cold RNase digest mix.

6. Incubate at 37 °C for 30 min. Add 50 µl RNase stop mix at 37 °C

7. Incubate at 37 °C for 30 min.

8. Extract once with 500 µl phenol:chloroform.

9. Transfer upper aqueous phase to a new tube. Add 8 µl 3 M sodium acetate. Add 1 ml 99 % ethanol.

10. Incubate at –20 °C for 30 min. Spin in a tabletop centrifuge for 30 min. Wash the pellet with one volume ice-cold 80 % ethanol. Dry the pellet.

11. Dissolve in 7 µl formamide loading buffer.

Preparation of undigested fragment (control)

1. Mix:

   0.2 µl of the labeled fragment (100 cps)

   9.8 µl formamide loading buffer

Gel electrophoresis.

1. Heat samples to 95 °C for 3 min.

2. Load 3.5 µl onto an 8 % denaturing sequencing type polyacrylamide gel (see "Polyacrylamide gel electrophoresis").

3. Run the gel at 50°–55 °C in 1x TBE buffer for 45–90 min.

4. Fix the gel in 10 % acetic acid for 15 min.

5. Transfer the gel to 3MM paper and dry the gel.

6. The gel is placed in a cassette, and exposed to a X-ray film with intensifying screen at –80 °C for 1–7 days.

Acknowledgements. The development of the methods was supported by the Danish Cancer Society, The Danish Biotechnology Programme, The European Commission and The Danish National University Hospital (Rigshospitalet).

References

Ausubel FM, Brent R, Kingston RE, Moore DD, Seidman JG, Smith JA, Struhl K (1995) Current Protocols in Molecular Biology, vol 3, John Wiley & Son, Inc

Davanloo P, Rosenberg AH, Dunn JJ, Studier FW (1984) Cloning and expression of the gene for bacteriophage T7 RNA polymerase. Proc Natl Acad Sci USA 81:2035–2039

Ferré F (1994) In: Mullis KB, Ferré F, Gibbs RA (eds) The Polymerase Chain Reaction, pp 67–86.

Liang P, Pardee AB (1992) Differential display of eukaryotic messenger RNA by means of the polymerase chain reaction. Science 257:967–971

Marzluff WF, Huang RCC (1985) In: Hames BD, Higgins SJ (eds) Transcription and Translation: A Practical Approach. IRL Press, Oxford, pp 89–129

Saccomanno CF, Chen JS, Nordstrom JL, Bordonaro M (1992) A faster ribonuclease protection assay. BioTechniques 13:847–849

Tabor S, Richardson CC (1985) A bacteriophage T7 polymerase/promotor system for controlled exclusive expression of specific genes. Proc Natl Acad Sci USA 82:1074–1078

Wang AM, Doyle MV, Mark DF (1989) Quantitation of mRNA by the polymerase chain reaction. Proc Natl Acad Sci USA:9717–9721

Chapter XXXIII

Direct Automated Sequencing of DDRT-PCR Fragments

C. Hansen, M. Jørgensen, M. Bévort, R. Hummel, M. Løfgreen, N. Pallisgaard, and H. Leffers

1
Introduction

Differential display is a powerful tool for comparing gene expression in different cell or tissue preparations (Liang and Pardee 1992). However, differential display does not give any indications regarding the identity of the differentially expressed genes. Thus, the next step is to identify the mRNAs that are represented in the differentially displayed bands and this is achieved by sequencing the DNA fragments. If the differential display is made from human cells, sequencing followed by searches in sequence databases will often identify the differentially expressed gene, since The Human Genome Project has resulted in the accumulation of a large number of human cDNA and gene sequences in sequence databases. In particular, the expressed sequence tag (EST) programs have produced a very large number of partial cDNA sequences consisting of "single read" sequences from the ends of randomly picked cDNAs (Adams et al. 1995). Since most differential display methods select and display fragments that originate from the 3'-ends of mRNAs, the sequences of the fragments will correspond to the 3'-EST sequences. Since the 3'-ESTs are linked to their corresponding 5'-ESTs which, in most cases, are derived from the coding region of the mRNAs, the 3'-ESTs will identify the gene. However, if the analysis is made on cells or tissues from other organisms, the sequence itself will often not reveal the identity of the gene, even though it may be known in, e. g., humans, because the sequence originates from the 3'-non translated part of the mRNA and not from the much more conserved coding region. Thus, in those cases it will be necessary to clone the corresponding cDNA and sequence parts of the coding region. If the gene is unknown the complete sequence of the cDNA should be determined. The procedure for using DDRT-PCR fragments as probes for the cloning of cDNAs is described in Chap. XXXV in this volume.

Sequencing of differential display bands has traditionally been done by reamplifying the bands followed by cloning the amplified fragments into a suitable vector. Since differential display normally is made using the *Taq* polymerase, the DNA fragments will have an additional A residue at their 3'-ends and this is often utilized in the cloning by using a vector with a 5' protruding T residue. Such vectors are available from several companies (e.g., Life Technologies). Alternatively, the amplified fragments can be treated or further amplified with the Pfu polymerase (Stratagene), or a similar polymerase, which results in blunt ended fragments that can be cloned into a blunt ended vector. Another possibility is to include a restriction site in the primers and use this for cloning of the fragments. Plasmid DNA is then prepared from mini preps and the insert is sequenced using primers that are complementary to the flanking vector sequences. However, cloning is a slow and laborious process that significantly reduces the number of differentially displayed bands that can be analyzed. Therefore it is of interest to bypass the cloning step and instead directly sequence the amplified PCR fragments.

2
Principles and Applications.

Below we describe two different protocols for direct sequencing of amplified DDRT-PCR fragments. One utilizes the new ThermoSequenase enzyme from Amersham/USB (Tabor and Richardson 1995) and this protocol can be used both for manual sequencing with ^{32}P end-labeled primers and for the ALF and ALFexpress sequenators (Pharmacia Biotech). Moreover, even the small 16-mer $HT_{11}V$ downstream primers can successfully be end-labeled and used for manual sequencing (not recommended for ALF/ALFexpress sequencing), although we recommend the amplification protocol described in Chap. XXVI and the end-labeled T7 primer. The other protocol utilizes *Taq* polymerase, however, we have not tested it in manual sequencing (kits are available from Pharmacia Biotech). Of these two methods, we recommend the ThermoSequenase protocol. There are alternative methods available for direct sequencing of DDRT-PCR fragments using incorporation of [^{33}P] or [^{35}S]αdATP (e.g., Xinkang and Feuerstein 1995), and a kit for manual sequencing using incorporation of [^{33}P] or [^{35}S]αdATP, the Taq-CS sequencing kit, is available from Perkin Elmer. However, superior results can be obtained by using ^{32}P end-labeled primers and the ThermoSequenase sequencing protocol (kits are available from Amersham/USB).

DDRT-PCR fragments can also be directly sequenced on the ABI 373/377 sequenators (Perkin Elmer) using the Dye Terminator Cycle sequencing kit from Perkin Elmer and the protocols that accompany the kit. Alternatively, sequencing on the ABI 373/377 can be performed using dye-labeled primers (four different dyes) as described below for manual and ALF/ALFexpress sequencing with ThermoSequenase; the four samples are then mixed and loaded onto the ABI 373/377 sequenators.

3
Materials

- tabletop centrifuge
- thermocycler for 0.1 ml tubes and with heated lid (Perkin Elmer 9600 or similar)
- DNA sequenator (ALF or ALFexpress, Pharmacia Biotech) or manual sequencing apparatus (e. g., SequiGen II from BioRad)

Equipment

- 0.1 ml PCR tubes
- for manual sequencing: X-ray film (e.g., Kodak BioMax MR)

Supplies

- for automated sequencing: CY5 (ALFexpress) 5' end-labeled primer or fluorescein (ALF) 5' end-labeled primer; 2.5 pmol/µl in water (for ThermoSequenase sequencing) or 5 pmol/µl (for *Taq* sequencing): 5'-CY5/fluorescein-TAATACGACTCACTATAGGGCC-3'
- for manual sequencing: T7-universal or "extended" T7 primer or $HT_{11}V$ (see Chap. XXVI): 10 pmol/µl in H_2O

Oligonucleotides

- for manual sequencing: T4 polynucleotide kinase: 10 U/µl (Amersham/USB)
- ThermoSequenase (TS): 32 U/µl (Amersham/USB)
- *Taq* polymerase: 5 U/µl (Perkin Elmer or similar)
- for manual sequencing: $\gamma[^{32}P]$ATP: 150–170 µCi/µl (6–7000 Ci/mmol crude γATP; Amersham or ICN)
- 0.1 M dATP, dCTP, dGTP and dTTP (Pharmacia)
- 5 mM ddATP, ddCTP, ddGTP and ddTTP (Pharmacia)
- DNA in water
- for manual sequencing: 10x T4 polynucleotide kinase buffer: 500 mM Tris-HCl (pH 7.5), 100 mM $MgCl_2$, 50 mM 1,4-dithiothreitol (DTT), 1 mM EDTA
- 10x TS buffer: 200 mM Tris-HCl (pH 9.3), 40 mM $MgCl_2$, 7.5 % Nonidet P40, 7.5 % Tween-20, 10 mM 1,4-dithiothreitol (DTT)

Reagents and solutions

- 10x *Taq* sequencing buffer: 500 mM Tris-HCl (pH 9.3), 25 mM MgCl$_2$, 40 % Triton X100
- TS-termination mixes: all TS-termination mixes contain 1 mM each of dATP; dCTP; dGTP and dTTP in 2.5 mM Tris-HCl (pH 9.3). In addition: A-termination mix: 5 µM ddATP; C-termination mix: 5 µM ddCTP; G-termination mix: 6 µM ddGTP; T-termination mix: 5 µM ddTTP
- stop mix (Alf/ALFexpress/Taq): 96 % deionized formamide, 10 mM NaOH, 10 mM EDTA (pH 9.5), 6 mg/ml blue dextran (Sigma)
- stop mix (manual): 96 % deionized formamide, 10 mM NaOH, 10 mM EDTA (pH 9.5), 0.04 % bromphenol blue, 0.04 % xylene cyanol
- *Taq*-termination mixes: all *Taq*-termination mixes contain 27.5 µM each of dATP; dCTP; 7-deaza-dGTP and dTTP in water. In addition: *Taq*-A-termination mix: 450 µM ddATP; *Taq*-C-termination mix: 300 µM ddCTP; *Taq*-G-termination mix: 30 µM ddGTP, *Taq*-T-termination mix: 900 µM ddTTP

4
Experimental Procedure

Preparation of DNA fragments

The amplified DNA fragments should be purified as described in Chap. XXVI. However, if only one PCR product is present in the samples, the gel purification step can be omitted and 1–4 µl of the amplification reaction can be used directly for sequencing.

[32]P 5'-End-Labeling of Sequencing Primer

Labeling

1. Mix (for 15–20 sequencing reactions):
      10 µl primer (10 pmol/µl)
      2 µl 10x kinase buffer
      0.5 µl γ[^{32}P]ATP (50–75 µCi)
      6.5 µl water
      1 µl T4 polynucleotide kinase (10 U/µl)

2. Incubate at 37 °C for 30 min.

3. Incubate at 70 °C for 5 min (inactivates the kinase).

4. Leave on ice or freeze at –20 °C. Use 1 µl per sequencing reaction.

ALF/ALFexpress and Manual Sequencing Using ThermoSequenase

The sequencing reactions for manual and ALF/ALFexpress sequencing are identical, except that the CY5 or fluorescein 5'-end-labeled primer used for sequencing on ALF/ALFexpress sequenators is replaced with 1 μl of the ^{32}P-end-labeled primer for manual sequencing, prepared as described above. Also, note that the stop solutions are different.

1. Prepare ThermoSequenase (TS) sequencing master mix. For one sequence: 2 μl 10x TS buffer; 16 μl[a] water. Mix.
   Add 0.2 μl TS (32 U/μl); mix and centrifuge.

2. For each sequence add to an Eppendorf tube:
   1 μl DNA[a] (10–60 ng DNA fragment[b]; 0.5–1 μg plasmid DNA; 0.35–0.5 μg M13 single-stranded DNA)
   1 μl primer[a] (2.5 pmol/μl if CY5- or fluorescein-labeled or 5 pmol/μl if ^{32}P-labeled)

3. Add 18 μl[a] master mix to each DNA/primer mix. Mix gently and centrifuge.

4. Dispense 2 μl of each termination mix into four 0.1 ml PCR tubes.

5. Dispense 4 μl of each DNA/primer/master mix into the four PCR tubes with the termination mixes. Centrifuge briefly.

6. Place the tubes in the PCR machine. Cycle conditions for T7 sequencing primer:
   1 cycle: 2 min at 96 °C
   20 cycles: 30 s at 96 °C; 40 s at 62 °C (or primer-specific annealing temperature)[c]
   1 cycle: 5 min at 62 °C (or primer-specific annealing temperature)

7. Add 5 μl stop mix (for ALF/ALFexpress or manual); mix and centrifuge briefly. The samples can be stored at −80 °C.

8. Heat to 95 °C for 10 min in the PCR machine, place on ice.

9. Load 2–3 μl per lane.

Manual sequencing

Note: [a] The volume of the DNA and primer can be between 2 and 5 μl, just adjust the water in the master mix. [b] The amount depends on the size of the fragment, if 200 bp use 20 ng, if 800 bp use 50–60 ng. [c] The annealing/extension temperature depends on the primer, 62 °C is for the T7 sequencing primer that is shown in the Materials section, **not** for the standard T7 universal primer, although this can be used since its sequence is present internally in the T7 primer that we recommend. Annealing temperature for the T7 universal primer is 50°–52 °C. If the 16-mer $HT_{11}V$ primers are used for TS sequencing, the cycle conditions should be changed to:

1 cycle: 2 min at 96 °C
20–40 cycles: 20 s at 96 °C; 30 s at 39 °C; 30 s at 60 °C
1 cycle: 5 min at 60 °C

ALF/ALFexpress Sequencing Using *Taq* Polymerase

Sequencing   (This is a modification of the protocol by Zimmermann et al. 1994).

1. Prepare *Taq* master mix. For one sequence: 4.5 μl primer (5 pmol/μl) (CY5- or fluorescein 5'-end-labeled) in water; 3 μl *Taq* sequencing buffer; 9.5 μl water. Mix. Add 1 μl (5 U) *Taq* polymerase.

2. Prepare Eppendorf tubes with 2 μl DNA (10–40 ng DNA fragment or 1 μg plasmid DNA)[b] in water.

3. Add 18 μl *Taq* master mix to the DNA; mix gently and centrifuge.

4. Dispense 2 μl of the A, C, G and T, *Taq* termination mixes into four 0.1 ml PCR tubes.

5. Dispense 4 μl of each DNA/primer/master mix into the four PCR tubes with the termination mixes. Centrifuge briefly.

6. Place the tubes in the thermocycler. The cycle conditions are:
    1 cycle: 2 min at 94 °C
    25cycles: 20 s at 94 °C; 20 s at 62 °C[c] (or primer-specific annealing temperature); 40 s at 72 °C
    1 cycle: 5 min at 72 °C

7. Add 5 μl stop solution; mix and centrifuge briefly. The samples can be stored at –80 °C.

8. Heat to 95 °C for 5–10 min in the thermocycler, place on ice.

9. Load 2–3 μl per lane.

Acknowledgements. Development of the methods was supported by The Danish Cancer Society, The Danish Biotechnology Programme, The European Commission and The Danish National University Hospital (Rigshospitalet).

References

Adams MD, Kerlavage AR, Fleischmann RD, Fuldner RA, Bult CJ, Lee NH, Kirknes EF, Weinstock KG, Gocayne JD, White O, et al. (1995) Initial assessment of human gene diversity and expression patterns based upon 83 million nucleotides of cDNA sequence. Nature 377:s3-s174

Liang P, Pardee AB (1992) Differential display of eukaryotic messenger RNA by means of the polymerase chain reaction. Science 257:967–971

Tabor S, Richardson CC (1995) A single residue in DNA polymerases of the *Escherichia coli* DNA polymerase I family is critical for distinguishing between deoxy- and dideoxyribonucleotides. Proc Natl Acad Sci USA 92:6339–6343.

Xinkang W, Feuerstein GZ (1995) Direct sequencing of DNA isolated from mRNA differential display. BioTechniques 18:448–452.

Zimmermann J, Wiemann S, Voss H, Schwager C, Ansorge W (1994) Improved Fluorescent cycle sequencing protocol allows reading nearly 1000 bases. BioTechniques 17:302–307

Chapter XXXIV

Ligation Linked PCR and Direct Sequencing of Differential Display Products

S. A. REEVES

1
Introduction

Novel and powerful techniques have been developed that should prove helpful in the identification of differentially or developmentally expressed genes. The technique of mRNA differential display, originally developed by Liang and Pardee (1992a, 1992b, 1993) and modified by others (Bauer et al. 1993; Mou et al. 1994), has provided a potential means of generating molecular fingerprints of developmentally expressed genes as well as genes that are regulated in response to extracellular signals.

However, identification of bona fide differentially expressed cDNAs generated using mRNA differential display can consume considerable amounts of time and effort. This is because, as in any PCR amplification-based scheme using arbitrary or degenerate primers, the potential for generating false positives is high (Bauer et al. 1993; Callard et al. 1994; Mou et al. 1994; Welsh et al. 1992).

In this chapter, the method of ligation linked PCR (LLR) (Reeves et al. 1995) is described. LLR facilitates the characterization of cDNAs generated with mRNA differential display in two ways: First, it optimizes the second round PCR amplification of mRNA differential display generated cDNAs. And second, because LLR incorporates a non-random and highly specific primer at the upstream or 5' end of the PCR amplified cDNAs, direct DNA sequencing can be performed on multiple cDNAs using a common and highly specific sequencing-primer. In this way, LLR can be used to rapidly characterize a large number of candidate differentially or developmentally expressed cDNAs.

Overview of LLR Method and Direct Sequencing

- Generate differentially expressed cDNAs as described in Chap. XXVI and fractionate PCR amplification products on a denaturing polyacrylamide gel.

- Excise and purify bands (cDNAs) unique to experimental sample, but not in reference sample.

- Transiently ligate purified cDNAs to T-overhang cloning vector.

- Re-PCR amplify ligated cDNAs and vector using arbitrary anchored oligo(dT) primer and T7 or SP6 primers.

- Perform direct DNA sequencing on re-PCR amplification products using T7 or SP6 primers

2
Materials

Equipment
- temperature controllable water bath
- programable thermal controller
- DNA sequencing apparatus

Supplies
- mineral oil
- Whatman 3MM filter paper

Kits
- CircumVent thermal cycle dideoxy DNA sequencing kit (New England Biolabs, Beverly, MA)

Oligonucleotides and vectors
- 10 µM arbitrary anchored oligo(dT) primer set (GenHunter Corp.): 5'-T_{12}MG-3', 5'-T_{12}MA-3', 5'-T_{12}MT-3', and 5'-T_{12}MC-3' (M represents G, A, or C)
- T7 or SP6 promoter primers or pCRII polylinker primer (5'-CGAATTGGGCCCTCTAGA)
- T-overhang cloning vector diluted to 10 ng/ml (e.g. pCRII, Invitrogen; pGEM-T, Promega)

Reagents and solutions
- 10x ligation buffer: 500 mM Tris-HCl (pH 7.5), 70 mM $MgCl_2$, 10 mM DTT
- 10x PCR buffer: 100 mM Tris-HCl (pH 9.0), 15 mM $MgCl_2$
- 10x T4 polynucleotide kinase buffer: 500 mM Tris-HCl (pH 7.6), 100 mM $MgCl_2$, 50 mM DTT
- stop/loading dye: 95 % formamide, 0.09 % (w/v) bromphenol blue, 0.09 % (w/v) xylene cyanol FF
- T4 DNA ligase (5 U/µl)
- *Taq* DNA polymerase (5 U/µl)
- T4 polynucleotide kinase (10000 U/ml)
- 10 mM rATP
- 25 µM dNTP mix
- 5 µCi/µl [γ^{32}P]rATP (3000 Ci/mmol)
- 3 M sodium acetate (pH 5.2)

- 7.5 M ammonium acetate
- 100 % and 70 % ethanol
- 100 % isopropanol
- 20 mg/ml glycogen
- 3 % Triton X-100
- 6 % denaturing polyacrylamide DNA sequencing gel

3
Experimental Procedure

Ligation Linked PCR

1. Using the orientation markers, align the X-ray film with the dried down gel. Use tape to firmly attach the X-ray film to the gel. Using a clean razor blade for each band representing a differentially expressed cDNA, cut through both the X-ray film and the dried down gel to remove the band. Place the excised gel fragment in a labeled microcentrifuge tube.   **Excision and elution of PCR amplified bands**

2. Add 100 µl distilled water to each tube and soak at room temperature for 10 min.

3. Incubate the tubes at 95 °C for 15 min.

4. Spin for 2 min on high in a microcentrifuge to clarify particulate material. Transfer the supernatant to a new microcentrifuge tube.

5. Precipitate the cDNAs by adding to each tube:
   - 10 µl 3 M sodium acetate (final 0.3 mM)
   - 5 µl of 10 mg/ml glycogen (5 µg)
   - 450 µl of 100 % ethanol

6. Incubate for 30 min at –80 °C. Microcentrifuge on high for 10 min at 4 °C and decant.

7. Add 200 µl ice-cold 70 % ethanol, briefly vortex, and microcentrifuge on high for 5 min at 4 °C.

8. Decant, air dry pellet, and dissolve in 10 µl distilled water.

9. Set up transient ligation reactions (10 µl total volume) for each excised and eluted cDNA as follows: **Transient ligation to pCRII**
   - 1 µl 10x ligation buffer (final 1x)
   - 1 µl 10 mM ATP (final 1 mM)
   - 2 µl T-overhang vector (20 ng)
   - 3 µl eluted cDNA

- 1 µl T4 ligase (final 4 U)
- 2 µl distilled water

10. Incubate at room temperature for 30 min (ligation), followed by heating at 65 °C for 5 min.

PCR amplification of transiently ligated cDNA (LLR cDNA)

11. Set up PCR reactions (40 µl total volume) for each of the ligation reactions as follows:
    - 4 µl 10x PCR buffer (final 1x)
    - 3.2 µl 25 µM dNTP mix (final 20 µM)
    - 4 µl 10 µM anchored oligo(dT) primer ($T_{12}MN$) (final 40 pmol)
    - T7 or SP6 primer (final 20 pmol)
    - 10 µl ligation reaction from above
    - distilled water to 39.8 µl
    - 0.2 µl *Taq* DNA polymerase (final 1 U)

    Note: It is important to use the same $T_{12}MN$ primer used in the reverse transcription for the PCR amplification.

12. Programable thermal controller conditions: 95 °C for 30 s, 40 °C for 2 min and 72 °C for 30 s with a total of 40 cycles.

13. Using agarose gel electrophoresis, examine 5 µl of each sample to confirm single PCR amplification product.

Direct DNA Sequencing of LLR cDNAs

Preparation for direct DNA sequencing

1. Precipitate each LLR cDNA as follows:
   - 10 µl PCR amplified LLR cDNA
   - 19 µl distilled water
   - 11 µl 7.5 M ammonium acetate (final 0.4 M)
   - 1 µl 20 mg/ml glycogen
   - 40 µl 100 % isopropanol

2. Mix well and incubate at room temperature for 10 min.

3. Spin on high in a microcentrifuge at room temperature for 10 min to pellet cDNA.

4. Decant and add 200 µl ice-cold 70 % ethanol, briefly vortex, and microcentrifuge on high for 5 min at 4 °C.

5. Repeat step 4.

6. Dry pellet and resuspend cDNA in 30 µl distilled water.

5' End labeling of sequencing primers

7. In a 0.5 ml microcentrifuge, add the following:
   - 2.5 µl 10x T4 polynucleotide kinase buffer (final 1x)

- T7 or SP6 promoter primer or pCRII primer (final 10.5 pmole)
- 7 µl [γ³²P]rATP (final 11.5 pmol)
- distilled water to a final volume of 24 µl
- 1 µl T4 polynucleotide kinase (final 10 U)

8. Incubate at 37 °C for 30 min.

9. Terminate the reaction by heating at 95 °C for 5 min. Microcentrifuge for 10 s on high to collect condensation. Labeled primers can be stored at –20 °C until use.

LLR cDNAs can be sequenced directly using reagents supplied with the CircumVent thermal cycle dideoxy DNA sequencing kit.

Direct DNA sequencing

10. Set up four microcentrifuge tubes labeled A, C, G, T.

11. Add 3 µl of the CircumVent mix A to tube A, and 3 µl of CircumVent mixtures C, G and T to tubes C, G and T, respectively. Mix by pipetting.

12. Combine cDNA with end-labeled sequencing primer in a separate 0.5 ml microcentrifuge tube as follows:
    - 3 µl cDNA
    - 3 µl [γ³²P]rATP-labeled sequencing primer (final 1.2 pmol)
    - 1.5 µl 10x sequencing buffer (final 1x)
    - 1 µl 3 % Triton X-100 (final 0.2 %)
    - 5.5 µl distilled water

13. Add 1 µl CircumVent DNA polymerase (2 U), mix by pipetting and distribute 3.2 µl to tubes A, C, G, and T.

14. Add 1 drop mineral oil to each tube.

15. Programable thermal controller conditions: 95 °C for 30 s, 40 °C for 2 min and 72 °C for 30 s with a total of 40 cycles.

16. After PCR amplification, add 4 µl stop/loading dye under the mineral oil of each sample.

17. Denature samples at 80 °C for 2 min and then place on ice. Briefly microcentrifuge to collect condensation.

18. Load samples directly onto a 6 % denaturing polyacrylamide gel and run at 60 Watts constant power for 3–4 h.

19. Transfer gel to 3MM Whatman paper, dry down and expose to X-ray film.

References

Bauer D, Müller H, Reich J, Reidel H, Ahrenkiel V, Warthoe P, Strauss M (1993) Identification of differentially expressed mRNA species by an improved display technique (DDRT-PCR). Nucleic Acids Res 21:4277–4280

Callard D, Lescure B, Mazzolini L (1994) A method for the elimination of false positives generated by the mRNA differential display technique. BioTechniques 16:1096–1103

Liang P, Pardee AB (1992a) Differential display of eukaryotic messenger RNA by means of the Polymerase Chain Reaction. Science 257:967–9871

Liang P, Pardee AB (1992b) Differential display and cloning of messenger RNAs from human breast cancer versus mammary epithelial cells. Cancer Res 52:6966–6968

Liang P, Pardee AB (1993) Distribution and cloning of eukaryotic mRNAs by means of differential display: refinements and optimization. Nucleic Acids Res 21:3269–3275

Mou L, Miller H, Li J, Wang E, Chalifour L (1994) Improvement of the differential display method for gene analysis. Biochem Biophys Res Comm 199:564–569

Reeves SA, Rubio M-P, Louis DN, (1995) A general method for PCR amplification and direct sequencing of mRNA differential display products. BioTechniques 18:18–20

Welsh J, Chada K, Dalal S, Cheng R, Ralph D, McClelland M (1992) Arbitrarily primed PCR fingerprinting of RNA. Nucleic Acids Res 20:4965–4970

Chapter XXXV

Differential Display PCR Fragments as Probes for cDNA Cloning

A. H. Nielsen, N. Pallisgaard, and H. Leffers

1
Introduction.

Differential display of expressed mRNAs (DDRT-PCR) (Liang and Pardee 1992) is rapidly becoming the principal method for comparing gene expression in different cells or tissues. Differentially expressed bands can be excised from the DDRT-PCR gels, reamplified and purified (see Chap. XXVI). The next step is to identify the genes (mRNAs) that correspond to the differentially displayed bands. As a first attempt, the DNA fragment is sequenced, as described in Chap. XXXIII in this volume, followed by searches in the DNA sequence databases, using the DNA sequence as query sequence. If the gene is known, this will identify the differentially expressed gene; however, unless if human samples have been analyzed, the chance that the gene is known is rather small since relatively few nonhuman genes have been sequenced. If human samples have been analyzed, there is a high probability that the gene is represented in the databases as an expressed sequence tag (EST), and this may then identify the gene (see Chap. XXXIII). If the sequence is not present in the database, or if it is a nonhuman sample, it is necessary to clone the corresponding cDNA and sequence parts of the coding region and then repeat the search in the database with coding region sequence as query sequence(s). If this still fails to identify the gene, it is probably an unknown gene and the complete sequence of the cDNA should be determined.

2
Principles and Applications

The following procedure describes how any DNA fragment can be radioactively labeled and used for the screening of cDNA (or genomic) libraries constructed in phage λ but essentially the same procedure can be used for screening of plasmid libraries. How to construct and screen both phage λ and plasmid libraries is described in Sambrook et al. (1989). Below we briefly describe the different steps and problems that are associated with these methods.

Labeling of Probes

When using DDRT-PCR fragments as probes, the main problem is that the fragments often are relatively small (<100 bp) which makes it difficult to perform an efficient labeling. In addition, random primed labeling or nick translation of a fragment of 100 bp will result in labeled fragments in the size range of 25–80 nucleotides and these are not optimal for hybridization since longer probes will result in stronger signals. However, we have random primed labeled fragments of 55–60 bp and successfully used these for high stringency screenings, with exposure times of 1–2 weeks (see Sect. 2.3). If the DNA fragment is very short (20–30 bp), it should be end-labeled (if possible). Alternatively, an oligonucleotide should be constructed, 5'-end labeled and used as probe. In that case, it may be necessary to change the hybridization protocol slightly as suggested by Sambrook et al. (1989). In the more extreme cases in which degenerated oligonucleotides are used, we recommend the protocol by Honoré et al. 1993.

Plating of Libraries

As mentioned above, we recommend using libraries that are constructed in phage λ, because: (1) they generally contain more independent recombinants and (2) they can be plated at a higher density. cDNA libraries can either be constructed in the laboratory (Sambrook et al. 1989), custom made by a company (Stratagene or Clontech) or bought as premade libraries (Clontech, Stratagene and other). We recommend using λ-ZAP vectors (Stratagene) because they greatly reduce the work "downstream" from plaque purification (i. e., preparation of DNA and analysis of the cDNA insert).

The library should be plated at a density of 500 plaque forming units (pfu) per cm^2 when using "normal" random primed labeled probes; this is a compromise between signal intensity and the number of plates to screen. Fewer pfu/cm^2 results in stronger signals (but more plates!), which may be necessary when using oligonucleotide probes. This is especially true if the probes are degenerated oligonucleotides, in which case we recommend using 300 pfu/cm^2 or less. The plaques are transferred to nylon membranes for hybridization which can present a problem, since the top agar occasionally sticks to the filter. This can partly be avoided by using moist filters, as described below; however, less DNA will be transferred to the filters resulting in lower signal intensity. For experienced users, it may be advantageous to use dry filters, especially when screening with degenerated oligonucleotides. The nylon filters can be used at least 10–15 times for screenings with DNA fragments, but the plates can only be stored (at 4 °C) for 3–4 weeks. Therefore we recommend freezing the plates at –80 °C and the titer will remain essentially unchanged for at least 2–3 years. The plates must remain frozen, positive plaques being excised directly from the frozen plates!

Generally there are no problems with the hybridization step, just be sure that the hybridization solution comes into contact with all the filters in the sandwich. This is achieved by inverting the tubes a few times during the prehybridization, hybridization and washing steps. For long exposures, it is essential to reduce the background. Therefore we use a relatively large amount of RNA as carrier (instead of sonicated DNA) because this results in less background. If there still is a high background, the RNA concentration in the prehybridization solution can be increased to 5–600 µg/ml.

Hybridization and Washing

The specificity of the screening is determined during the washing steps. Low stringency washing will identify clones with more than 60 %–80 % sequence identity to the probe; high stringency washing identifies clones with more than 90 %–95 % identity (depending on the length of the probe). When using high stringency washing, only a few positive clones per filter are to be expected. Thus, if autoradiography reveals a large number of positive clones, it may be caused by the presence of a repetitive sequence (e.g., an Alu or CpA repeat) in the DNA fragment (this should be checked by sequencing, see Chap. XXXIII). If so, it may be impossible to use the DNA fragment as probe; however, even repetitive sequences include minor differences and it may therefore be possible to solve the problem by increasing the stringency during the washing step (e.g., to 70 °C in 0.1x SSC or 65°–70 °C in 0.05x SSC), and then pick the clones that exhibit the strongest signals. Note that it is possible to first do a low stringency wash and an exposure and then remove the filters from the plastic bags and do a high stringency wash, without an additional hybridization step.

3
Materials

Equipment – hybridization oven (e.g., from Hybaid)
– tabletop centrifuge
– waterbath
– 37 °C incubator
– temperature regulated shaker
– UV transilluminator (e.g., Stratalinker from Stratagene)
– plastic bag sealer

Supplies – 10 and 50 ml plastic tubes (NUNC, Denmark)
– 22×22 cm screening dishes (NUNC, Denmark)
– 9 cm petri dishes (NUNC, Denmark)
– nylon membranes (e.g., Hybond-N from Amersham)
– plastic bags
– X-ray film and exposure cassette with intensifying screen
– Mesh nylon sheet (e.g., Scrynel from ZBF AG, Rüschlikon, Switzerland)

Media – LB media: 5 g yeast extract, 10 g peptone, 5 g NaCl, water to 1 l
– LBMM media: 5 g yeast extract, 10 g peptone, 5 g NaCl, 2 g $MgCl_2$, 2 g maltose, water to 1 l
– agar for plates: 2 % agar in LB media
– top agar: 0.8 % agar in LBMM media

Reagents and solutions – total yeast RNA (Boehringer Mannheim)
– 0.1 M dCTP, dGTP and dTTP (Pharmacia)
– $[\alpha^{32}P]$dATP: 10 µCi/µl (3000 Ci/mmol; Amersham, NEN or ICN)
– phenol (molecular biology grade) saturated with 0.25 M sodium acetate (pH 6.0)
– chloroform
– ethanol (99 % and 80 %)
– 0.5 M EDTA (pH 8.0)
– 3 M sodium acetate (pH 6.0)
– TE buffer: 10 mM Tris-HCl (pH 7.5), 0.1 mM EDTA
– 10x random priming buffer: 500 mM Tris-HCl (pH 7.5), 50 mM $MgCl_2$, 100 mM β-mercaptoethanol
– 10x A-labeling nucleotide mix: 2.5 mM each of dGTP, dCTP and dTTP
– 100x Denhart: 2 % bovine serum albumin (Boehringer Mannheim), 2 % Ficoll 400 (Sigma), 2 % polyvinylpyrrolidone (Sigma)
– 20x SSC: 3 M NaCl, 0.3 M sodium citrate
– 10 % SDS

- SM buffer: 100 mM NaCl, 10 mM MgCl$_2$, 50 mM Tris-HCl (pH 7.5)
- denaturation buffer: 0.5 M NaOH, 1.5 M NaCl
- renaturation buffer (neutralization solution): 0.5 M Tris-HCl (pH 7.5), 1.5 M NaCl
- prehybridization solution: 5x SSC, 5x Denharts, 0.5 % SDS, 200 µg carrier RNA per ml
- hybridization solution: 5x SSC, 5x Denharts, 0.5 % SDS, 100 µg carrier RNA per ml
- wash buffer-1: 2x SSC, 0.1 % SDS, 0.5 mM EDTA
- wash buffer-2: 1x SSC, 0.1 % SDS, 0.5 mM EDTA
- wash buffer-3: 0.1x SSC, 0.1 % SDS, 0.5 mM EDTA
- random 8-mer primer (N$_8$): 0.25 µg/µl in water
- amplified DDRT-PCR DNA fragment in water or TE
- Klenow fragment enzyme: 5 U/µl (Amersham/USB; Pharmacia Biotech; Boehringer Mannheim or others)

4
Experimental Procedure

1. Dissolve total yeast RNA at 100 µg/µl in 0.25 M sodium acetate (pH 6.0).

    Preparation of carrier RNA

2. Add equal volume of phenol saturated with 0.25 M sodium acetate (pH 6.0). Shake vigorously for 2–3 min.

3. Centrifuge for 5 min at room temperature. Transfer water phase to a new tube (note that the phases may be inverted!).

4. Repeat steps 2 and 3.

5. Add equal volume of chloroform. Shake vigorously for 2–3 min.

6. Centrifuge for 5 min at room temperature. Transfer water phase to a new tube.

7. Repeat steps 5 and 6.

8. Add 2 volumes of 99 % ethanol, vortex vigorously, incubate at –20 °C for 10 min.

9. Centrifuge at 8000 rpm for 10 min at room temperature.

10. Remove supernatant and wash the pellet (vortex vigorously) with 80 % ethanol, centrifuge and dry the pellet carefully.

11. Dissolve in original volume of 0.25 M sodium acetate (pH 6.0).

12. Repeat steps 8–10.

13. Dissolve in water at about 150 µg/µl. Measure concentration, adjust to 100 µg/µl with water. Freeze in aliquots at –20 °C.

Preparation of labeled probe

1. Prepare random labeling mix. For one labeling, mix in an Eppendorf tube: 2.5 µl 10x random priming buffer; 2.5 µl 10x A-labeling nucleotide mix; 4–6 µl [α^{32}P]dATP (40–60 µCi); water to 18 µl. Leave at room temperature.

2. Mix 4 µl DNA fragment (20–100 ng) and 3 µl random 8-mer primer (0.25 µg/µl).

3. Transfer the DNA/primer mix to a glass capillary, seal the ends carefully (Bunsen burner).

4. Heat in boiling water for 3 min. Cool on ice for 30 s.

5. Add DNA/primer mix to random labeling-mix (break off the ends of the capillary).

6. Add 1 µl Klenow fragment enzyme (5 U), mix gently and centrifuge. Incubate at room temperature for 1 h.

7. Add 75 µl TE buffer, 3 µl 0.5 M EDTA, 10 µl 3 M sodium acetate. Vortex.

8. Add 250 µl ethanol, vortex and incubate for 10 min at –70 °C.

9. Centrifuge at 14000 rpm for 10 min. Remove supernatant.

10. Wash the pellet with 0.5 ml 80 % ethanol, dry and dissolve in 0.5 ml hybridization solution.

Plating of cDNA library

The following protocol is for plating on large 22×22 cm dishes. If smaller dishes are used the amount of phages should be reduced. The aim is to get about 500 pfu/cm^2.

1. Prepare agar dishes. For a 22×22 cm dish, use 250 ml of 2 % agar in LB media. Leave plates at 4 °C overnight. Prewarm and dry the plates upside down without lids for 1–3 h at 37 °C just before use.

2. Grow an overnight culture of the appropriate bacterial strain in LBMM media.

3. Dilute bacteria 1:10 in LBMM media. Grow for 3 h at 37 °C with vigorous shaking.

4. Pipette phages corresponding to 150000 pfu into a 50 ml NUNC tube.

5. Add 5 ml of diluted bacteria, incubate at 37 °C (or temperature appropriate for the particular phage strain) for 20 min.

6. Add 35 ml 0.8 % top agar (at 42 °C) to each tube, mix carefully by inverting the tubes.

7. Pour the top agar with phages and bacteria onto the prewarmed (37 °C) dishes. Allow the top agar to harden at room temperature. If the plates are wet, they should be incubated without lids until they are dry. Incubate overnight at 37 °C (or temperature appropriate for the particular phage strain).

8. The next day, cool the plates at 4 °C for 1–2 h.

1. Premoisten filter on a piece of moist 3MM paper. The filter should just be moist, not wet!   **Plaque lift**

2. Carefully place the filter on the dish, mark the filter and the plate carefully! (Use a water resistant pen).

3. Carefully lift off the filter, it should be on the dish for as short a time as possible (i.e., the time it takes to mark the plate and the filter).

4. Place filter "plaque side up" on a piece of 3MM paper soaked with denaturation solution; leave filter for 5 min.

5. Transfer filter to a piece of 3MM paper soaked with neutralization solution; leave filter for 3 min.

6. Repeat step 5.

7. Transfer the filter to a dish containing 200 ml 2x SSC. Gently wash off bacterial debris and agar pieces. Leave the filter plaque side up on a piece of benchtop paper and let it dry.

8. Place a second filter on the plate, mark the filter according to the marks on the plate (from step 2). Leave the filter on the plate for 3 min. This is a replica to avoid artifacts later in the procedure.

9. Repeat steps 4–7.

10. Cross-link the DNA to the filter by UV irradiation for 90 s using an UV transilluminator. Invert the filter and repeat UV irradiation for 1 min.

11. Store filters dry at room temperature or proceed with hybridization. The filters can be used for at least 10–15 hybridizations.

12. The plates can be stored at –80 °C simply by placing them in an –80 °C freezer. The phage titer will remain essentially unchanged for at least 2–3 years, although ice crystals will grow on the plates; however, this does not affect the viability of the phages.

Note: Never thaw the plates! Remove plaques while the plates are still frozen!

Hybridization   (This is for four filters in one hybridization tube. For fewer filters, reduce the volumes).

It is essential that the hybridization and washing solutions come into contact with each filter, which can be achieved by inverting the tubes. This results in the filters coiling up when the tube is placed in one direction and rolling out when the tubes are inverted. During overnight hybridization, the filters should not be coiled up.

1. Wet the filters and mesh sheets in 5x SSC.

2. Sandwich the filters and mesh sheets (e.g., for four filters in one hybridization tube: mesh-filter-mesh-filter-mesh-mesh-filter-mesh-filter-mesh).

3. Wash with 100 ml 5x SSC.

4. Remove the 5x SSC carefully!

5. Add 35 ml prehybridization solution (without labeled probe).

6. Incubate in hybridization oven for 2 h, invert the tubes a few times.

7. Transfer labeled probe to a plastic tube containing 20 ml hybridization solution. Boil for 10 min in a waterbath, quickly cool to 65 °C.

8. Carefully remove the prehybridization solution from the filters.

9. Add hybridization solution with the labeled probe to the filters.

10. Incubate in the hybridization oven at 65 °C for 18–36 h. Invert the tubes a few times in the beginning of the incubation period.

Low and high stringency washing   (All incubations are done in the hybridization oven)

1. Remove the hybridization solution. Store at –20 °C. It can be reused 2–3 times within a 2 week period (e.g., for purification of positive plaques and verification of the cDNA clone, see below).

2. Add 100 ml prewarmed (60 °C) wash buffer-1. Incubate at 60 °C for 15 min; invert the tubes a few times.

3. Remove wash buffer-1 and repeat step 2.

4. Remove wash buffer-1 carefully and add 100 ml prewarmed wash buffer-2. Incubate at 60 °C for 15 min; invert the tubes a few times.

5. Repeat step 4.
   This corresponds to a low stringency wash. If related cDNAs are of interest, go to step 8. For high stringency, continue with step 6.

6. Remove wash buffer-2 and add 100 ml prewarmed (65 °C) wash buffer-3. Incubate at 65 °C for 15 min; invert the tubes a few times.

7. Repeat step 6 twice.

8. Remove the filters from the hybridization tube. Seal the filters in plastic bags.

9. Mark the plastic bags with radioactive ink according to the markings on the filters.

10. Expose to an X-ray film at −70 °C using an intensifying screen. Exposure time is from 12 h to 2 weeks, depending on the specific activity, amount and length of the probe and the quality of the filters. After autoradiography, the filters are stored at −20 °C in the sealed plastic bags.

Positive plaques will appear on both filter 1 and the replica filter. If the signal only appears on one of the filters, it is probably an artifact.

If the plates are frozen, it is essential that they are kept frozen during the following procedure.

Purification of positive plaques

1. Prepare small (9 cm) agar plates, each containing 25 ml 2 % agar in LB media. Three plates are needed for each purification.

2. Align the autoradiograph on the bottom of the original plate using the markers that were made on the plates and filters (use filter 1, since it generally is better aligned with the markers on the plates than the replica).

3. Mark the position of the positive plaque(s) on the bottom of the plate using a thick black pen. The marking should be visible through the frozen agar from the top of the plate. If it is difficult to see the marking, use a light box.

4. Excise an area corresponding to about 1 cm^2 around the positive plaque. Ideally only the top agar should be taken from the plate; however, a few mm of the bottom agar will not interfere with the procedure.

5. Transfer the agar plug to an Eppendorf tube containing 200 μl SM buffer.

6. Crush the agar plug carefully using a glass rod. Shake for 15 min at room temperature.

7. Add 30 μl chloroform and shake for 5 min. Centrifuge briefly. Store at 4 °C. This is the stock for this plaque; it should be kept until it has been verified that the cDNA insert in the phage clone present in the purified plaque corresponds to the desired cDNA (see below).

8. Make a dilution in SM buffer. The dilution factor depends on the quality of the plates, the amount of phage DNA that was transferred to the filters and the phage strain. For λ-ZAPII, we generally dilute 1 μl into 1 ml SM buffer for frozen plates and 1 μl into 2 ml for fresh plates; however this can vary considerably!

9. Transfer 1, 2 and 3 μl of the diluted phage stock to three 10 ml tubes, respectively.

10. Add 150 μl diluted bacteria (steps 2 and 3 in "Plating of cDNA library"). Incubate at 37 °C (or as appropriate for the particular phage strain) for 20 min.

11. Add 3 ml 0.8 % top agar in LBMM media (prewarmed to 42 °C); mix by inverting the tubes.

12. Pour the phage-top agar-bacteria onto the prewarmed (37 °C) small petri dish. Allow the top agar to harden at room temperature. Incubate overnight at 37 °C(or temperature appropriate for the particular phage strain).

13. The next day, cool the plates at 4 °C for 1 h.

The plates are now ready for plaque lift, as described in "Plaque lift". Select a plate with 200–700 plaques, fewer plaques may result in losing the "positive" plaque, and if there are too many plaques on the plate it will be necessary to repeat the purification step. Moreover, for the small plates it is not necessary to prewet the filters and replica filters are normally not required since it should be possible to align a positive plaque (on an autoradiograph) with a plaque on the plate. The labeled probe that was used for the original screening can be reused in the purification step. When the plaque(s) have been purified, DNA should be prepared either from the λ phages or as plasmid mini-prep (for λ-ZAP) as described by Sambrook et al., (1989).

We strongly recommend verifying that the probe actually hybridizes with the isolated cDNA as described below.

1. Transfer 1–5 µl of the DNA solutions (20–1000 ng) to a piece of nylon filter; allow it to dry. Include a sample with only vector DNA.

2. Denature and renature the DNA and cross-link it to the filter as described in "Plaque lift", steps 4–7.

3. Prehybridize, hybridize and wash the filter as described in "Hybridization" and "Low and high stringency washing".

The old labeled probe can often still be used for the verification, since very few counts are required because of the (comparatively) large amount of DNA that is applied to the filters.

This procedure applies to four 22×22 cm filters. Nylon filters can be stripped and reused at least 10–15 times.

1. Wash filters in 1 l water in a large beaker; remove water. Do not allow the filters to dry.

2. Boil 2 l water.

3. Add 20 ml 10 % SDS to the filters.

4. Immediately add the boiling water to the filters.

5. Leave until the water has cooled to 40°–50 °C.

The filters are now ready for prehybridization as described in "Hybridization".

Acknowledgements. Development of the methods was supported by The Danish Cancer Society, The Danish Biotechnology Programme, The European Commission and The Danish National University Hospital (Rigshospitalet).

References

Honoré B, Madsen P, Leffers H (1993). The tetramethylammonium chloride method for screening of cDNA libraries using highly degenerate oligonucleotides obtained by backtranslation of amino acid sequences. J Biochem Biophys Methods 27: 9–48
Liang P, Pardee AB (1992) Differential display of eukaryotic messenger RNA by means of the polymerase chain reaction. Science 257:967–971
Sambrook J, Fritsch EF, Maniatis T. (1989) Molecular cloning, a laboratory manual. Cold Spring Harbor Laboratory Press, Cold Spring Harbor

Chapter XXXVI

Targeted RNA Fingerprinting

B. Stone

1
Introduction

The wealth of sequence data generated in the last 20 years has revealed the existence of many multigene families (MacIntyre 1994; Ohta 1991). Genes within a family typically share short stretches of conserved primary sequence, usually as part of conserved functional domains. Genes within a family may also play similar or overlapping roles in diverse tissues and processes. In some cases, individual genes within a family are differentially expressed, with up-regulation coinciding with a requirement for the gene product in a specific biological process. Examples include the differential expression of homeobox containing genes during development (Lawrence and Morata 1994; Scott and Weiner 1984) and altered abundance of members of the cyclin gene family during various phases of the cell cycle (Pines 1995; Hunt 1991). Thus, for processes in which specific gene families are suspected to play a role, an assay that allows for the rapid detection and cloning of differentially expressed family members would be of great value. Targeted RNA fingerprinting (TRF) was developed to respond to such a need.

TRF is a PCR based technique which can be used to clone differentially expressed members of any multigene family of interest (Stone and Wharton 1994). TRF can be used with almost any gene family during any biological process. Both previously identified and novel members of the gene family under investigation can be detected and cloned using this protocol.

Two recent protocols describe the cloning of low to medium abundance cDNA fragments which are differentially expressed among multiple samples of RNA (Liang and Pardee 1992; Welsh et al. 1992). However, due to the exclusive use of arbitrary primers, these techniques are unable to efficiently detect differential expression within specific gene families of interest. In contrast, TRF uses arbitrary primers in the reverse transcription reaction, then, during the amplifications, adds a set of degenerate primers corresponding to a conserved coding region to favor amplification of a specific gene family. This allows the rapid detection of differentially expressed members of known gene families expressed at levels as low as 100 copies/cell.

Yoshikawa et al. have recently described a variation of TRF which also targets multigene families (Yoshikawa et al. 1995). In their protocol, two primers are designed which correspond to two distinct, conserved coding regions within the gene family of interest, much as in conventional degenerate PCR. Their protocol requires these two coding regions to be separated by a variable distance in the individual family members. While this approach may prove useful for some sets of sequences, many gene families may not contain conserved coding domains that are separated in this manner. In addition, the extra sequence information required may limit the success of this approach when the researcher is attempting to detect novel family members.

One major advantage to using a targeted approach is the ease with which a particular cDNA fragment can be determined to be of interest. In TRF, candidate fragments are cloned and sequenced. Following sequencing, only cDNA fragments with sequences that are consistent with the gene family under investigation are subjected to further analysis. This rapidly reduces the pool of potentially interesting clones to a manageable number. Fragments that are unrelated to the targeted gene family, yet appear to be differentially expressed may also be saved and investigated. In contrast, in differential display, candidate fragments typically consist of only 3' untranslated sequences, thus their significance remains unknown following sequencing. Clones isolated with this technique must be subjected to further cloning and sequencing. This can be a time consuming process since large numbers of candidate fragments can be generated.

The goal of this chapter is to provide detailed protocols for targeted RNA fingerprinting. All unique or critical procedures are described. However, details of common molecular techniques, such as running sequencing gels and cloning into plasmid vectors are omitted.

2
Experimental Strategy

Overview of Procedure

Targeted RNA fingerprinting begins with the isolation of RNA representing two or more different physiological states. The RNA pools are then reverse transcribed using a primer with an arbitrary sequence on the 3' end. The reverse transcription primers are designed such that priming of RNA occurs about every 1000 bases. This allows amplification of cDNA fragments from within the coding region of messenger RNA. The cDNA samples are then split into several aliquots for amplification. Separate, parallel radiolabeled amplifications are carried out using the reverse transcription primer alone and in combination with a primer corresponding to a conserved coding region in the gene family of interest (the targeting primer). The targeting primer initiates amplification of cDNA fragments which contain the conserved coding sequence. These fragments are identified through comparison with amplification products amplified in the absence of the targeting primer. Four to six different reverse transcription primers and one to six similar targeting primers are used.

Following amplification, products are resolved on a 4 % sequencing gel and autoradiographed. Fragments that are unique to, or more abundant in one RNA pool are then eluted, reamplified, cloned and sequenced. CDNAs with sequences that are consistent with the gene family under investigation can then be used as probes to confirm differential expression using conventional techniques.

Primer Design

In general, all primers should have about 50 % G+C content where possible and end in G or C. Three types of primers are used. Reverse transcription (RT) primers are arbitrary in sequence on their 3' end and are used in reverse transcription and amplification reactions. Targeted (TA) primers are based on a conserved coding domain within a gene family and are used during amplification steps to favor amplification of members of the gene family under investigation. Amplification (AM) primers are used to amplify candidate cDNA fragments of interest following gel electrophoresis. The use of different restriction sites on the RT primers and TA primers allows efficient, directional cloning and also provides selection against products that are amplified with only one species of primer.

The RT primers are typically 20 bp in length, with one position of fourfold degeneracy within the last six bases on the 3' end. This limited degeneracy allows random priming to occur approximately every kilobase. The 5' end has a *Nhe*I site with a four base GA repeat as a clamp. Six different RT primers are suggested to ensure initiation of reverse transcription within 300 bases of most coding sequences.

The TA primers should ideally be comprised of five codons of conserved coding sequence in the sense direction. Codons with limited degeneracy should be positioned in the 3' end of the primer if possible. In positions of complete degeneracy, inosine may be used. The third position of the last codon is usually omitted.

Gene families with a limited number of members may be amplified with one or two degenerate TA primers. In this case, these primers should be designed to include all possible coding sequences for the last three codons present at the 3' end, while the two codons on the 5' end should include only the most common codons used in the species being studied (for human codon bias, see Lathe 1985). No more than 24-fold degeneracy should be used for any one of these primers. For gene families with a large number of members, or with repeated conserved motifs, a separate primer should be made for each permutation of coding sequence for the three codons on the 3' end of the conserved sequence. The remaining two codons in the 5' end are designed as described above. Immediately 5' to the conserved coding sequence, each TA primer has a C, followed by an *Eco*RI restriction site and two GTC sequences as a clamp.

AM primers are synthesized which anneal to the invariant 5' end of each RT primer and each targeted primer. This allows for efficient reamplification of excised fragments with identical primers in all cases.

Suggested RT and AM primer sequences are listed below. An example of a set of TA primers used to amplify members of the zinc finger gene family is also shown.

Primer examples

RT primers:
- clamp *Nhe*I arbitrary sequence
- GAGAGCTAGCTTAGGNTCAGA
- GAGAGCTAGCTTATCNGGACC

TA primers for messages coding for conserved H/C link region (HTGEKP) present in zinc finger containing peptides:
- clamp *Eco*RI   H   T   G   E   K   P
- GTCGTCGAATTCCA(C/T)AC(A/C/T)GG(A/G)GAAAAACC
- GTCGTCGAATTCCA(C/T)AC(A/C/T)GG(A/G)GAAAAGCC

- GTCGTCGAATTCCA(C/T)AC(A/C/T)GG(A/G)GAGAAACC
- GTCGTCGAATTCCA(C/T)AC(A/C/T)GG(A/G)GAGAAGCC

AM primers:
- AM 1: GGAACTGAGAGCTAGCTT
  (amplifies the RT primed cDNA end)
- AM 2: TGGTTGTCGTCGAATTCC
  (amplifies the TA primed cDNA end)

3
Materials

- microcentrifuge   **Equipment**
- UV spectrophotometer
- water bath (cycling temperature block can be used as substitute)
- cycling temperature block
- sequencing gel apparatus
- gel drying apparatus large enough for sequencing gels

- prepackaged 200 µl and 1 ml pipette tips   **Supplies**
- rubber policeman (tissue scrapers)
- RNase-free 1.5 ml microcentrifuge tubes
- 0.5 ml and 1.5 ml microcentrifuge tubes (not thin-walled)
- 30S104>40 cm 3MM paper
- plastic wrap
- X-ray film and exposure cassette
- plasmid cloning vector with restriction sites compatible with those present in TA and RT primers
- plasmid miniprep columns (Qiagen or equivalent)
- sequencing kit with labeled nucleotide

- RT, TA and AM primers (see primer design section). Primers may   **Oligonucleo-** be purified in any manner that is compatible with PCR. This   **tides** includes simple butanol extractions followed by precipitation (Sambrook et al. 1989). Gel purification is not necessary. Primers should be suspended in TE at a final concentration of 20 µM.

All chemicals should be molecular biology grade.   **Reagents**
- phosphate buffered saline (PBS) solution: 137 mM NaCl, 2.7 mM   **and solutions** KCl, 4.3 mM Na_2HPO_4, 1.4 mM KH_2PO_4 (pH 7.3)
- guanidinium solution: 4 M guanidinium thiocyanate, 25 mM sodium citrate (pH 7), 0.5 % sarcosyl, 0.1 M 2-mercaptoethanol. This solution does not have to be made in diethylpyrocarbonate (DEPC) treated water but should be made in double distilled water.

- DEPC treated water. Mix 0.5 % DEPC in double distilled water and hold overnight at 37 °C. Autoclave 1 hr.
- 2 M sodium acetate (pH 4) made in DEPC treated water. Dissolve the sodium acetate in a minimum of water and adjust pH with acetic acid.
- phenol, water saturated
- chloroform
- isopropanol
- 80 % ethanol made with DEPC treated water
- 10 mM Tris (pH 7.5) made in DEPC treated water
- RNase inhibitor
- 2.5 M sodium acetate (pH 5.2)
- absolute ethanol
- 5x reverse transcriptase buffer: 250 mM Tris (pH 8.3), 250 mM KCl, 20 mM DTT, 20 mM $MgCl_2$
- dNTPs mix: 2.5 mM each
- MMLV reverse transcriptase
- 10x amplification buffer: 250 mM Tris (pH 8.3), 250 mM KCl, 25 mM $MgCl_2$
- 3000 Ci/mmol [αP^{32}]dCTP
- *Taq* DNA polymerase
- 4 % sequencing gel
- 2x gel loading buffer: deionized formamide with 0.05 % bromphenol blue and 0.05 % xylene cyanol
- gel elution buffer: 0.5 M ammonium acetate, 10 mM magnesium acetate, 1 mM EDTA, 0.1 % SDS
- TE buffer: 10 mM Tris (pH 8.0), 0.1 mM EDTA
- phenol (Tris pH 7.0 equilibrated)
- chloroform:isoamyl alcohol 24:1 (v/v)
- glycogen
- restriction enzymes and buffers which recognize TA and RT sites (*Eco*RI, *Nhe*I and *Xba*I)

4
Experimental Procedure

RNA Isolation

For each RNA isolation, cells should be as homogenous as possible. Cultured cells work very well but any source of RNA is suitable as long as the resulting RNA is of high quality and is representative of the physiological state being studied.

In general, any method that results in the isolation of high quality RNA is suitable. Poly-A selection is unnecessary. We have used the acid guanidinium thiocyanate/phenol:chloroform extraction (APGC) method of Chomczynski and Sacchi (1987) for our experiments beginning with cultured cells. Superior results are obtained when freshly isolated RNA is used immediately in reverse transcriptions. The APGC method is outlined below.

1. For tissue culture cells, rinse cells 1x with cold PBS solution. Add 1.8 ml of guanidinium solution to a single 100 mm tissue culture plate, tip plate up on one side and scrape viscous cell lysate down.

2. Pipette 0.5 ml of lysate into 1.5 ml microcentrifuge tubes.

3. Add 100 µl of 2 M sodium acetate (pH 4) and vortex 10 s.

4. Add 0.5 ml water saturated (not Tris buffered) phenol and vortex 10 s.

5. Add 200 µl of chloroform and vortex 10 s.

6. Hold on ice for 10 min and centrifuge at maximum speed for 20 min at 4 °C.

7. Being careful not to disturb the interface, remove and save the upper aqueous layer in an RNase-free 1.5 ml tube (as much as 20 % may be left covering the organic layer; if the RNA is extremely valuable, back extract with guanidinium solution, and repeat the extraction). The organic (lower) layer can be discarded.

8. Precipitate the RNA by adding 0.5 ml isopropanol to the aqueous layer and briefly vortex.

9. Centrifuge at maximum speed for 20 min at 4 °C.

10. Dissolve the pellet in 0.5 ml guanidinium solution and reprecipitate by adding 0.5 ml isopropanol and hold on ice for 1 h.

11. Centrifuge at maximum speed for 20 min at 4 °C.

12. A pellet should be obvious. Remove and discard the fluid.

13. Wash the pellet by adding 300 µl of ice cold 80 % ethanol and centrifuging for 1 min. Repeat once.

14. Resuspend the pellet in 10 mM Tris (pH 7.5) made in DEPC treated water by carefully pipetting up and down several times. If RNA does not go into solution quickly, heat at 70 °C for 5 min, hold on ice for 5 min and then gently pipette several times.

15. Quantitation is determined by 260/280 absorption. Label 1.5 ml microcentrifuge tubes for blank and for each RNA sample. Add 500 µl of water to the blank and 495 µl water to each sample tube.

16. Take 5 µl of each RNA sample and add to 495 µl water in the appropriate tube and vortex.

17. Add 1 µl of RNase inhibitor to each stock of RNA.

18. Determine the 260/280 absorbence for each diluted RNA sample. Concentration is determined by multiplying the OD_{260} by the dilution factor (100 in this case) and by the conversion factor of 40 µg/ml for each OD.

19. RNA samples should be reverse transcribed immediately. If this is not possible, RNA can be stored at −80 °C following precipitation with on tenth volume DEPC treated 2.5 M sodium acetate and 2 volumes absolute ethanol. It is suggested that RNA quality be tested through electrophoresis of 1 µg on a **fresh** (used within 10 min of solidification) 1 % agarose mini-gel. This gel can be run in TAE or TBE buffer. Sharp ribosomal bands should be evident after 20 min of electrophoresis.

Reverse Transcription

These reactions generate cDNA from two or more samples of RNA prior to amplification. One microgram of each RNA sample is reverse transcribed separately with each RT primer in 50 µl reactions. In addition, for each RNA pool, negative controls consisting of parallel reactions without the addition of MMLV reverse transcriptase should be included. These negative controls need to be done only once using one species of RT primer for each RNA isolation. The negative controls will be amplified to check for DNA contamination in the RNA preparations (in the absence of reverse transcription, very few amplification products should be observed).

The following is an example of reverse transcription of two different RNA pools (RNA 1 and RNA 2). The resulting cDNAs will be used to compare targeted amplification products using two TA primers in the next section. These reactions should be conducted using barrier pipette tips.

1. Mark four tubes with the following designations, 1+, 1−, 2+ and 2−. In step 2, the tubes marked 1 will receive RNA 1 and the other tubes RNA 2. The tubes marked with a + will receive MMLV reverse transcriptase while those with a − will not receive enzyme.

2. In each tube, mix:
   - x μl (x=1 μg) of RNA
   - 33 − x μl DEPC treated water
   - 1 μl (50 μM) stock RT primer

3. Heat the RT primer/RNA mixtures at 65 °C for 5 min, ice 5 min.

4. Mix on ice a master mix for five reactions (four reaction tubes plus some extra volume to ease pipetting).

reagent	vol×5	volume added to mix
5x RT buffer	10 μl×5	50 μl
dNTPs mix	4 μl×5	20 μl
RNase inhibitor	1 μl×5	5 μl

Mix well with pipette tip.

5. Add 15 μl of master mix to each RNA tube.

6. Add 1 μl (200 U) of MMLV reverse transcriptase to those tubes marked + only. The tubes marked − are the negative controls and do not receive enzyme.

7. Hold reactions at 37 °C for 30 min. Heat 70 °C for 10 min. Reactions should be quick frozen on dry ice and stored at −80 °C.

Amplifications

The following amplification protocol utilizes the cDNA samples generated in the previous section. The cDNA samples are amplified with two different TA primers, TA 1 and TA 2. The negative controls are included, though these need to be done only once for each RNA isolation. It is probably best to begin with an experiment of this size. As the researcher gains experience, additional TA primers can be included. A grid describing the numbering and contents of each reaction tube is presented below to assist in organization. The reverse transcription reaction number used as template for each amplification is listed at the top of each column, while the TA primers included for each amplification are listed at the beginning of each row. Amplification tube numbers are designated with a # sign.

RT reactions	1+	2+	1−	2−
primer TA 1	#1	#2	#7	#8
primer TA 2	#3	#4		
no TA primer	#5	#6		

1. Label eight 0.5 ml amplification tubes 1–8 as in the above grid.

2. Program a cycling temperature block to perform the following two linked profiles.
    1 cycle: 94 °C for 5 min, 50 °C for 5 min, 72 °C for 5 min
    linked to 20 cycles of: 94 °C for 30 s, 60 °C for 30 s, 72 °C for 1 min
    hold at 4 °C

3. Set up a master mix for nine amplification reactions (eight reactions plus one for ease in pipetting).

reagent	vol/reaction×9	vol in mix
10x amplification buffer	4.0 µl×9	36.0 µl
dNTPs mix	0.4 µl×9	3.6 µl
RT 1 primer 50 µM stock	1.0 µl×9	9.0 µl
3000 Ci/mmol [αP^{32}]dCTP	0.5 µl×9	4.5 µl
water	21.1 µl×9	189.9 µl

4. Add 5 µl of the appropriate RT reactions to the amplification tubes numbered 1–8 (see grid). Refreeze the cDNA on dry ice immediately.

5. Add 1 µl TA primer 1 to tubes 1, 2, 7, 8. Add 1 µl TA primer 2 to tubes 3 and 4. Do not add TA primers to tubes 5 and 6.

These last tubes are necessary to compare the products that are amplified by the RT primer alone with those amplified with the RT and TA primers.

6. Transfer 29 µl of master mix to each of the amplification tubes.

7. Overlay reactions with two drops of mineral oil.

8. Dilute *Taq* polymerase (5 U/µl stock) 1:50 in 1x amplification buffer, mixing well by pipetting gently.

9. Place reaction tubes in cycling temperature block and begin cycling program described in step 1.

10. When the program reaches the first annealing temperature (50 °C), pause the program and add 1 µl diluted *Taq* polymerase to each tube. Make sure that the pipette tip goes through the oil layer before expelling the diluted polymerase.

11. Cancel the pause in the program to allow cycling to resume.

12. When the final cycles are finished, hold samples at 4 °C. Dilute *Taq* polymerase 1:24 in 1x amplification buffer (final concentration 5 µl = 1 U).

13. Add 5 µl diluted *Taq* (1 U) to each amplification tube and repeat the 20 cycles with high stringency annealing (60 °C).

14. Amplification products can be stored at 4 °C and must be run on a gel within 36 h of amplification.

Gel Electrophoresis and Autoradiography.

Products are resolved on conventional 4 % sequencing gels. Protocols for pouring and running these types of gels are common so only those steps important to TRF will be mentioned here. All products amplified with the same RT and TA primers should be run in adjacent lanes. (In the example presented above, lanes should be in consecutive order 1–8.)

1. Dilute 5 µl of each sample in 5 µl 2x loading buffer.

2. Heat samples to 70 °C for 3 min and hold on ice.

3. Load 3 µl of each sample in each lane.

4. Run gels at 65 Watts of constant current until the xylene cyanol dye is 3/4 of the way down the gel.

5. Separate gel plates, do **not** fix, transfer gel to paper backing and dry 1 h at 80 °C.

6. Take 1 µl of an amplification reaction and dilute in 9 µl of loading buffer. Use this diluted solution and a 200 µl pipette tip to make some indexing marks on the paper next to the dried gel. The dye in the loading buffer will stain the paper backing while the radioactive label will leave a corresponding mark on the autoradiogram. This will allow precise alignment of the gel with the autorad prior to band excision.

7. The gels can be exposed to film from 6–12 h and then developed. At least two exposures should be done. One exposure will be a permanent record and the other will be cut as a template for band excision. Save the dried gel for band excision.

Evaluation of Results

The first lanes that should be examined are lanes 7 and 8 (see example in Fig. 1). These are negative controls (no reverse transcriptase was used). There should be significantly fewer bands present in lanes 7 and 8 than in the other lanes. In addition, bands should not be similar when lane 7 is compared with lane 8 (no more than one or two bands in common). If either of these criteria is not met, the RNA samples are contaminated by DNA and an additional DNase treatment step prior to reverse transcription is advised (see Troubleshooting).

If there is no evidence of DNA contamination, the other lanes can be examined. Lanes that were amplified with matching primer sets should be compared (compare lane 1 with lane 2 and lane 3 with lane 4, etc.). Bands that are present only in lanes with TA primers and seem to be unique to or more abundant in one RNA sample are considered as candidate differentially expressed, targeted cDNAs (bands marked with arrowhead in Fig. 1). Bands that vary in intensity only

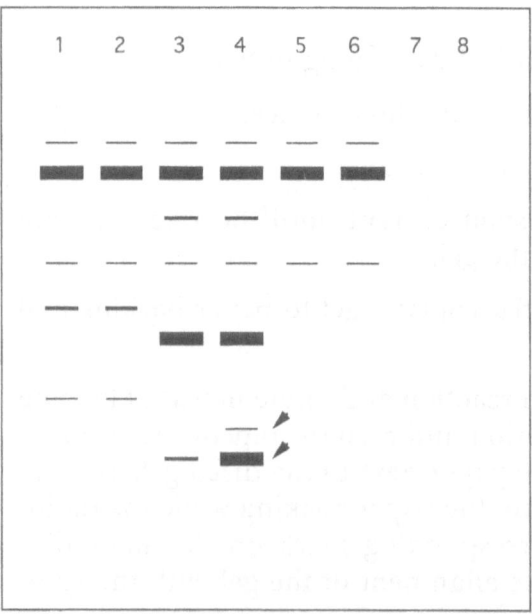

Fig. 1. This is an idealized representation of the (TRF) targeted RNA fingerprinting autoradiogram produced in the example given in the text. The amplification reaction numbers are listed above the appropriate lanes. For further details, see text

slightly between the RNA pools may be of interest, especially if other bands above and below appear to be very uniform. Bands that exhibit differing intensities in those lanes amplified with RT primers only may be differentially expressed, but are unlikely to be members of the targeted gene family.

The number of fragments dependent upon TA primers may vary considerably, depending on the number of distinct genes within the family, their relative levels of expression and the size of the coding sequence used to design the TA primers. Thus the number of bands that should be analyzed is difficult to predict and must be determined empirically for each set of targeting primers. It is often observed that the average size of amplification products produced with RT combined with TA primers is smaller than those amplified with RT primers alone.

Band Excision and Reamplification.

1. Bands identified as described in the "Evaluation of Results" section should be marked lightly with a sharp pointed, permanent marker on both sets of autorads. Include some sort of identification number that is unique to each band.

2. Mark two tubes with the unique identifier for each band to be excised.

3. On one set of autoradiograms only, the desired bands should be cut out with a single-edged razor-blade.

4. Lay the gel and the "cut out" autoradiogram over a flat cutting board. The cut out autoradiogram should be oriented over the gel by aligning the blue dye marks on the paper with the corresponding signal on the film. Heavy weights can be set on three corners to maintain the orientation.

5. Use a single-edge razor-blade to cut the gel and paper backing through the windows of the autoradiogram. If the gel is covered by plastic wrap, this should be taken off each fragment. The cut out fragments are placed in the tube marked with the appropriate identification number. A separate, written record of the fragment number and the primers that were used to generate it should also be kept.

6. Rehydrate the gel slice with 200 μl of gel elution buffer.

7. Take the gel slice out using clean forceps and place on a piece of plastic wrap. Separate the gel slices from the paper backing using

forceps and 18 gauge syringe needles (it is impossible to remove all of the paper). Place the gel in a fresh tube with 200 µl of elution buffer.

8. Rotate at room temperature overnight.

9. Centrifuge at maximum speed and transfer the supernatant to a fresh Eppendorf tube. Add 0.2 ml of ethanol. Hold at –20 °C for 4 h.

10. Centrifuge at maximum speed in a microfuge for 20 min at 4 °C.

11. Wash the pellet once with 100 % ice-cold ethanol.

12. Resuspend in 10 µl of TE.

13. Reamplify the fragments using the following recipe for each reaction:
    - 3.0 µl eluted fragment
    - 5.0 µl 10x amplification buffer
    - 0.5 µl AM primer 1
    - 0.5 µl AM primer 2
    - 2.0 µl dNTPs mix
    - 38.0 µl water

14. Dilute *Taq* 1:4 in 1x amplification buffer. Dilute enough *Taq* such that there is 1 µl of diluted *Taq* (1 µl = 1 U) for each reamplification reaction.

15. Place reactions in cycling temperature block and program the 30 cycles with the following profile: 94 °C for 30 s, 53 °C for 30 s, 72 °C for 1 min.

16. Begin cycling, pause program at the first annealing step and add 1 µl of diluted *Taq* to each reaction. Resume cycling.

17. Products should be checked on a 1 % agarose gel. The relative sizes of each fragment should be preserved (compare the sizes of the fragments seen on the original fingerprinting acrylamide gel with those observed on the agarose gel).

Cloning and Sequencing.

Fragments should be cloned directionally into plasmid vectors using conventional techniques (Sambrook et al. 1989). The following procedure is a workup of each reamplified fragment prior to cloning. Protocols describing restriction digestion, ligation into a vector and transformation are common and are not described. Direct sequencing of PCR products has not been tested with this protocol but may work for some fragments.

1. Remove 30 µl of each successful reamplification reaction to a fresh tube.

2. Extract once with Tris (pH 7.0) buffered phenol:chloroform.

3. Extract once with chloroform:isoamyl alcohol.

4. Add 1 µl of glycogen, 3 µl of 2.5 M sodium acetate and 60 µl of absolute ethanol. Vortex lightly.

5. Hold at –20 °C for 1 h.

6. Centrifuge in microfuge at 4 °C at maximum speed for 15 min.

7. Wash pellet once with 100 µl of ice cold 80 % ethanol.

8. Resuspend in 20 µl of TE buffer.

9. The fragment can now be digested with *Eco*RI and *Nhe*I and cloned into a plasmid vector (*Nhe*I and *Xba*I have compatible overhangs).

10. Following cloning, two or three colonies from each transformation should be picked, streaked and checked for insert using a conventional miniprep or by PCR.

11. Insert-positive clones can be used to generate sequencing templates. There are many commercial mini-prep kits that yield high quality plasmid DNA suitable for sequencing.

12. Sequence clones manually using dideoxy sequencing or automated sequencing protocols. Automated sequencing is suggested for large numbers of clones (20 or more). Sequencing in the sense direction only is usually enough to identify sequences with homologies to the targeted gene family.

Database Comparisons.

Compare fragment sequences with the known members of the targeted gene family. For information about running a blast search of GenBank and other databases, send an e-mail "help" message to the blast server at blast--40--ncbi.nlm.nih.gov . More options for database searching can be found at the Baylor College of Medicine search launcher. The URL for this service is http://kiwi.imgen.bcm.tmc.edu.

Fragments which do not correspond to known gene family members should be translated and checked for additional amino acid residues typically conserved in the gene family being targeted. These additional residues should be outside of the conserved region used to design the TA primers. Matches here may identify previously unidentified family members. A useful DOS based shareware program for this type of manipulation is Seqaid by Douglas D Rhoads and Donald J Roufa (1988) and is available at ftp://iubio.bio.indiana.edu/molbio/ibmpc .

Confirmation of differential expression

Known or novel members of the targeted gene family should be analyzed through northern blots or RNase protection assays to confirm differential expression (for detailed northern blot and RNase protection protocols see Chaps. XXXI and XXXII respectively).

5
Troubleshooting

If RNA samples are contaminated by DNA, perform a DNase digestion as described below.

1. Following RNA extraction, resuspend RNA at 1 µg/µl.

2. Mix for each RNA sample:
   - 10 µl (10 µg) RNA
   - 1 µl RNase inhibitor
   - 29 µl DEPC treated water
   - 5 µl 10x DNase I buffer
   - 5 µl RNase-free DNase I (10 U, diluted in 1x DNaseI buffer)

3. Hold at 37 °C for 1 h.

4. Extract once with Tris buffered phenol:chloroform.

5. Extract once with chloroform:isoamyl alcohol.

6. Precipitate with 5 µl 2.5 M sodium acetate (pH 5.2) and 100 µl ethanol, vortex and hold at –20 °C for 1 h.

References

Chomczynski P, Sacchi N (1987) Single-Step Method of RNA Isolation by Acid Guanidinium Thiocyanate-Phenol-Chloroform Extraction. Anal Biochem 162:156–159
Hunt T (1991) Cyclins and their partners: from a simple idea to complicated reality. Seminars in Cell Biology 2:213–222
Lathe R (1985) Synthetic Oligonucleotide Probes Deduced from Amino Acid Sequence Data. J Mol Biol 183:1–12

Lawrence P A, Morata G (1994) Homeobox Genes: Their Function in *Drosophila* Segmentation and Pattern Formation. Cell 78:181–189

Liang P, Pardee A B (1992) Differential Display of Eukaryotic Messenger RNA by Means of the Polymerase Chain Reaction. Science 257:967–971

MacIntyre R J (1994) Molecular evolution: codes, clocks, genes and genomes. BioEssays 16(9):699–703

Ohta T (1991) Multigene families and the evolution of complexity. J Mol Evol 33(1):34–41

Pines J (1995) Cyclins and cyclin-dependent kinases: theme and variations. Advances in Cancer Research 66:181–212

Sambrook J, Fritsch E F, Maniatis T (1989) Molecular Cloning: A Laboratory Manual Cold Spring Harbor Laboratory Press, Cold Spring Harbor

Scott M P, Weiner A J (1984) Structural relationships among genes that control development: Sequence homology between the *Antennapedia*, *Ultrabithorax*, and *fushi tarazu* loci of *Drosophila*. Proc Natl Acad Sci USA 81:4115–4119

Stone B, Wharton W (1994) Targeted RNA fingerprinting: the cloning of differentially-expressed cDNA fragments enriched for members of the zinc finger gene family. Nucleic Acids Res 22:2612–2618

Welsh J, Chada K, Dalal S S, Cheng R, Ralph D, McClelland M (1992) Arbitrarily primed PCR fingerprinting of RNA. Nucleic Acids Res 20:4965–4970

Yoshikawa T, Xing G, Detera-Wadleigh S D (1995) Detection, simultaneous display and direct sequencing of multiple nuclear hormone receptor genes using bilaterally targeted RNA fingerprinting. Biochim Biophys Acta 1264:63–71

Chapter XXXVII

Generation Of Tumor-Specific DEST Catalogues

M. A. WATSON and T. P. FLEMING

1
Introduction

The differential display polymerase chain reaction (DDPCR) and arbitrarily primed polymerase chain reaction (APPCR) are two techniques that have been implemented with increasing frequency to identify novel, differentially expressed mRNAs (Liang and Pardee 1995; McClelland et al. 1995). In most applications of these techniques, two populations of cell lines (e.g., tumor cell line vs corresponding nonmalignant primary cells) are compared and PCR fragments that are differentially present in one or another population are used as probes to isolate corresponding full-length cDNAs. We have made three basic modifications to this approach to study abnormal gene expression in human breast neoplasia (Watson and Fleming 1994).

The first modification involves the use of ^{32}P end-labeled oligonucleotides rather than ^{35}S deoxynucleotides to generate labeled PCR fragments. This approach eliminates the hazards of thermal cycler contamination by aerosolized ^{35}S (Trentmann et al. 1995). Furthermore, PCR fragments generated by symmetric amplification of the arbitrary 10-mer oligonucleotide used in the reaction are not visualized; only fragments generated by incorporation of a longer, labeled poly-T oligonucleotide are seen. While this approach may overlook potentially interesting sequences generated solely by the 10-mer oligonucleotide, it should also promote amplification of authentic 3' poly-A ends of mRNAs rather than random amplification of contaminating ribosomal RNA and multiple upstream regions of mRNA. As a result, the "display" produced by this method is less complex and may result in a higher yield of truly differentially expressed fragments.

A second modification involves the use of patient tumor biopsy specimens rather than cell lines as a source of RNA. We believe that genetic alterations identified in carcinoma cell lines may have limited value as clinical markers or therapeutic targets if they do not accurately represent events that occur during the natural course of disease. Identification of tumor-specific changes in gene expression that occur in vivo may ultimately provide more useful markers for patient management. However, because tumor biopsy specimens do not contain homogeneous cell populations, a large number of isolated DDPCR fragments may represent differences in gene expression due to tissue heterogeneity between specimens rather than malignancy-associated changes. Furthermore, contamination of tumor specimens with nonmalignant tissue may obscure any cancer-specific loss of gene expression. This problem may be ameliorated in part by the use of microdissected tumor specimens (Jensen and Page 1994), but will always be an important caveat whenever working with primary tumors rather than cell lines.

The most significant modification of this protocol is its adaptation to a high-throughput system to catalogue sequences that are abnormally expressed in human tumors. The creation of expressed sequence tag libraries is a rapid and powerful method to identify or "tag" sequences that are expressed in specific tissues or cell lines (Adams et al. 1995). The advantage of this methodology, compared to isolating and sequencing individual cDNAs, is that a large number of sequences can be "catalogued" with small amounts of sequencing data. We have therefore combined the concepts of differential display and expressed sequence tag libraries to create a tumor-specific differentially expressed sequence tag (DEST) catalogue. In this approach, fragments isolated by DDPCR are sequenced directly and their tumor-specific differential expression is confirmed without the time- and labor-intensive steps of subcloning, library screening, and individual cDNA sequencing. By analyzing a large number of patient specimens with this protocol and correlating tumor DESTs with clinical and pathological data, it should be possible to identify sequence tags that have prognostic value for disease parameters such as treatment response, relapse interval, and overall patient survival. An overview of this approach is illustrated in Fig. 1.

2
Materials

Equipment   – tissue homogenizer
– thermal cycler
– sequencing gel electrophoresis apparatus

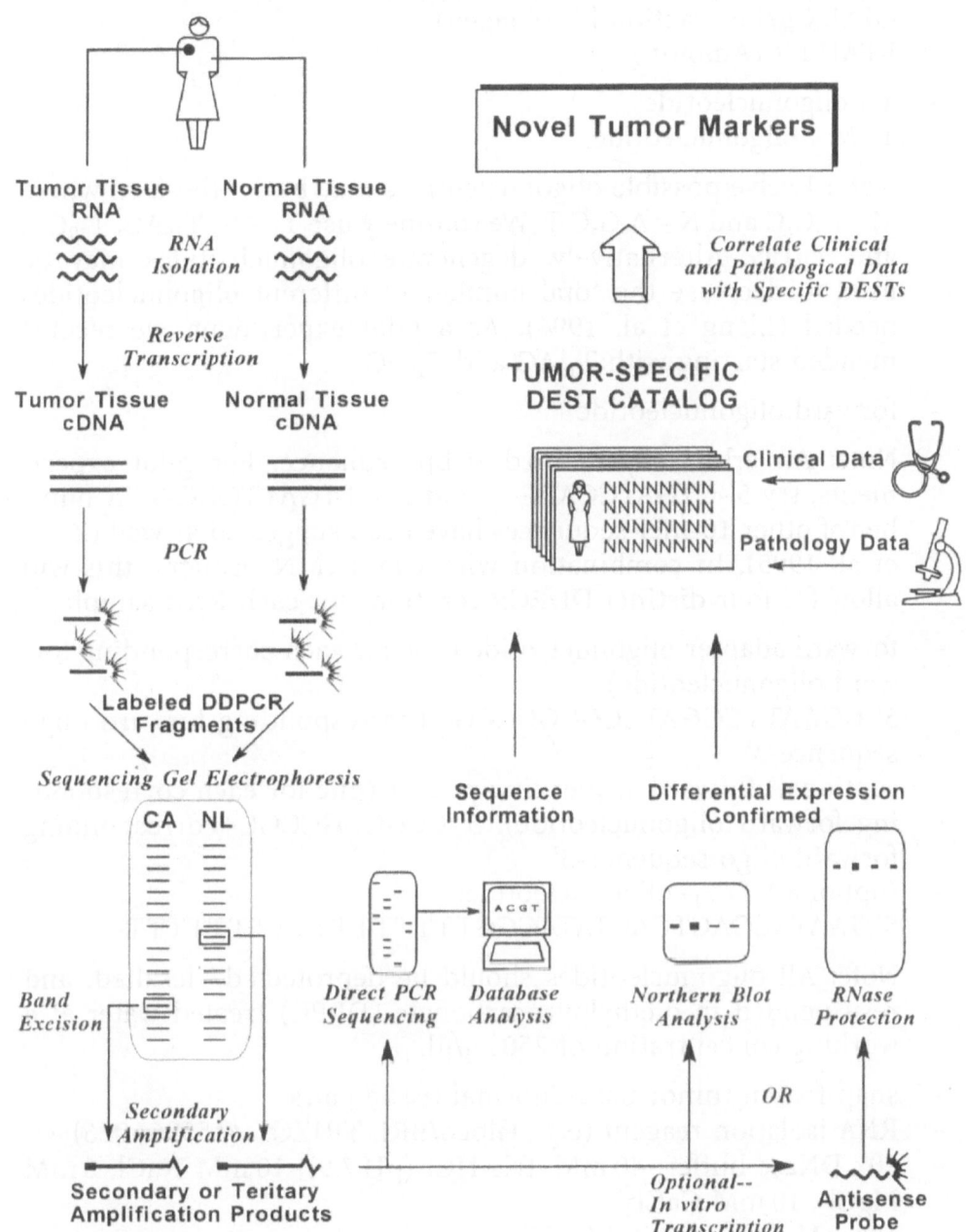

Fig. 1. Outline of a procedure for creating a tumor-specific DEST catalogue

Kits	– QIAEX gel extraction kit (Qiagen)
	– RPAII kit (Ambion)
Oligonucleo-	– T_{21} oligonucleotide
tides	– $T_{19}MN$ oligonucleotide

Note: Twelve possible oligonucleotides may be synthesized where $M=A,G,C$ and $N=A,G,C,T$. We routinely use $T_{19}AG$, $T_{19}AC$, $T_{19}CC$, and $T_{19}CG$. Alternatively, degenerate oligonucleotides may be used to decrease the total number of different oligonucleotides needed (Liang et al. 1994). As a pilot experiment, we recommended starting with $T_{19}AG$ and $T_{19}AC$.

– forward oligonucleotide

Note: An arbitrarily defined 10 bp sequence. For pilot experiments, try 5'-CTGATCCATG-3' and 5'-CTTGATTGCC-3'. A number of other 10-mer sequences have been suggested as well (Zhao et al. 1995). In combination with two $T_{19}MN$ primers, this will allow for four distinct DDPCR reactions for each RNA sample.

– forward adapter oligonucleotide (one for each corresponding forward oligonucleotide):
5'-GGAATTCGGATCCGCGGCCGC+corresponding forward oligo sequence-3'
– (optional) forward sequencing primer (one for each corresponding forward oligonucleotide): 5'-CCGCGGCCGC+corresponding forward oligo sequence-3'
– (optional) $T7T_{17}$ oligonucleotide:
5'-TAATACGACTCACTATAGGGTTTTTTTTTTTTTTTTT-3'

Note: All oligonucleotides should be deprotected, desalted, and resuspended in diethylpyrocarbonate (DEPC) treated water at a working concentration of 250 ng/μl.

Reagents	– snap-frozen tumor tissue/normal tissue pairs
and solutions	– RNA isolation reagent (e.g., Gibco/BRL TRIZOL #15596–026)
	– 10x DNase buffer: 40 mM Tris-HCl (pH 7.9), 10 mM NaCl, 6 mM $MgCl_2$, 10 mM $CaCl_2$
	– 100 mM dithiothreitol (DTT)
	– RNase inhibitor (e.g., Promega RNasin #N2511)
	– RNase-free DNase (e.g., Promega RQ1-DNase #M6101)
	– phenol:chloroform
	– 3 M sodium acetate
	– ethanol
	– DEPC treated water
	– 5x reverse transcriptase buffer (supplied with enzyme)

- 10 mM dNTP solution: 10 mM each dATP, dCTP, dGTP, TTP in DEPC treated water
- Superscript II MMLV-modified reverse transcriptase (Gibco/BRL #18064–014)
- $[\gamma^{32}P]ATP$ (10 mCi/ml; 3000 Ci/mmol)
- 10x T4 polynucleotide kinase buffer (supplied with enzyme)
- T4 polynucleotide kinase
- 10x *Taq* DNA polymerase buffer (supplied with enzyme)
- 25 mM $MgCl_2$
- *Taq* DNA polymerase
- mineral oil
- 10 mg/ml glycogen
- agarose
- acrylamide and buffers for sequencing gel electrophoresis

3
Experimental Procedure

Several tumors of interest and corresponding normal tissue from the same patient should be obtained. For a pilot experiment, two to three pairs of tumor/normal specimens are sufficient to generate a modest number of sequence tags. These specimens may be obtained at random from cancer patients, or may be selected based on specific clinical or pathological parameters (e.g., hormone receptor status, family history, stage of disease). We believe that it is important to obtain patient-matched normal tissue to control for interindividual variation in gene expression (e.g., patient hormonal status in breast and ovarian tumors). The tumor and surrounding normal tissue biopsy specimens should be well circumscribed and free of extraneous tissue such as muscle and fat. Obviously, grossly dissected tumor tissue will be composed of tumor tissue, normal tissue, and stroma. This is an inevitable drawback to this in vivo approach, but we have nevertheless been successful in isolating tumor-specific sequence tags using such specimens. Alternatively, microdissected frozen sections may be used for isolating RNA (Jensen and Page 1994).

Tissues should be rapidly dissected from surgical specimens and snap-frozen. Autopsy material or biopsy specimens that have not been rapidly processed fail to provide quality RNA. If surgical specimens are not available locally, the Cooperative Human Tissue Network (LiVolsi et al. 1993) is an excellent source for tumor specimens. We have found that 70 %–80 % of samples obtained from this source yield high quality RNA. Specimens are also accompanied by reports that provide useful pathological information.

Tumor procurement

RNA preparation

RNA can be isolated using any number of commercially available reagents (e.g., Gibco/BRL TRIZOL reagent) or other standard methodology. Isolation of poly-A mRNA is not necessary. Although poly-A mRNA theoretically provides a more specific template for amplification by removing the large excess of ribosomal RNA, we have found little difference in message display between poly-A and total RNA using this protocol.

Prior to first-strand synthesis, contaminating genomic DNA should be removed from RNA samples. This step is not necessary if RNA is poly-A selected.

1. Mix the following:
   - 13 μl   10 μg RNA sample
   - 2 μl   10x DNase buffer
   - 1 μl   100 mM DTT
   - 1 μl   (~20 U) RNasin
   - 3 μl   (~3 U) RNase-free DNase

2. Incubate at 37 °C for 30 min.

3. Phenol:chloroform extract the RNA and precipitate with ethanol.

4. Resuspend the RNA in DEPC water at 0.1–0.5 μg/μl.

5. Examine a small fraction (0.5 μg) of RNA by gel electrophoresis to ascertain integrity. RNA is now ready for first-strand synthesis.

Reverse transcription of RNA

1. For each RNA sample to be displayed, perform three reverse transcriptase (RT) reactions: two duplicate reactions and one reaction without enzyme as a negative control.

2. For each reaction, add 0.1–1 μg of total RNA and 250 ng (35 pmoles; Fc=1.8 μM) of T_{21} oligonucleotide to a 0.5 or 1.5 ml tube. Add DEPC treated water to a final volume of 10 μl. Heat the RNA/oligonucleotide at 68 °C for 10 min.

3. While RNA samples are heating, prepare a reaction mix (e.g., for 10 reactions). For each reaction add:
   - 4 μl   5x RT buffer   40 μl
   - 2 μl   100mM DTT   20 μl
   - 1 μl   10 mM dNTPs   10 μl
   - 0.5 μl   RNasin (~20 U)   5 μl
   - 1.5 μl   DEPC water   15 μl

   Note: Make sufficient reaction mix for one or two more reactions than planned.

4. Incubate the RNA/oligonucleotide at 45 °C for 10 min.

5. Add 10 µl of reaction mix (without enzyme) to negative control RNAs.

6. Add 1 µl (200 U) of RT enzyme per reaction to the reaction mix. Mix briefly.

7. Add 10 µl of reaction mix (+enzyme) to remaining RNAs.

8. Incubate all reactions at 45 °C for 60 min.

9. After the reactions are complete, add 80 µl of DEPC water to each reaction.

10. Incubate at 95 °C for 10 min to denature cDNA/RNA hybrids.

11. These first-strand reactions may now be stored for at least 1 year at −20 °C.

Oligonucleotide labeling

1. Add the following to a 0.5 ml tube:
   - 1 µl   250 ng (36 pmol) $T_{19}MN$ oligonucleotide
   - 3 µl $[\gamma^{32}P]ATP$ (3000 Ci/mmol; 10 mCi/ml)
   - 4 µl   water
   - 1 µl   10x T4 polynucleotide kinase reaction buffer
   - 1 µl   (5–10 U) T4 polynucleotide kinase

2. Incubate at 37 °C for 30 min.

3. Heat at 90 °C for 2 min to inactivate enzyme. Spin briefly.

Differential display PCR

1. Make a master reaction mix. For each PCR reaction to be performed add:
   - 40 µl water
   - 5   µl 10x *Taq* polymerase reaction buffer
   - 4   µl 25 mM $MgCl_2$ (Fc=2 mM)
   - 0.1 µl 10 mM dNTPs (Fc=20 µM)
   - 0.2 µl 50 ng (7 pmol) $T_{19}MN$ oligonucleotide (Fc=0.15 µM)
   - 0.2 µl 50 ng (15 pmol) forward oligonucleotide (Fc=0.3 µM)
   - 0.3 µl 7.5 ng (1.1 pmol) labeled $T_{19}MN$ oligonucleotide (Fc=23.5 nM)
   - 0.3 µl (1.5 U) *Taq* DNA polymerase

   Note: Make sufficient reaction mix for one or two more reactions than planned.

2. Aliquot 50 µl of reaction mix to appropriately sized PCR tubes.

3. Add 1 µl of each first-strand cDNA to PCR reaction tube.

4. Perform PCR reactions with the following program:
   1 cycle: 94 °C for 2 min.
   40 cycles: 94 °C for 45 s, 40 °C for 2 min, 72 °C for 30 s
   1 cycle: 72 °C for 2 min

5. After reactions are complete, remove from mineral oil (if necessary) and precipitate nucleic acid with 1/10 volume (5 μl) 3 M sodium acetate and 2 volumes (100 μl) ethanol.

6. Resuspend each dried pellet in 8 μl of standard formamide sequencing dye.

7. Heat reactions for 5 min at 90 °C and load ~3 μl on a standard 6 % polyacrylamide/7.5 M urea sequencing gel. Also load a sequencing reaction or other labeled molecular weight ladder to identify the size of displayed fragments.

8. Samples should be run to resolve fragments in the 100–300 nucleotide range. This usually corresponds to running the second (xylene cyanol) tracking dye to the bottom of the sequencing gel.

9. Gel is dried (**but not fixed**) and exposed to film for 24–48 h. Avoid the use of intensifying screens as this will decrease band resolution on the film. Be certain to add a sufficient number of orientation marks on the gel so that the gel and film can be accurately aligned for band excision (Kim et al. 1995).

Isolation of differentially expressed bands

1. Figure 2 demonstrates typical results of a differential display using this protocol. Several features are indicative of a successful experiment:
   - Between 10 and 30 intense, discreet bands should be seen in each lane.
   - The majority (>95 %) of bands should appear identical between duplicate RT samples.
   - Most bands (>80 %) should appear identical between individual tumor/normal samples.
   - Bands appearing in the negative control (no RT enzyme) reactions should be sparse, irreproducible, and completely different from experimental lanes. Theoretically, there should be no bands present in the negative control lanes, but in practice this is rarely the case. These bands probably represent trace nucleic acid contamination present in reagents or (more likely) enzyme preparations.

2. Based on examination of the autoradiograph, pick PCR bands for further analysis. Use the following criteria to choose bands which will most likely represent differentially expressed mRNAs:

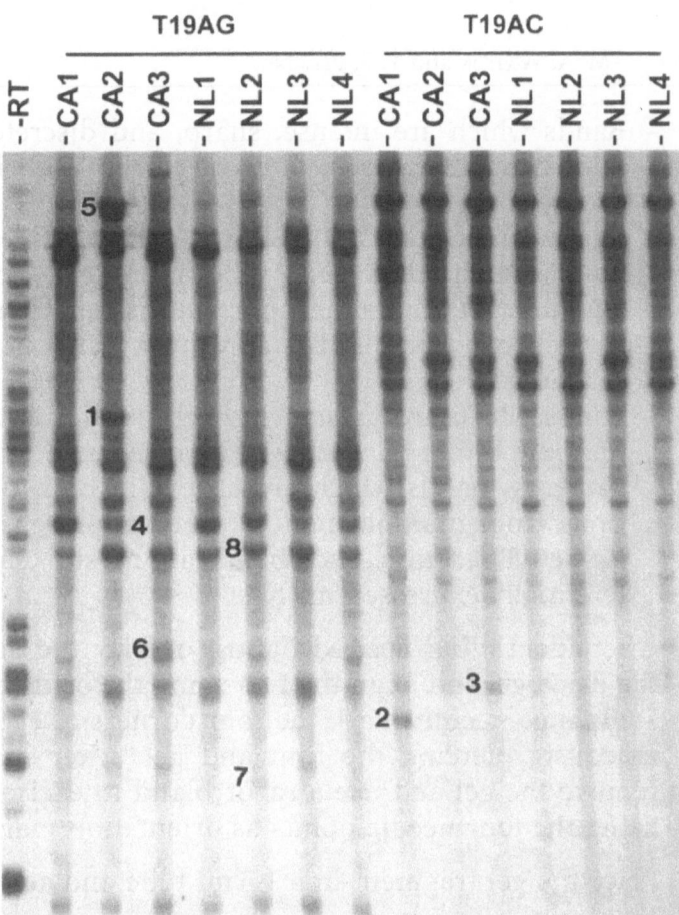

Fig. 2. Representative results of a differential display experiment. RNAs from three breast tumor biopsy specimens (CA1, CA2, CA3) and four normal breast tissue controls (NL1, NL2, NL3, NL4) were subjected to the DDPCR protocol described with the forward oligonucleotide 5'-CTGATCCATG-3' and either $T_{19}AG$ or $T_{19}AC$. Note that for a given oligonucleotide combination, the majority of displayed bands are identical between tumor and normal tissue reactions. However, use of different $T_{19}MN$ oligonucleotides produces a strikingly different display pattern. Theoretically, the negative (without RT) control reaction should not display any fragments. However, we often observe a banding pattern that is irreproducible and exhibits no similarities to other experimental samples. Discreet, intense bands that are present in tumor samples but not normal tissue controls (*1, 2, 3*) are good candidates for differentially expressed sequences. Note that if duplicate RT reactions had been run for these samples (as recommended in the protocol) and these bands had been present in both reactions, this would increase the likelihood of their true differential expression. Occasionally, we observe bands which are absent from a particular tumor sample but are present in other tumors and normal tissue controls (*4*). High molecular weight bands (*5*), fuzzy bands (*6*), and very faint bands (*7*) often do not reamplify or are not truly differentially expressed in target tissues. It may be helpful to excise an intense, sharp band that is present in all samples (*8*) as a control. Such a band may represent a housekeeping gene whose identity can be immediately confirmed by direct sequencing of the secondary amplification product and comparison to established sequence databanks (Watson and Fleming 1994)

- Bands which are intense, sharp, and discrete. Avoid fuzzy or faint bands as these will more likely result in amplified PCR artifacts.
- Bands larger than 100 bp. Avoid small doublet bands.
- Bands which are present in both duplicate tumor RT reactions and present in neither normal tissue RT reaction (or vice versa). Avoid bands that appear in only one of two duplicate reactions.
- Bands which are present in both duplicate RT reactions from multiple tumors and absent in all normal tissue RT reactions; similarly, bands which are absent in both duplicate RT reactions from multiple tumors, but are present in every normal tissue RT reaction. Such bands are most likely to represent differentially expressed mRNAs.

3. After identifying desired bands, overlay the autoradiograph on the dried gel and align the two using the orientation marks. With a 20 gauge needle, mark the four corners of each band by simultaneously piercing the film and gel. After marking all bands, remove the gel and use a razor bland to excise the gel fragment using the four needle points as orientation marks

4. Place the gel fragment in a 1.5 ml tube and add 100 µl of water.

5. Vortex tube for 1 h at room temperature to rehydrate gel slice.

6. Heat tube at 95 °C for 15 min.

7. Vortex tube for 1 h at room temperature.

8. Spin tube in a microcentrifuge for 5 min to pellet gel debris.

9. Carefully remove ~80 µl of supernatant without touching acrylamide or paper debris.

10. Precipitate DNA:
    80 µl   DNA
    4 µl   10 mg/ml glycogen (Fc=0.1 mg/ml)
    9 µl   3 M sodium acetate
    250 µl   ethanol
    Incubate at –20 °C for 1 h

11. Resuspend pellet in 20 µl water.

Secondary amplification

1a. Amplify the eluted PCR product in a secondary PCR reaction:
    30 µl   water
    5 µl   10x *Taq* polymerase reaction buffer
    4 µl   25 mM $MgCl_2$

  1 µl 10 mM dNTPs (Fc=200 µM)
  2 µl 500 ng (70 pmol) $T_{19}MN$ oligonucleotide (Fc=1.5 µM)
  2 µl 500 ng (50 pmol) forward adapter oligonucleotide
        (Fc=1.0 µM)
0.3 µl (1.5 U) *Taq* DNA polymerase
  5 µl eluted PCR band

1b. Perform PCR reactions as follows:
    1 cycle: 94 °C for 2 min
    5 cycles: 94 °C for 45 s, 45 °C for 1 min, 72 °C for 30 s
  40 cycles: 94 °C for 45 s, 55 °C for 1 min, 72 °C for 30 s,
    1 cycle: 72 °C for 2 min

2. Analyze reactions on a 2 % agarose gel. The secondarily amplified PCR product should be a sharp, discrete band of roughly the same size as the primary DDPCR band excised. If the PCR product is a smear, multiple bands, or an inappropriately sized band, it will very seldom represent the desired DDPCR product originally identified.

3. Excise the PCR product from the agarose gel and purify using any commercially available products (e.g., Qiagen QIAEX gel extraction kit). The DDPCR fragment is now ready for additional analysis.

There are several rapid approaches to determine the significance of the isolated DDPCR product without the need for subcloning.

Analysis of DDPCR product

1. The DDPCR product may be sequenced directly using the forward sequencing primer or even the forward adapter oligonucleotide. A number of kits for the direct sequencing of PCR products are commercially available. We routinely use the *fmol* sequencing system (Promega) with ^{32}P end-labeled sequencing primer.

In our experience, approximately 30 % of DDPCR products generated by this method are amenable to direct sequencing. The remaining 70 % of products contain duplicated ends ($T_{19}MN$ sequence or forward primer sequence at each end) and therefore produce either no sequence or two overlapping sequences when primed with forward sequencing or forward adapter primers. Interestingly, those products that can be sequenced directly are usually differentially expressed in the target tissues. The remaining products often are not.

    Sequence obtained can be directly compared to various sequence databases at the NCBI (Benson et al. 1994) or TIGR (Adams et al. 1995) via Internet access. For more information on these resources contact:

NCBI: http://www.ncbi.nlm.nih.gov/
TIGR: http://www.tigr.org/

2. The differential expression pattern of the DDPCR product should be confirmed by traditional methods. The secondary amplification product may be used directly in a ^{32}P random primer labeling reaction to generate a radiolabeled cDNA probe. Original tumor and normally tissue RNAs used for the DDPCR experiment are then subjected to northern blot analysis using standard methodology (for a detailed northern blotting protocol see Chap. XXXI).

3. Often, there is insufficient tumor RNA for northern Analysis. Alternatively, the mRNA represented by the DDPCR product is too rare to be detected by this method. In these cases, the secondarily amplified product may be used to generate anti-sense RNA probes for RNase protection experiments.

Note: This method is only useful if the DDPCR product does not contain identical ends (i.e., is amenable to direct sequence analysis).

Perform the following tertiary PCR amplification:
34   μl   water
 5   μl   10x *Taq* polymerase reaction buffer
 4   μl   25 mm $MgCl_2$
 1   μl   10 mM dNTPs (Fc=200 μM)
 2   μl   500 ng (40 pmol) $T7T_{17}$ oligonucleotide (Fc=0.7 μM)
 2   μl   500 ng (50 pmol) forward adapter oligonucleotide (Fc=1.0 μM)
0.3 μl   (1.5 U) *Taq* DNA polymerase
 1   μl   secondary PCR product

Perform PCR reactions as follows:
1 cycle: 94 °C for 2 min
35 cycles: 94 °C for 45 s, 55 °C for 1 min, 72 °C for 30 s
1 cycle: 72 °C for 2 min

The PCR product is electrophoresed on a 2 % agarose gel, excised, and purified as described above for the secondary amplification product. Although the $T7T_{17}$ primer has no anchoring 3' nucleotide, we have found that under the above conditions, a discrete, specific PCR product can be produced from secondary products generated from any $T_{19}MN$ oligonucleotide.

This PCR product can now be used as a template for T7 RNA polymerase to generate anti-sense probes for RNase protection experiments. Many commercially available kits can be used for this methodology (e.g., Ambion, RPAII kit).

4. Once differential expression of the DDPCR fragment is confirmed and sequence information is obtained, sequence data are entered into a database along with all available clinical and pathological parameters available for the tumor specimen from which the DDPCR fragment was obtained. Each successive sequence tag obtained is then compared to this database. Sequence tags that are frequently isolated from independent tumor specimens (particularly if those tumors share a common pathological or clinical variable) are used as probes to isolate full length cDNAs for future study (Fig. 1).

4
Troubleshooting

- The primary PCR does not produce an acceptable differential display
    - Check the integrity of the first-strand reactions by using them in a standard PCR reaction with control primers for housekeeping genes such as β-actin.
    - There are several components in the first-strand reaction that decrease the efficiency of the PCR reaction. Try using a further 1:10 dilution of the first-strand reaction in the PCR reaction. While this may select against rare differentially expressed messages, it often improves the quality of the display.
    - Perform the differential display PCR with a plasmid cDNA library template ($\sim$1–10 ng of DNA) rather than first-strand cDNA reactions. This will serve as an excellent positive control for the PCR reaction and is a good template to use for optimizing PCR parameters. Once an acceptable display is obtained with the cDNA library, return to the use of first-strand RT reactions.
    - Depending on the thermal cycler employed, the annealing segment of the PCR program may be varied from 35 °C to 45 °C. We have not found that any other parameter in the PCR program significantly affects the resulting display.
    - Try other combinations of $T_{19}MN$ and forward primers. The combination of primers employed has the greatest affect on the quality of the display. We have not found that the concentration of either primer used (over a tenfold range) has a noticeable effect on the display.
    - If a radiolabeled smear and few discrete bands are seen, try decreasing the concentration of dNTPs in the reaction.

- Increasing or decreasing the MgCl$_2$ concentration in the PCR reaction will have a significant effect on the display. However, it should not be necessary to change this parameter to obtain a high quality banding pattern.
- Consider trying different suppliers of *Taq* polymerase, particularly if a large number of bands are seen in the negative control (no RT enzyme) reaction. We have seen some variation in display quality based on the supplier *Taq* polymerase used. Some authors recommend using sequencing grade *Taq* polymerase which contains very low 5'-3' exonuclease activity (Hadman et al. 1995). We have not found this to be necessary.

- An eluted DDPCR fails to reamplify
  - Approximately 10 % of eluted bands fail to reamplify. Usually, these are faint, irreproducible bands on the primary display. Choose bands that are dark, discrete, greater than 100 nucleotides in length, and can be seen in duplicate RT sources.
  - Changing PCR program parameters for the secondary amplification seldom improves results. We have found that secondarily amplified bands obtained by heroic measures (e.g., 60 cycle amplification, two rounds of amplification) are uniformly artifacts.
  - If all secondary amplifications result in single bands of identical size, consider contamination of reagents by a previously amplified product.

References

Adams MD, Kerlavage AR, Fleischmann RD, Fuldner RA, Bult CJ, Lee NH, Kirknes EF, Weinstock KG, Gocayne JD, White O, et al. (1995) Initial assessment of human gene diversity and expression patterns based upon 83 million nucleotides of cDNA sequence. Nature 377:s3-s174

Benson DA, Boguski M, Lipman DJ, Ostell J (1994) GenBank. Nucleic Acids Res 22:3441–3444

Hadman M, Adam BL, Wright GL Jr, Bos TJ (1995) Modifications to the differential display technique reduce background and increase sensitivity. Anal Biochem 226:383–386

Jensen RA, Page DL, (1994) Identification of genes expressed in premalignant breast disease by microscopy-directed cloning. Proc Natl Acad Sci USA 91:9257–9261

Kim A, Rofflertarlov S, Lin CS (1995) New technique for precise alignment of an RNA differential display gel with its film image. BioTechniques 19:346

Liang P, Pardee AB (1995) Recent advances in differential display. Curr Opin Immunol 7:274–280

Liang P, Zhu W, Zhang X, Guo Z, O'Connell RP, Averboukh L, Wang F, Pardee AB (1994) Differential display using one-base anchored oligo-dT primers. Nucleic Acids Res 22:5763–5764

LiVolsi VA, Clausen KP, Grizzle W, Newton W, Pretlow TG 2d, Aamodt R (1993) The Cooperative Human Tissue Network. An update. Cancer 71:1391–1394

McClelland M, Mathieu-Daude, F, Welsh J (1995) RNA fingerprinting and differential display using arbitrarily primed PCR. Trends Genet 11:242–246

Trentmann SM, van der Knaap E, Kende H (1995) Alternatives to [35] S as a label for the differential display of eukaryotic messenger RNA. Science 267:1186–11877

Watson MA, Fleming, TP (1994) Isolation of differentially expressed sequence tags from human breast cancer. Cancer Research 54:4598–4602

Zhao S, Ooi SL, Pardee AB (1995) New primer strategy improves precision of differential display. BioTechniques 18:842

Analysis of Gene Expression in the Preimplantation Mouse Embryo Using mRNA Differential Display

J. W. ZIMMERMANN and R. M. SCHULTZ

1
Introduction

Preimplantation development is characterized by cell proliferation and differentiation. For example, the outer cells of the morula stage embryo differentiate into the trophectoderm of the blastocyst. The trophectoderm, which is a fluid-transporting epithelium, is the first differentiated cell type present in the preimplantation embryo and ultimately gives rise to extraembryonic tissues. In contrast, the inner cells of the morula generate the inner cell mass in the blastocyst. These cells remain totipotent/pluripotent and give rise to the embryo proper. Although changes in gene expression must underlie these differences in cell lineage, the paucity of biological material has severely hampered efforts to analyze changes in gene expression during preimplantation development. The generation of cDNA libraries to preimplantation embryos at different stages of development, coupled with subtraction hybridization methods (Rothstein et al. 1992), has made some inroads into the systematic identification of genes that are transiently expressed during the two-cell stage. The mRNA differential display method (Liang and Pardee 1992), as adapted for preimplantation mouse embryos (Zimmermann and Schultz 1994), provides an ideal approach to analyze changes in the temporal and spatial pattern of gene expression in the preimplantation embryo.

The major advantages of mRNA differential display in analyzing gene expression, as opposed to using conventional cDNA libraries generated from preimplantation embryos at different stages of development, are that the mRNA differential display method requires readily obtainable numbers of embryos and permits analysis of gene expression during very narrow developmental windows, e.g., embryos that are only a few hours apart in developmental time. Because the mRNA differential display method generates amplified cDNAs that tend to be small and at the 3' untranslated region, these cDNAs frequently do not contain informative sequence information. The marriage of mRNA differential display with either the cDNA libraries generated from preimplantation embryos or 5' RACE methods circumvents this problem and permits the isolation and characterization of longer or full-length cDNAs.

We describe here two methods for analyzing gene expression in the preimplantation mouse embryo by mRNA differential display. In the first method, total RNA is isolated from the embryos and then subjected to reverse transcription. In the second method, the embryos are directly lysed in a tube and subjected to reverse transcription with no prior isolation of the RNA. In both methods, an aliquot is removed for PCR, and the products (amplicons) resolved on a sequencing gel (Fig. 1).

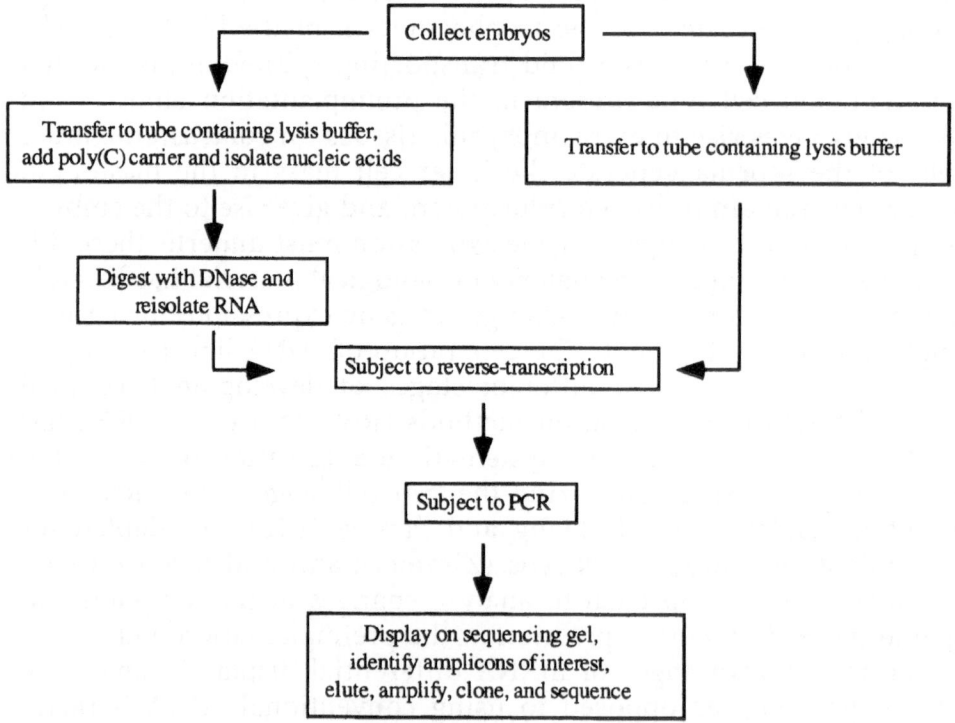

Fig. 1. Schematic flowsheet demonstrating the two methods of reverse transcription and PCR to analyze gene expression in preimplantation mouse embryos

2
Materials

Equipment

- model 9600 thermocycler (Perkin-Elmer Cetus)
- sequencing gel electrophoresis apparatus
- gel dryer
- agarose mini-gel electrophoresis apparatus

Supplies

- 0.5 ml polypropylene tubes
- thin-walled PCR tubes
- mineral oil
- X-ray film (XAR-5, Kodak)

Reagents
and solutions

For "Collection and culture of embryos:"
- MEM/PVP: bicarbonate-free minimal essential medium (Earl's salts) supplemented with 100 µg/ml pyruvate, 10 µg/ml gentamycin, 3 mg/ml polyvinylpyrrolidone, and 25 mM Hepes (pH 7.2).
- KSOM: 95 mM NaCl, 2.5 mM KCl, 0.35 mM KH_2PO_4, 0.2 mM $MgSO_4$, 10 mM sodium lactate, 0.2 mM glucose, 0.2 mM sodium pyruvate, 25 mM sodium bicarbonate, 1.71 mM $CaCl_2$, 1 mM glutamine, 0.01 mM EDTA, 1 mg/ml bovine serum albumin (BSA), 10 µg/ml gentamycin
  For "poly(C) RNA/cDNA preparation:"
- DEPC-treated water: water treated in the standard manner with 0.1 % diethylpyrocarbonate
- poly(C) at 8 µg per ml in DEPC- treated water(Boehringer Mannheim)
- lysis buffer: 4 M guanidine thiocyanate, 1 M 2-mercaptoethanol, 0.1 M Tris·HCl (pH 7.4)
- 1 M acetic acid (in DEPC-treated water)
- 2 M potassium acetate (in DEPC-treated water)
  75 % ethanol (in DEPC-treated water)
- Resuspension buffer: 40 mM Tris·HCl (pH 7.9), 10 mM NaCl, 6 mM $MgCl_2$ in DEPC-treated water
- RQ1 DNase (Promega)
- RS-saturated phenol: saturated phenol (Amresco) stored under an equal volume of resuspension buffer
- 3 M potassium acetate (pH 5.2) (in DEPC-treated water)
- 25 µM 3' primer (a 13-mer oligonucleotide: either $T_{11}MG$, $T_{11}MC$, $T_{11}MA$, or $T_{11}MT$, where "M" signifies any base but T) in DEPC-treated water
- reverse transcription (RT) master mix: four parts 250 mM Tris-HCl (pH 8.3), 375 mM KCl, 15 mM $MgCl_2$; two parts 100 mM

dithiothreitol; 1 part RNasin (at 40 U/µl; Promega); and two parts
200 µM dATP, 200 µM dTTP, 200 µM dCTP, 200 µM dGTP
- Superscript II reverse transcriptase (GIBCO/BRL)
  For "Direct-lysis RNA/cDNA preparation" (alternate protocol):
- Direct-lysis RT master mix: 75 mM Tris-HCl (pH 8.3), 112.5 mM
  KCl, 4.5 mM MgCl$_2$, 10 mM dithiothreitol, 30 µM dATP, 30 µM
  dTTP, 30 µM dCTP, 30 µM dGTP, 2.5 µM 3' primer, RNasin to a
  final concentration of 2 U/µl, and 0.5 % Nonidet P-40.
- Superscript II reverse transcriptase (GIBCO/BRL)
  For "Differential display PCR":
- PCR master-mix: (per 18 µl aliquot) 2 µl of 500 mM KCl, 100 mM
  Tris-HCl (pH 8.3), 2 µl of 25 mM MgCl$_2$, 2 µl of 5 µM 5' primer,
  2 µl of 25 µM 3' primer, 0.2 µl of 200 µM dATP, 200 µM dTTP,
  200 µM dCTP, 200 µM dGTP, 8.3 µl of DEPC-treated water, 1 µl
  10 µCi of ^{35}S- labeled dATP (specific activity >1000 Ci/mmol;
  Amersham), and 0.5 µl (2.5 U) of AmpliTaq DNA polymerase
  (Perkin-Elmer Cetus). The KCl, Tris, and MgCl$_2$ may be contri-
  buted by the buffer(s) supplied with Perkin-Elmer's AmpliTaq.
  The primers may be suspended in DEPC-treated water.

Note: Under some conditions DEPC may inhibit PCR amplification
(although this laboratory has never had such problems with its differential display PCRs).

For "Visualization and isolation of amplified cDNAs:"
- standard 6 % DNA sequencing gel
- DEPC-treated water
- 3 M potassium acetate
- glycogen (20 mg/ml)
- 100 % ethanol
- agarose gel

3
Experimental Procedure

**Collection
and culture
of embryos**

The following is given as an example. Collection conditions will vary
according to the stage of interest.

1. Collect blastocysts via flushing the uterus with MEM/PVP.

2. Culture the embryos approximately 1 h in KSOM (under 5 % CO$_2$
   at 37 °C) to allow them to recover from the collection.

Poly(C) RNA/cDNA preparation

This is the method as originally described (Zimmermann and Schultz 1994). Its main advantage is that it allows a single proven RNA preparation to serve as the basis for a large number of differential displays. Its main disadvantage is that the efficiency of the RNA recovery can be variable; recoveries range from 20 % to 60 %.

1. Prepare a 0.5 ml polypropylene tube with 100 µl of lysis buffer and 20 µg of poly(C) (2.5 µl of the stock solution). Vortex well. Keep this tube on ice.

2. Add the embryos (preferably >100) in as small a volume of medium as possible (preferably <1 µl, but up to 5 µl should not be a problem) to the tube. Vortex well.

3. Add 8 µl of 1 M acetic acid, 5 µl of 2 M potassium acetate, and 300 µl of 100 % ethanol. Vortex well.

4. Precipitate the nucleic acid by incubating the tube at –20 °C overnight.

5. Pellet the nucleic acids by centrifugation at 10000×g for 15 min at 4 °C.

6. Wash the pellet with 300 µl of ice-cold 75 % ethanol, then carefully remove the supernatant such that approximately 0.5 µl of fluid remains over the pellet.

   Note: If the poly(C) pellets are allowed to dry at all, they become virtually insoluble.

7. Resuspend the pellet in 20 µl of resuspension buffer.

8. Degrade DNA by adding 1 U of RQ1 DNase and incubating at 37 °C for 1 h.

9. Add 50 µl of DEPC-treated water and 80 µl of RS-saturated phenol. Vortex well.

10. Centrifuge for 8 min at 10000×g.

11. Transfer the aqueous phase to a new 0.5 ml polyproplyene tube.

12. Add 8 µl of 3 M potassium acetate (pH 5.2) and 300 µl of 100 % ethanol. Vortex well.

13. Precipitate the RNA overnight at –20 °C .

14. Pellet the RNA by centrifugation at 10000×g for 15 min at 4 °C .

15. Wash the pellet as described in step 6.

16. After carefully removing the supernatant, resuspend the pellet in 10 μl of DEPC-treated water. The efficiency of the recovery should be calculated by determining the A_{260} of an aliquot and comparing this to that of 20 μg of poly(C). Store the RNA at −20 °C . The RNA can now be reverse transcribed with the 3' primer of your choice.

17. Mix 2 μl of the 3' primer stock in a thin-walled PCR tube with an appropriate amount of the RNA prep. Note that this amount will vary according to your stage of interest and your purposes; 10 blastocyst-equivalents worth of RNA works well. Add sufficient DEPC-treated water such that the total volume is brought to 9.5 μl.

18. Incubate the sample at 70 °C for 10 min in the thermocycler.

19. Drop the temperature to 45 °C and then add 9 μl of RT master mix. Agitate the tube with your fingertip. Allow the sample to equilibrate to 45 °C for approximately 1 min.

20. Add 1.5 μl (300 units) of Superscript II reverse transcriptase. Again agitate the tube.

21. Incubate this reaction mixture at 45 °C for 1 h.

22. Disable the reverse transcriptase by incubating at 95 °C for 5 min.

23. Chill the tube on ice, centrifuge briefly, and then vortex lightly. Store the cDNA at −20 °C.

Direct-lysis RNA/cDNA preparation

This is an alternate method of RNA preparation/reverse transcription. Its main advantage is that it allows one to conduct the reverse transcription and PCR reactions quickly and simply with low background from a specific, selected, small group of embryos. Its main disadvantage is the danger of genomic DNA contamination of the cDNA prep.

1. Add a small number of embryos in a minimal volume of medium (with care, this can be kept below 0.5 ml) to 9 μl of direct-lysis RT master mix in a thin-walled PCR tube. (Note that using three blastocysts is clearly superior to using five blastocysts; too much cellular debris possibly begins to inhibit the reverse transcription.) Agitate the tube with the fingertip.

2. Incubate at 70 °C for 1 min in the thermocycler.

3. Remove the tubes from the thermocycler and incubate for 3 min at room temperature on the benchtop.

4. Add 1 µl (200 U) of Superscript II reverse transcriptase. Agitate the tube.

5. Return the tubes to the thermocycler. Incubate at 37 °C for 10 min.

6. Raise the temperature to 45 °C for 20 min.

7. Disable the reverse transcriptase by raising the temperature to 95 °C for 5 min.

8. Chill the tube on ice, centrifuge briefly, and then vortex lightly. Store the cDNA at –20 °C.

cDNA prepared by either of the two above-described methods is used as template for PCR amplification.

Differential display PCR

1. Prepare PCR master mix and distribute 18 µl into thin-walled PCR tubes on ice.

2. Add 2 µl of the cDNA prep to the tube. Vortex well.

3. Overlay with 50 µl of ice-cold mineral oil.

4. Subject the samples to 40 cycles of PCR amplification using the following 3-step cycle parameters: 94 °C for 30 s, 42 °C for 60 s, and then 72 °C for 30 s.

5. Follow the amplification with a 5 min extension at 72 °C.

6. Store the PCR samples at –20 °C .

The following is an overview of the process leading up to the sequencing of differential display amplicons. Nonstandard techniques used in the process are detailed.

Visualization and isolation of amplified cDNAs

1. Resolve 5 µl of the PCR sample on a standard 6 % DNA sequencing gel. It is recommended that three separate amplifications are run for each stage of interest. Run stages to be compared in adjacent lanes; shark's-tooth combs are not recommended.

2. Dry the gel onto filter paper **without** fixation.

3. Visualize the bands by autoradiography using Kodak XAR-5 X-ray film at –80 °C for 1–5 days.

4. Locate the band of interest. Apply phosporescent ink to the gel; the resulting autoradiogram functions as a template to allow the band to be precisely cut from the gel.

5. Soak the excised gel fragment in 100 µl of DEPC-treated water for 10 min in a 0.5 ml polypropylene tube.

6. Incubate the tube in boiling water for 15 min.

7. Pellet debris by centrifugation at 10000×g for 2 min, and transfer the supernatant to a fresh tube.

8. Add 8 µl of 3 M potassium acetate and 2.5 µl of glycogen (20 mg/ml). Vortex well. Add 300 µl of 100 % ethanol. Vortex well.

9. Precipitate the glycogen and nucleic acid overnight at –20 °C.

10. Pellet by centrifugation at 10000×g for 15 min at 4 °C, wash with 200 µl of ice-cold 85 % ethanol, and then resuspend in 10 µl of DEPC-treated water.

11. Use 5 µl for reamplification via PCR as described above, with the following modifications: use a 40 µl reaction volume, use the dNTPs at 20 µM, and use no radionucleotide. Visualize the amplified products on an agarose mini-gel.

This procedure will almost invariably result in clearly reamplified bands. These amplicons can then be cloned (the Invitrogen TA cloning kit is recommended) and sequenced using standard techniques (see Chaps. XVIII and XIX).

4
Results and Comments

We have used the poly(C) carrier method to analyze gene expression in preimplantation embryos (Zimmermann and Schultz 1994) and to identify genes whose expression is modulated by growth factors in the blastocyst (Babalola and Schultz 1995). In addition, we have used this method to identify genes (e.g., translation factor eIF-4C) that are transiently expressed during the two-cell stage by comparing the amplicon profiles for one-cell, mid-two-cell and late two-cell/early four-cell embryos (Davis, De Sousa and Schultz, unpublished results).

To make valid comparisons of the differences in amplicon intensity for embryos at different stages of preimplantation development, it is necessary to subject equivalent amounts of poly(A)$^+$ RNA to reverse transcription. Thus, we correct for the developmental changes in the amount of poly(A)$^+$ RNA in the embryos. For example, the amount of poly(A)$^+$ in the blastocyst is about 2.5 and five times that present in the eight-cell and two-cell embryos, respectively. Thus, for every blastocyst equivalent, 2.5 eight-cell embryo equivalents or five two-cell embryo equivalents are subjected to the reverse transcription reaction. In addition, is critical to run triplicate samples to insure reproducibility with respect to both the presence and intensity of the amplicon bands.

We have had good success with the more direct alternative method in which the embryos are directly subjected to the reverse transcription reaction without any isolation of RNA. To insure that DNA contamination is not a problem, initial experiments should be conducted to demonstrate that in the absence of the reverse transcription reaction essentially no amplicons are detected following the PCR and that the pattern of those that are detected differs markedly from the pattern obtained when the reverse transcription reaction is conducted. To date, we have not observed that DNA contamination is a problem.

We have also noted that the alternative method gives higher reproducibility of the amplicon band pattern and, moreover, it is probably a fairer representation of the reproducibility of the method, since each sample is subjected to reverse transcription and then each sample is subjected to PCR. In contrast, when poly(C) is used as the carrier, a single sample is subjected to reverse transcription and then this sample is subjected to multiple PCRs. We have used this direct method to identify genes whose expression is preferentially restricted to either the inner cells of the blastocyst or the trophectoderm. In this analysis three inner cell masses are used for each blastocyst, since the number of cells in the inner cell mass is about one third that of the intact blastocyst. Genes that are preferentially expressed in the trophectoderm will yield amplicons whose intensity is much greater in the intact blastocysts than in the inner cell mass cells, and genes that are preferentially expressed in the inner cell mass will reveal amplicons whose intensity is greater than the corresponding amplicon obtained from the intact blastocysts. A representative region of such a gel is seen in Fig. 2.

Sequence analysis of a number of amplicons reveals that the 3' primer primes internal to the junction between the end of the 3' untranslated region and poly(A) tail at a frequency of about 10%–15%. In these cases, the 3' primer appears to prime as a hexamer and useful sequence information is obtained, e.g., one can identify the gene if the sequence is already in the data bases. Last, it must be emphasized that an independent method should be employed to confirm the expression pattern obtained with the mRNA differential display method. We use the DNA sequence obtained to generate gene-specific primers and a reverse transcription/PCR assay that permits determining relative changes in the abundance of the mRNA under investigation (Temeles et al. 1994).

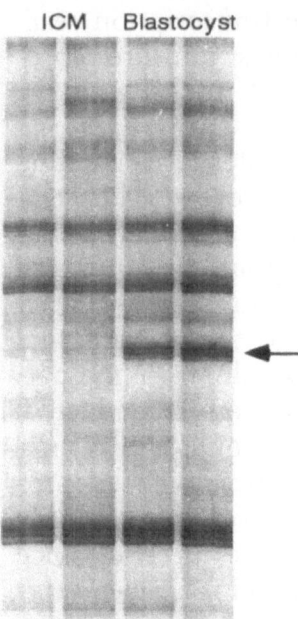

ICM   Blastocyst

Fig. 2. Region of autoradiogram showing amplicon profile obtained from inner cell masses (ICM, *first two lanes*) and intact *blastocysts (third and fourth lanes)*. The *arrow* points to an amplicon(s) that are preferentially expressed in the trophectoderm. These multiple bands (doublets and triplets) are due to the additional deoxyadenosine that is added by *Taq* polymerase and the small differences in the electrophoretic mobility of the two complementary DNA strands in the DNA sequencing gel

5
Troubleshooting

As described above, it is critical not to let the pellets completely dry when using the poly(C) RNA carrier method. This will result in essentially no RNA recovery, and hence few, if any, bands will be observed following reverse transcription and PCR. When bands are not observed, the reverse transcription reaction, rather than the PCR, is usually at fault.

References

Babalola GO, Schultz RM (1995) Effect of TGF-αand TGF-β on gene expression in the preimplantation mouse embryo. Mol Reprod Dev 41:133–139

Liang P and Pardee AB (1992) Differential display of eukaryotic messenger RNA by means of the polymerase chain reaction. Science 257:967–971

Rothstein J., Johnson D, DeLoia J, Skowronski J, Solter D, Knowles B (1992) Gene expression during preimplantation mouse development. Genes Develop 6:1190–1201

Temeles GL, Ram PT, Rothstein JL, Schultz RM (1994) Expression patterns of novel genes during mouse preimplantation embryogenesis. Mol Reprod Dev 37:121–129

Zimmermann JW, Schultz RM (1994) Analysis of gene expression in the preimplantation mouse embryo: Use of mRNA differential display. Proc Natl Acad Sci USA 91:5456–5460

Chapter XXXIX

Rapid Identification of Differentially Expressed Genes in Trypanosomes

N. B. MURPHY and R. PELLÉ

1
Introduction

Members of the genus *Trypanosoma*, within the order Kinetoplastida, are unicellular parasitic organisms which are a major threat to human health and livestock productivity in large areas of the developing world. Despite their relatively small genome size, these primitive organisms undergo complex life cycles, are capable of controlling their rates of proliferation at different stages, can modulate the immune system of their mammalian hosts, and, more recently, have been shown to undergo programmed cell death (see Ameisen et al. 1995; Vickerman 1985; Zilberstein and Shapira 1994). Many of the molecular features and processes of these organisms provide paradigms for eukaryotic biology (see Adler and Hajduk 1994; Bonen 1993; Gonners-Ampt and Borst 1995; Pays et al. 1994; Simpson and Maslov 1994; Tschudi and Ullu 1994). To gain a better understanding of the molecular mechanisms involved in the control of differentiation of these intriguing and deadly organisms we have developed a differential display PCR method, dubbed randomly amplified differentially expressed sequences (RADES), for the rapid identification of differentially expressed trypanosome or leishmania genes (Murphy and Pellé 1994).

2
Principles

The RADES-PCR method (an outline of the whole experimental procedure is shown in Fig. 1) differs from other differential display methods in that the template for the PCR fingerprinting step is amplified double-stranded cDNA which can be generated from very small quantities of starting material. Fingerprints are visualized by agarose gel electrophoresis, which simplifies the purification, cloning and

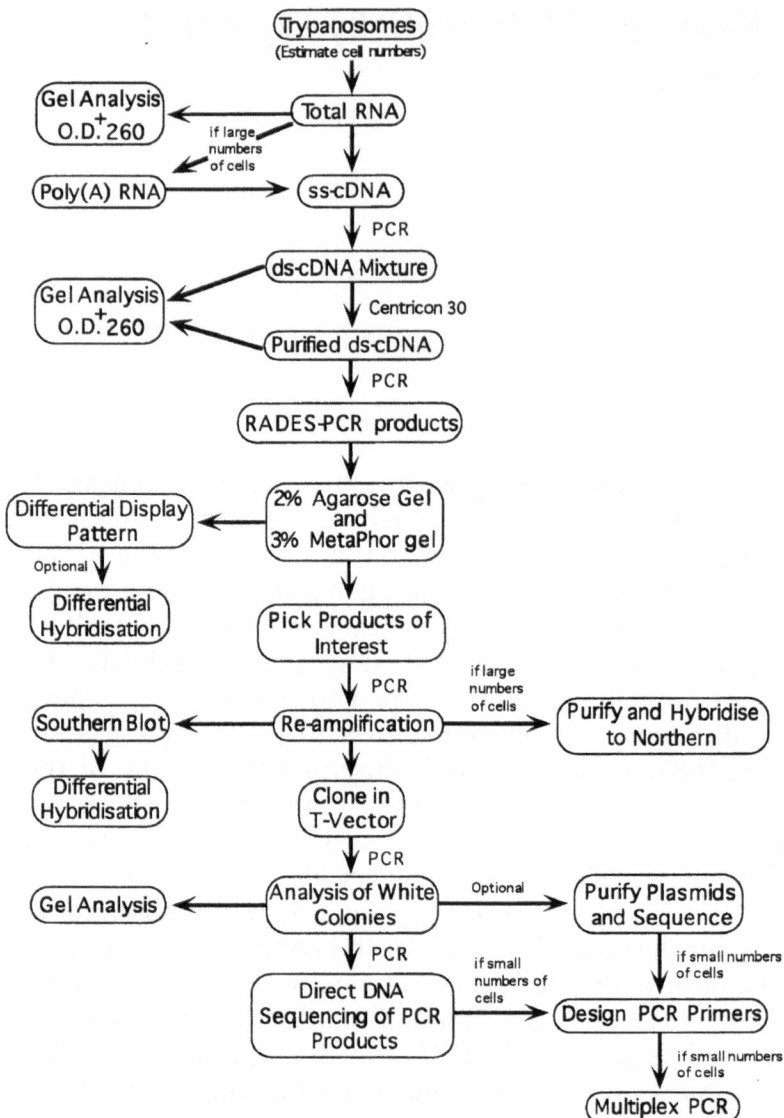

Fig. 1. Schematic flow-sheet of the different steps in the RADES-PCR procedure from the isolation of RNA from trypanosomes to the DNA sequencing and analysis of the differential display products

analysis steps. The method exploits the fixed 5' and 3' sequences of all trypanosome mRNA sequences characterized to date (De Lange et al. 1984; Parsons et al. 1984) and therefore can be carried out with samples which are contaminated with host material, a feature which is unique to kinetoplastids. This feature is particularly important for intracellular and extracellular mammalian forms in which lengthy purification procedures can themselves alter gene expression. It also overcomes potential problems of nonspecific fingerprints from contaminating genomic DNA in RNA samples and is particularly useful for samples in which cell numbers are limiting. Although the protocols focus on African trypanosomes, the method is applicable to all kinetoplastids.

Following the generation of first-strand cDNA, trypanosome-specific cDNAs are selectively amplified in the PCR, even from mixtures of host and parasite material, using 3'-specific oligo(dT) and 5'-specific miniexon primers. The RADES-PCR differential display products are visualized in agarose gels, which are more convenient and easier to handle than acrylamide gels. This analysis also avoids the requirement for using radioisotopes to detect the differential display products. Products can be transferred directly from the agarose gels to nylon membranes and hybridized with labeled cDNA from the different trypanosome populations being analysed, thus reducing the number of northern blot hybridizations required to confirm the differential expression of amplified products.

3
Materials

– spectrophotometer   **Equipment**
– peristaltic pump
– submarine gel electrophoresis apparatus
– UV transilluminator
– photographic equipment
– thermocycler (e.g., MJ Research)

– standard agarose   **Supplies**
– MetaPhor agarose (FMC BioProducts #50182)
– oligo(dT)-cellulose (Collaborative Biomedical Products #20003)
– Centricon 30 columns (Amicon #4208)
– Thermowell plates (Corning Costar Corp. #6520)
– mineral oil (Sigma #8042–47–5)
– Nytran nylon membranes (Schleicher and Schuell #414596)

– random priming kit: Multiprime kit (Amersham #RPN 1601Y) or   **Kits**
  Prime-It (Stratagene)
– Wizard PCR preps (Promega #A7170)
– GeneClean (BIO 101 #3106) (optional)
– *fmol* DNA sequencing system (Promega #Q4100)

Oligonucleo- tides and vectors	– anchored oligo(dT) primer (e.g., TAGGCGCGCC(T)$_{20}$) – miniexon primer: TAGGCGCGCCTAGAACAGTTTCTGTACTATATTG – arbitrary 10-mer primers – universal forward and reverse sequencing primers (oligonucleotides can be purchased from Genosys Biotechnologies, Inc.) – pGEM-T plasmid vector (Promega #A3600)

Reagents and solutions
- sterile RNase-free water
- denaturation solution (solution D) for RNA: 4 M guanidinium thiocyanate (Fluka #50980), 25 mM sodium citrate (pH 7.0), 0.5 % (w/v) sarkosyl (IBI), 0.1 M β-mercaptoethanol.

Note: Add β-mercaptoethanol just prior to use (72 µl/10 ml solution D).

- 2 M sodium acetate (pH 4.0)
- water saturated phenol
- chloroform
- isopropanol
- ethanol
- 60 mM sodium phosphate buffer (pH 6.8). Autoclave before use.
- ethidium bromide stock solution (10 mg/ml)
- 6x loading buffer for RNA electrophoresis: 0.25 % (w/v) bromphenol blue, 0.25 % (w/v) xylene cyanol, 30 % (v/v) glycerol, 1.2 % SDS
- RNA size markers (e.g., GIBCO-BRL)
- 0.1 N NaOH
- 2x loading buffer for purification of poly(A)+ RNA: 40 mM Tris-HCl (pH 7.6), 1 M NaCl, 2 mM EDTA, 0.2 % (w/v) SDS
- 3 M sodium acetate (pH 5.2)
- 5x first-strand cDNA buffer: 250 mM Tris-HCl (pH 7.6), 350 mM KCl, 50 mM MgCl$_2$, 20 mM dithiothreitol
- dNTPs stock solution: 5 mM of each
- MMLV reverse transcriptase (Promega, Stratagene or GIBCO-BRL)
- 1 M MgCl$_2$ stock solution
- 10x *Taq* polymerase buffer: 100 mM Tris-HCl (pH 8.3), 500 mM KCl, 0.5 % (v/v) NP40, 0.5 % (v/v) Tween-20
- *Taq* DNA polymerase (e.g., Promega)
- DNA size markers
- TE: 10 mM Tris, 1 mM EDTA
- [^{32}P]dCTP (Amersham #PB 10205)
- 50 % formamide
- 20x SSC (pH 7.0): 175.3 g NaCl, 88.2 g sodium citrate in 1 l of water

4

Experimental Procedure

Pleomorphic forms of *Trypanosoma brucei* display distinct morphological differences in the bloodstream of an infected host represented by actively dividing long slender forms and nondividing short stumpy forms. Trypanosome parasites can be isolated from whole blood by DEAE-cellulose chromatography (Lanham and Godfrey 1970). This procedure exposes the parasites to alkaline pH and buffers which lack essential nutrients; factors which can influence gene expression. In cases where these factors present problems, the *T. brucei* parasites may be rapidly isolated from whole blood, together with host lymphocytes, by centrifugation and isolation of the resultant buffy coat.

Isolation of trypanosomes

The RADES-PCR method is also particularly applicable to studies on differential gene expression in in vitro propagated trypanosomes, even when cell numbers of particular life cycle stages are limiting. Recent improvements for the in vitro propagation of different life cycle stages of trypanosomes (Bienen et al. 1991; Evans 1993; Fish et al. 1989; Hirumi and Hirumi 1991; Sanchezmoreno et al. 1995), in conjunction with RADES-PCR, will allow for the analysis of alterations in gene expression during differentiation or in response to external factors. For all propagation methods, isolated parasites are pelleted, rapidly frozen and stored at –70 °C until required.

The success of every step in this procedure is dependent on the quality of the materials and reagents. When attempting the procedure for the first time you should try it on an abundant source of material so that each stage can be properly checked (i.e., the quality of the RNA, cDNA, etc.). Once this has been achieved, a control tube of this material can be included when handling more difficult and less abundant material, which will test the efficiency of each step.

RNA purification

The single step RNA isolation method by Chomczynski and Sacchi (1987), with improvements by Puissant and Houdebine (1991), is rapid and simple and results in RNA which is essentially free of DNA. Several manufacturers offer kits for isolating RNA and purifying mRNA (e.g., Clontech, Promega, Invitrogen). The procedure outlined below generates good quality RNA for the subsequent steps of the RADES-PCR method.

1. In an 1.5 ml Eppendorf tube containing a cell pellet of approximately 5×10^8 cells add 500 µl solution D and resuspend by pipetting up and down and vortexing.

2. Add 50 µl 2 M sodium acetate (pH 4.0), 500 µl water saturated phenol and 100 µl chloroform, sequentially, and mix tube after each addition.

3. Vortex for 20 s and leave mix on ice for 10 min.

4. Centrifuge in a microfuge (14000 rpm) for 10 min.

5. Transfer upper aqueous layer containing RNA to a new Eppendorf tube, carefully avoiding disturbing the white interface containing the DNA and proteins.

6. Add an equal volume (550 µl) of isopropanol, mix and leave on dry ice or at –70 °C for 10 min.

7. Centrifuge for 10 min in a microfuge.

8. Remove the supernatant either by pouring off or with a pipette tip, taking care not to dislodge the pellet.

9. Redissolve the pellet in 400 µl of solution D. The pellet is difficult to dissolve and should be broken up by pipetting up and down through a micropipette tip. Placing the tube at 50 °C between pipettings will help to dissolve the pellet more easily.

10. After redissolving the RNA, add 2.5 volumes (1 ml) of cold (–20 °C) ethanol, mix and leave on dry ice or at –70 °C for 10 min.

11. Centrifuge as before and discard supernatant, taking care not to disturb the RNA pellet.

12. Wash the RNA pellet with cold (–20 °C) 70 % ethanol and dry under vacuum or in a 50 °C oven.

13. Dissolve the RNA pellet in 200 µl of sterile RNase-free water. The pellet should contain approximately 200–300 µg of total trypanosome RNA.

14. Check the quality and approximate concentration of RNA by electrophoresis in an RNA gel (for 10^7 cells or more). If sufficient numbers of cells have been used (10^8 or more), the concentration of the RNA can be determined by diluting a small amount of the sample 100-fold in sterile water and determining its absorbance at 260 nm in a UV spectrophotometer. One absorbance unit is equivalent to approximately 40 µg/ml of RNA. Pure RNA will exhibit an A_{260}/A_{280} ratio of 2.0 to 2.2.

Note: Under certain conditions, the number of trypanosomes of a particular life cycle stage, or treated in a particular way, may be limiting (10^6 or less). For RNA isolations using small numbers of cells, the volume of solution D should be adjusted to 50 µl or 100 µl and the other components adjusted accordingly. The quantity of RNA that can be isolated from such small numbers of cells is not sufficient to determine its integrity by electrophoresis, its concentration in a spectrophotometer or to purify poly(A)-enriched mRNA. In these cases, the overall integrity of the RNA can only be surmised following amplification of total cDNA and analysis of the RADES-PCR products produced. In general, this has not caused problems, provided the cells have been properly prepared, stored and the different steps in the RNA and cDNA preparation have been adhered to.

For northern blot analysis, RNA is generally denatured with glyoxal or formaldehyde prior to electrophoresis in agarose gels (Sambrook et al. 1989). A rapid and simple method which requires just a 5 min denaturation of RNA samples in loading buffer and allows the experimenter to monitor the integrity and migration of major RNA species and size markers during electrophoresis has recently been described (Pellé and Murphy 1993).

Analysis of RNA by gel electrophoresis

Although RNA is more labile than DNA and ribonucleases are difficult to inactivate in comparison to DNases, excessive care is not required in handling RNA for analysis. It is only important to ensure that gel formers, combs and tanks are well cleaned and rinsed and that gloves are worn at all times during cleaning to prevent transfer of ribonucleases from the surface of the skin. Decon 90 (Decon Laboratories Ltd.), or a similar phosphate-free detergent, is used to wash the gel tank, comb and gel former followed by rinsing with deionized water. A peristaltic pump is required to recirculate the buffer during electrophoresis, and the tubing used should be well rinsed with sterile distilled water.

1. A 1.4 % (w/v) agarose gel is made up by boiling in 10 mM sodium phosphate buffer (pH 6.8) containing 0.1 µg/ml of ethidium bromide (1 µl of a 10 mg/ml stock solution per 100 ml of buffer), then cooled to 60 °C and poured.

2. In a sterile Eppendorf tube, 10 µl of the RNA (1–10 µg) dissolved in sterile water is mixed with 2 µl of sterile 6x loading buffer. The mixture is incubated at 75 °C for 5 min followed by immediate loading of the sample onto a submarine gel. Commercially available RNA size markers, to verify the sizes of the major ribosomal bands, are similarly treated prior to electrophoresis. When analyzing many samples, the denatured RNA can be placed on ice before loading the gel.

3. The gel is electrophoresed at 3–7 V/cm in 10 mM sodium phosphate buffer (pH 6.8) containing 0.1 µg/ml ethidium bromide. Because the buffering capacity of the electrophoretic buffer is relatively weak due to its low ionic strength, constant recirculation of the buffer is maintained to prevent the formation of an undesirable pH gradient which can lead to degradation of the RNA during electrophoresis. The electrophoresis can be interrupted at any time and the migrating RNA in the gel visualized with medium-wave UV light to verify the migration and integrity of the RNA. For *T. brucei* RNA, three distinct ribosomal RNA bands (approximately 2.2, 1.7 and 1.5 kb) and smaller RNA transcripts, including tRNA molecules, should be clearly visible.

Purification of poly(A)-enriched RNA

For RADES-PCR analysis it is not necessary to use poly(A)-enriched RNA for first-strand cDNA synthesis, although results with poly(A)-enriched RNA can be slightly cleaner than with total RNA. If the quantity of cells and subsequent RNA is not limiting, it is recommended to enrich for poly(A)$^+$ RNA. Several methods for the enrichment of poly(A)$^+$ RNA are available (Sambrook et al. 1989) and a number of commercial companies supply easy to use kits. The following method is the standard one used for affinity purification of RNA by oligo(dT)-cellulose chromatography.

1. Oligo(dT)-cellulose is weighed to obtain 100 mg of resin per 1 mg of total RNA.

2. Suspend the oligo(dT)-cellulose in 0.1 N NaOH and pour into a 1 ml sterile syringe barrel plugged with siliconized sterile glass wool. Wash the column with sterile water.

3. Wash the column with sterile 1x loading buffer until the pH has equilibrated to less than 8.0.

4. Dissolve the RNA in sterile water, heat to 65 °C for 5 min, add an equal volume of 2x loading buffer, cool to room temperature and apply to the column.

5. Collect the eluate from the column in a sterile tube, reheat to 65 °C for 5 min, cool to room temperature and reapply to the column. Collect the eluate in a sterile tube as previously.

6. Wash the column with approximately 10 column volumes of 1x loading buffer, collecting fractions in sterile tubes. The A_{260} nm of the early fractions will be high as these will contain most of the nonpolyadenylated RNA, whereas the latter fractions should have a negligible absorbance.

7. Elute the poly(A)$^+$ RNA with 2–3 column volumes of sterile water heated to 50 °C. Pool the poly(A)$^+$ fractions, add 3 M sodium acetate (pH 5.2) to a final concentration of 0.3 M, mix and add 2.5 volumes of –20 °C ethanol. Place on dry ice or at –70 °C for 10 min and recover the RNA by centrifugation in a microfuge for 10 min. Carefully discard the supernatant and wash the pellet with cold 70 % ethanol. Recentrifuge briefly, carefully discard the supernatant and allow the pellet to dry.

8. Resuspend the RNA in a small volume of sterile water to give an approximate concentration of 0.–1 µg/µl (about 100 µl for each milligram of total RNA applied to the column). The concentration of the poly(A)-enriched RNA can be estimated by gel electrophoresis and by measuring its absorbance at 260 nm.

9. The poly(A)-enriched RNA will still contain nonpolyadenylated RNA and can be purified further by repeating the affinity purification procedure from step 4 above. The column is regenerated prior to loading the RNA by washing with 3–5 column volumes of 0.1 N NaOH followed by sterile water and then 1x loading buffer until the pH of the eluate is below 8.0.

Note: For large numbers of samples, it is more efficient to use a batch purification procedure in which the oligo(dT)-cellulose is added to the denatured RNA in step 4 above. Detailed protocols for batch mRNA are provided by manufacturers of oligo(dT)-cellulose and can be found in Sambrook et al. (1989).

Preparation of first-strand cDNA

The conversion of poly(A)$^+$ RNA to cDNA in the RADES-PCR method is achieved through an oligo(dT) primer and reverse transcriptase, MMLV (this reverse transcriptase is preferable to avian reverse transcriptase). The RADES-PCR differential display method differs from other differential display methods from this point, since, following first-strand cDNA synthesis, double-stranded cDNA is generated using specific primers for the fixed 5' and 3' ends of mRNAs of trypanosomes. Following removal of primers and dNTPs, this double-stranded cDNA serves as the template for arbitrary primers in the subsequent PCR fingerprinting reactions. First-strand cDNA synthesis is carried out in a final volume of 20 µl containing between 50 and 100 ng of poly(A)-enriched RNA. If the starting sample is limiting (10^6 trypanosomes or less) and poly(A)$^+$ RNA has not been prepared, one half of the total RNA sample is used. It is not necessary to adjust components in the reaction mixture to accommodate for smaller quantities of RNA. However, when dealing with several samples, the relative quantities of RNA added to each reaction should be similar. If the amount of RNA is too small to quantitate, estimates should be calculated on the number of trypanosomes used to prepare the RNA sample.

1. To sterile 500 µl microfuge tubes add the RNA samples in 12 µl of water and 1 µl of an anchored oligo(dT) primer (100 ng/µl). The primer TAG GCG CGC C(T)$_{20}$ was designed on the basis of its similar melting point to a primer for the miniexon sequence and, similar to this miniexon primer, contains an *Asc*I site for the cloning of full length cDNA in the plasmid vector pNEB193 (New England Biolabs, Beverly, MA). The mixture is heated to 70 °C for 2 min and placed back on ice.

2. Add 4 µl of 5x first-strand buffer, 2 µl of dNTPs (from a stock containing 5 mM of each) and 1 µl of MMLV reverse transcriptase (GIBCO-BRL, 200000 U/ml)

3. Mix, spin briefly and place the tubes at 37 °C for 1 h.

4. The first-strand cDNAs can be used directly for PCR amplification to generate full length double-stranded cDNA, or stored at either –20 °C or –70 °C until used. If the cDNAs are to be stored for long periods, it is advisable to ethanol precipitate the nucleic acids in the reactions and store at –70 °C in 70 % ethanol.

Generation of double-stranded cDNA by PCR amplification

Most suppliers of *Taq* DNA polymerase (e.g., Promega) also supply buffer with the enzyme. We have found our own buffer works well with a variety of different commercial preparations of the enzyme. MgCl$_2$ is added separately to the reaction to a final concentration of 2 mM for most reactions. If there are problems with the PCRs, test reactions can be carried out varying the MgCl$_2$ concentration between 1 mM and 3 mM final concentration.

1. Take 5 µl, or one quarter, of the first-strand cDNA reactions and place in sterile 500 µl microfuge tubes (GeneAmp thin-walled tubes from Perkin Elmer Cetus can be used, if preferred). Add 10 µl of 10x *Taq* buffer, 4 µl dNTPs (from a stock containing 5 mM of each), 1 µl each of the oligo(dT) primer and miniexon primer TAG GCG CGC CTA GAA CAG TTT CTG TAC TAT ATT G (100 ng/µl each).

2. Add 2.5 U of *Taq* DNA polymerase per reaction and adjust the mixtures to 100 µl with water. For multiple samples it is more convenient to make a 2x master mix containing buffer, dNTPs and *Taq* polymerase. The reaction tubes contain first-strand cDNAs and primers in 50 µl of water, to which is added 50 µl of the 2x

master mix. Overlay the reactions with mineral oil , and place in a thermal cycler (for machines with a heated lid, the mineral oil can be omitted).

3. Cycling conditions for the PCR are 94 °C for 1 min, 55 °C for 1 min, 72 °C for 2 min for 40 cycles followed by a 5 min extension at 72 °C.

4. Examine 5 µl of the PCR products on an agarose gel with DNA size markers. The products should appear as a smear, with the majority of products migrating close to 1 kb. The total amount of product generated in the 100 µl PCR should be about 10 µg of double-stranded cDNA, and can be estimated from the sample loaded on the gel.

5. Removal of buffer, primers and dNTPs are achieved by ultrafiltration in a Centricon 30 column. Other methods can be also used, such as the Wizard PCR purification system or chromatography through a Sephadex G50 spun column (Sambrook et al. 1989), but we have found the Centricon 30 columns to give the best results. Add each cDNA solution with 1.5 ml of water to the sample reservoir of the Centricon 30 filtration units.

6. Centrifuge at 7000 rpm for 7 min in a fixed angle rotor (e.g., an SS-34 rotor of an RC-5B Sorvall centrifuge) with adapter. The volume remaining in the sample reservoir should be approximately 100 µl. Add a further 1.5 ml of water to the sample reservoirs and repeat the centrifugation step.

7. The washing step is repeated a further two times, removing excess flow through accumulating in the filtrate cup. The final centrifugation is carried out for 10 min, or until the volume remaining in the sample reservoirs is about 100 µl.

8. The cDNAs are collected by spinning into the retentate cups of the filtration units, and are diluted to 20 ng/µl in TE buffer, based on the estimated DNA concentration from the gel analysis of the PCR products (for 10 µg of sample this is equivalent to a 500 µl final volume).

9. Dispense the cDNAs in 50 µl aliquots into microfuge tubes. Place the working tubes at 4 °C and store the others at –20 °C. Excessive freezing and thawing of the DNA can cause alterations in the PCR fingerprinting patterns and the DNA is stable at 4 °C for several months. Dividing the sample into aliquots also helps to overcome problems of losing them through accidental contamination. Furthermore, at each stage material is retained, so it is possible to repeat any steps to produce the final double-stranded cDNA samples.

RADES-PCR differential display analysis of cDNA

Once good quality cDNAs are generated, the RADES-PCR is carried out essentially the same as arbitrarily primed PCR (AP-PCR) of genomic DNA (Welsh and McClelland 1990; Williams et al. 1990). The RADES-PCR reactions can be carried out either in microfuge tubes or in 96-well Thermowell plates. We find the 96-well plates convenient for the analysis of multiple samples and for storage purposes. The PCR reactions are carried out at two template concentrations to determine the robustness of each arbitrary primer to generate reproducible fingerprints and reduce the likelihood of picking artefactual PCR products that do not represent products of differentially expressed sequences. It is worthwhile including a genomic DNA sample which produces reliable AP-PCR fingerprints as a positive control.

1. Aliquot 1 μl of the target cDNAs (20 ng) and 1 μl of a one in ten dilution of the cDNAs (2 ng) in separate tubes or wells of the 96-well plate. Add 1 μl of an arbitrary 10-mer primer (20 ng/μl) to each sample and bring the volume to 10 μl with water.

2. In a separate sterile microfuge tube prepare a 2x reaction buffer mix containing 20 mM Tris-HCl (pH 8.3), 100 mM KCl, 4 mM $MgCl_2$, 0.1 % NP40, 0.1 % Tween-20 and 125 U/ml of *Taq* DNA polymerase. Prepare slightly more (about 10 %) 2x reaction buffer mix than required, as the detergent in the mix causes foaming and it is difficult to accurately pipette the final samples.

3. Mix 10 μl of the 2x reaction buffer mix with the cDNA-primer samples and place the plate or tubes in a thermocycler.

4. Cycling conditions for the PCR are 94 °C for 45 s, 40 °C for 1 min, 72 °C for 1 min for 40 cycles followed by a 2 min extension at 72 °C.

5a. Following the PCR, products are analyzed by electrophoresis of 7 μl of sample in a 2 % (w/v) agarose gel. Place sets of samples of different dilutions alongside each other for easy visualization of any differences in the fingerprint profiles due to template concentration. PCR reactions which do not generate reproducible products in the different samples and at the different template concentrations are usually discarded. However, it is still possible to identify differentially expressed sequences in such reactions by transferring the agarose-separated products to a nylon membrane and hybridizing with labeled cDNA from the different samples being compared.

5b. For a finer separation of the PCR products, 10 µl of each sample set generating clear and reproducible fingerprints is electrophoresed in a 3 % (w/v) MetaPhor agarose gel.

6. Recover DNA from products displaying differential amplification as outlined below.

DNA from bands of interest displaying differential amplification can be rapidly recovered from both the 2 % standard and 3 % MetaPhor agarose gels. Low melting point agarose is not required. The recovered DNA is then reamplified for cloning and characterization. The procedure is rapid, simple and, like all the steps in the procedures above, sufficient material remains for repeating experiments if they fail at the first attempt.

Recovery of DNA products of interest from agarose gels

1. After photographing the gels containing the RADES-PCR products, mark on the photograph the bands of interest to be purified.

2. Label one sterile 500 µl microfuge tube for each band of interest and pipette 50 µl of water into it.

3. For each band of interest take a 200 µl wide-bottomed micropipette tip or cut about 2 mm from the base of narrow-bottomed micropipette tips with a sterile blade or sharp scissors.

4. Place the ethidium bromide-stained gel on a long-wave UV transilluminator to visualize the DNA bands. If there are multiple bands to be purified, it is advisable to cut the gel with a sterile blade into sections containing each set of PCR products for each arbitrary primer, as extended exposure to UV will result in fading of the gel and difficulty in visualizing the bands of interest. Decreased exposure also reduces the generation of thymidine dimers by the UV light source. If the gel does fade, it is possible to restain it with ethidium bromide (1 µg/ml) in water.

5. Use one micropipette tip for each band and remove a core of the gel containing DNA from the band of interest. It is possible to embed the tip into the gel several times to capture several core pieces in the tip, but it is not necessary to recover all of the DNA band and speed of recovery is more important than DNA quantity.

6. Place the pipette tip containing the core gel piece into the 50 µl of water of one of the microfuge tubes. Repeat these steps for each of the bands of interest until all are recovered.

7. Place the open microfuge tubes with the micropipette tips containing the core gel pieces in a thermocycler, and program the machine to heat to 95 °C.

8. As the gel cores melt in the micropipette tips, the agarose will mix with the 50 μl of water. Use a micropipette to push out any remaining agarose and mix the solution.

9. The tubes can be stored indefinitely at 4 °C until the reamplification step is carried out. The solution may gel at this temperature, particularly if a large core has been taken or if the percentage of the gel was greater than 3 %. However, samples that have gelled can be remelted prior to dispensing them for the reamplification step.

Reamplification and verification of products of interest

Products of interest are reamplified with the same primer and under the same conditions as the RADES-PCR differential display reaction. The procedure below is for a 100 μl PCR to generate the maximum amount of product, but can be carried out in a smaller volume by adjusting all the components of the reaction accordingly. The products are transferred to a nylon membrane and tested for differential expression by hybridization with labeled double-stranded cDNA from the different samples. Although not all products produce a hybridization signal, this step reduces wasted effort on some differentially amplified products which are not from differentially expressed genes and the number of northern hybridizations that need to be carried out.

1. Take 5 μl of the recovered DNA sample and place in a sterile 500 μl microfuge tube. If the sample has solidified, heat it to between 90 °C and 95 °C first.

2. Add 100 ng of the appropriate arbitrary primer and adjust the volume to 50 μl with water.

3. In a separate tube prepare a 2x reaction buffer mix, as outlined above for the RADES-PCR. Add 50 μl of the 2x reaction buffer mix with the cDNA-primer samples.

4. Overlay the samples with mineral oil and place the plate or tubes in a thermocycler. Use the same cycling conditions as for the RADES-PCR.

5. Following the PCR, take 5 μl of each sample for analysis by agarose gel electrophoresis in a 2 % (w/v) agarose gel. Generally multiple products are produced in each reaction, and the product of interest is not necessarily the most abundant in a reaction. It is therefore important to accurately estimate the sizes of the products and correctly identify those that correspond to the expected sizes. Note also that about 5 %–10 % of products do not reprodu-

cibly amplify for undetermined reasons. A repeat experiment can be attempted for such products, but if they still do not amplify they are discarded.

6. DNA products from the agarose gel are transferred to a nylon membrane by standard techniques for blotting or electrotransfer, and fixed to the filter by a short UV light exposure or NaOH treatment (as recommended by the suppliers). The blot is hybridized with labeled double-stranded cDNA from the different samples being compared. Labeling of the cDNA is achieved using a commercial random priming kit and [^{32}P]dCTP.

7. Following each hybridization and exposure of the filter, the probe is removed by immersing the membrane in 50 % formamide in 2x SSC for 1 h at 65 °C followed by washing with 0.2x SSC at room temperature. The blot is then re-hybridized with the next cDNA probe.

Purification, cloning and analysis of PCR products

The steps for the purification and cloning of PCR products of interest are not given in detail, as many of these procedures are routinely carried out in many laboratories and there are a large variety of kits available from different suppliers of molecular biology products (for a detailed procedure see Chap. XVIII).

Only half of each sample is used for the purification of DNA fragments of interest, unless the products are not very abundant. For reactions that display more than one amplified PCR product, the product of interest is purified by preparative agarose gel electrophoresis, whereas those reactions that display a single product can be cleaned using, for example, a Wizard PCR purification system. There are many different procedures for purifying DNA fragments from agarose gels which usually depend on the preferences of the experimenter. The GeneClean kit has become a popular method for the purification of DNA from agarose gels due to its speed and simplicity. However, we have found that DNA, either purified from low melting point agarose gels by phenol extraction or electro-eluted from standard agarose gels (Sambrook et al. 1989), is more amenable to cloning.

Generally we use the pGEM-T vector for cloning the purified PCR products, and similar vectors are available from other commercial suppliers (e.g., Novagen). These vectors are relatively expensive, and it is possible to generate a large amount of T-vector relatively cheaply following the procedures outlined in Marchuk et al. (1991).

Bacteria harboring plasmids with cloned inserts are identified by the blue-white X-Gal selection procedure. We find variable numbers of false-positive white colonies, depending on the ligation efficiency, and materials and time can be wasted in their analysis. PCR amplification directly from a white colony with the universal forward and reverse plasmid sequencing primers not only reduces the cost and speed of analysis, but also allows the size of the cloned insert to be estimated followed by direct sequencing of the PCR product. Clones can also be grown in liquid culture, and if a large number are to be processed, a 96-well flat-bottomed microtiter plate can be used. The procedure for colonies and liquid cultures is the same except for how the cells are generated.

1. With a sterile micropipette tip take a small amount of a white colony from an agar plate incubated overnight at 37 °C. Colonies more than several days old do not give good results. Avoid taking any agar as this inhibits the PCR. Resuspend each colony sample in 50 µl of water in a sterile 500 µl microfuge tube. For liquid cultures take 5 µl of an overnight culture and place in 45 µl of water.

2. Place the tubes in a boiling water bath or in a thermocycler heating block at between 95 °C and 100 °C for 5 min (the boiling temperature depends on your altitude and is approximately 95 °C in Nairobi).

3. Remove 2 µl of each sample and aliquot into a 20 µl final volume PCR reaction containing 20 ng of both the universal forward and reverse sequencing primers. The cycling conditions are 94 °C for 1 min, 50 °C for 1 min, 72 °C for 1 min for 40 cycles, followed by a 2 min extension at 72 °C.

4. Analyze the PCR products on a 2 % (w/v) agarose gel with DNA size markers. The PCR product size of the generated from a blue colony containing pGEM-T with the above primers is about 270 bp, so the size of a cloned insert can be estimated from the total size of the PCR product generated minus 270 bp.

5. Select tubes containing a single clear PCR product of the correct estimated size for DNA sequence analysis. Some white colonies will produce more than one PCR product, whereas others will generate a product of only 270 bp, showing that they do not contain a cloned insert.

6. For each PCR product of interest take 1 µl and add to 9 µl of water. Only 1–2 µl of this dilution is required for DNA sequence analysis using the *fmol* DNA sequencing system. It is not necessary to clean the PCR products or remove primers and dNTPs, as the dilution step reduces these to trace levels. Since the products are double-stranded DNA, sequence information can be derived for

both strands, and the sequence information is equally as clean as for purified plasmid samples. In some cases we have resolved ambiguities in sequences from plasmids by this procedure.

Cloned PCR products that have not displayed hybridization signals with labeled cDNA above can be used for northern blot analysis to confirm their differential expression. Hybridization probes can be generated by random priming of the purified cloned insert or incorporating a labeled nucleotide in a PCR reaction with the universal forward and reverse primers. For more sensitive detection, riboprobes can be generated from the T7 and SP6 promoters in pGEM-T vector and protocols plus reagents are available in kits from the manufacturer (Promega). Unless the direction of transcription is known, both T7 and SP6 probes are required for the northern blot analysis. RNA gels are run as described above and directly transferred to a nylon membrane as for the Southern blotting procedure. Improved sensitivity in the hybridization procedure can be achieved by treating the gel with 7% (v/v) formamide for 10 min prior to blotting (Pellé and Murphy 1993).

Differential expression analysis

If RNA is limiting due to cell numbers, specific PCR primers can be designed from the derived DNA sequence and tested, together with primers for a constitutively expressed control gene, such as actin, in a semiquantitative multiplex PCR assay on single-stranded cDNA from stages of interest (Murphy et al. 1990).

5
Results and Comments

The number of PCR products produced by the RADES-PCR method is primer dependent; some primers produce many products (Fig. 2, primer 1061) whereas others produce relatively few products (Fig. 2, primer 508). Some primers produce a different fingerprint patterns every time they are used (Fig. 2, primer 878), hence the importance of conducting the analysis at two template concentrations. Primers which generate reproducible fingerprints at the different template concentrations (Fig. 2, primers 1059 and 1061) are chosen. It is important to note that any primers you might have, even the universal sequencing primers, constitute arbitrary primers when used under the PCR conditions outlined above. It is therefore not necessary to have primers specifically designed for this methodology. When using primers greater than 15 nucleotides, a two-step PCR should be carried out; the first two to three rounds are carried out at a low primer annealing temperature followed by 40 rounds at a more

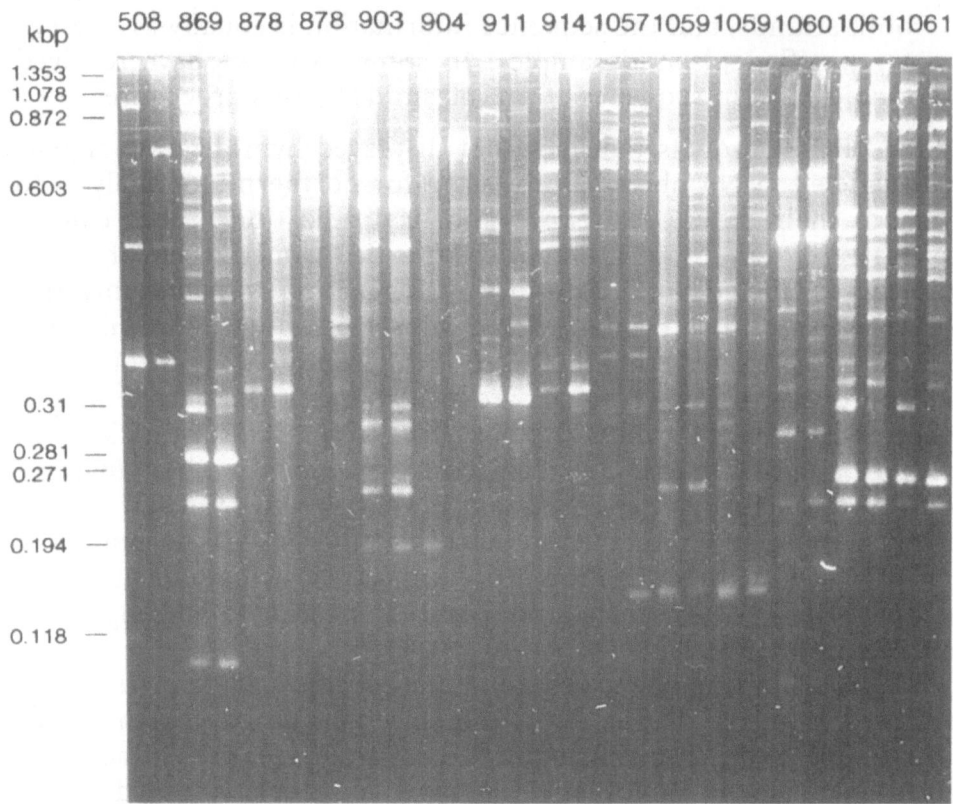

Fig. 2. Separation of RADES-PCR products generated from bloodstream and metacyclic forms of *T. congolense* on a 3% (w/v) MetaPhor agarose gel. *Numbers* above represent the oligodeoxynucleotide number, and each pair of tracks represents the two forms of the parasite used. Examples of reproducible (primers 1059, 1061) and nonreproducible (primer 878) fingerprints for two different template concentrations are displayed

stringent annealing temperature (Welsh and McClelland 1990; Welsh et al. 1992). In all cases, products chosen for further analysis are those that clearly and reproducibly differ in intensity between the samples being analyzed at the different template concentrations (see Fig. 2). In general, the percentage of differentially amplified products screened in this manner that represent differentially expressed genes is about 70%. Further checking by hybridization analysis increases the percentage of correctly identified products.

Combinations of primers can be used in the RADES-PCR method, thus extending the number of fingerprints that can be obtained. We have used combinations of arbitrary primers or an arbitrary primer together with a miniexon-specific primer with good results. However, the reamplification of products of interest with combinations of primers is more difficult, and a number of products never re-amplify or always produce products of a different size than expected.

6
Troubleshooting

The most important consideration in these procedures is to ensure the quality of the materials and reagents by checking each step. RNA preparation has been simplified, and any degradation observed is generally due to poor storage conditions of cells or RNA. Once first-strand cDNA has been generated, this material is stable and the only problems to overcome are defining the correct conditions for the PCR. Differences between thermocyclers and buffer components used in different laboratories will generate different results. If no fingerprint pattern is observed, try increasing the $MgCl_2$ concentration; we generally try a range between 1 mM and 3 mM. Another possibility is that the ramping rates between the different temperatures are too rapid and the tube contents do not have sufficient time to equilibrate. This can be corrected by slowing down the ramping rates or extending the times at the different temperatures. For many, the first attempt at generating a fingerprint pattern is disappointing, and it is not always clear why a problem has occurred. By repeating the procedure under different conditions, one condition is usually identified which generates a good fingerprint and all subsequent experiments then miraculously begin to work. We have found inexplicable differences between experimenters; using the same materials, buffers, primers and thermocycler one person in the laboratory finds that only 3 mM $MgCl_2$ works reproducibly whereas others find that only 2 mM produces reproducible results.

References

Adler BK, Hajduk SL (1994) Mechanisms and origins of RNA editing. Curr Opin Genet Dev 4:316–322

Ameisen JC, Idziorek T, Billaut-Mulot O, Loyens M, Tissier JP, Potentier A, Ouaissi A (1995) Apoptosis in a unicellular eukaryote (*Trypanosoma cruzi*): implications for the evolutionary origin and role of programmed cell death in the control of cell proliferation, differentiation and survival. Cell Death Diff 2:285–300

Bienen EJ, Webster P, Fish WR (1991) Trypanosoma (*Nannomomas*) *congolense*. Changes in respiratory metabolism during the life cycle. Exp Parasitol 73:403–412

Bonen L (1993) Trans-splicing of pre-mRNA in plants, animals and protists. FASEB J 7:40–46

Chomczynski P, Sacchi N (1987) Single step RNA isolation by acid guanidinium thiocyanate-phenol-chloroform extraction. Anal Biochem 162:156–159

De Lange T, Berkvens TM, Veerman HJ, Frasch AC, Barry JD, Borst P (1984) Comparison of the genes coding for the common 5' terminal sequence of messenger RNAs in three trypanosome species. Nucleic Acids Res 12:4431–4443

Evans DA (1993) *In vitro* cultivation and biological cloning of *Leishmania*. Methods Mol Biol 21:29–41

Fish WR, Muriuki C, Muthiani AM, Grab DJ, Lonsdale-Eccles JD (1989) Disulfide bond involvement in the maintenance of the cryptic nature of the cross-reacting determinant of metacyclic forms of *Trypanosoma congolense*. Biochemistry 28:5415–5421

Gommers-Ampt JH, Borst P (1995) Hypermodified bases in DNA. FASEB J 9:1034–1042

Hirumi H, Hirumi K (1991) *In vitro* cultivation of *Trypanosoma congolense* bloodstream forms in the absence of feeder cell layers. Parasitology 102:225–236

Lanham SM, Godfrey DG (1970) Isolation of salivarian trypanosomes from man and other mammals using DEAE-cellulose. Exp Parasitol 28:521–534

Marchuk D, Drumm M, Saulino A, Collins FS (1991) Construction of T-vectors, a rapid and general system for direct cloning of unmodified PCR products. Nucleic Acids Res 19:1154

Murphy LD, Herzog CE, Rudick JB, Fojo AT, Bates SE (1990) Use of the polymerase chain reaction in the quantitation of *mdr*-1 gene expression. Biochemistry 29:10351–10356

Murphy NB, Pellé R (1994) The use of arbitrary primers and the RADES method for the rapid identification of developmentally regulated genes in trypanosomes. Gene 141:53–61

Parsons M, Nelson RG, Watkins KP, Agabian N (1984) Trypanosome mRNAs share a common 5' spliced leader sequence. Cell 38:309–316

Pays E, Vanhamme L, Berberof M (1994) Genetic controls for the expression of surface antigens in African trypanosomes. Annu Rev Microbiol 48:25–52

Pellé R, Murphy NB (1993) Northern hybridization: rapid and simple electrophoretic conditions. Nucleic Acids Res 21:2783–2784

Puissant C, Houdebine L (1991) An improvement of the single-step method of RNA isolation by acid guanidinium-thiocyanate-phenol-chloroform extraction. BioTechniques 8:148–149

Sambrook J, Fritsch EF, Maniatis T (1989) Molecular Cloning: A Laboratory Manual. Cold Spring Harbor Laboratory Press, Cold Spring Harbor, N.Y.

Sanchezmoreno M, Fernandezbecerra MC, Castillacalvente JJ, Osuna A (1995) Metabolic studies by H-1 NMR of different forms of *Trypanosoma cruzi* as obtained by in vitro culture. FEMS Microbiol Letts 133:119–125

Simpson L, Maslov DA (1994) Ancient origin of RNA editing in kinetoplastid protozoa. Curr Opin Genet Dev 4:887–894

Tschudi C, Ullu E (1994) Trypanosomatid protozoa provide paradigms of eukaryotic biology. Infect Agents Dis 3:181–186

Vickerman K (1985) Developmental cycles and biology of pathogenic trypanosomes. Brit Med Bull 41:105–114

Welsh J, McClelland M (1990) Fingerprinting genomes using PCR with arbitrary primers. Nucleic Acids Res 18:7213–7218

Welsh J, Chada K, Dalai SS, Cheng R, Ralph D, McClelland M (1992) Arbitrarily primed PCR fingerprinting of RNA. Nucleic Acids Res 20:4965–4970

Williams JGK, Kubelik AR, Livak KJ, Rafalski JA, Tingey SV (1990) DNA polymorphisms amplified by arbitrary primers are useful as genetic markers. Nucleic Acids Res 18:6531–6535

Zilberstein D, Shapira M (1994) The role of pH in the development of *Leishmania* parasites. Annu Rev Microbiol 48:449–470

Subject Index